Programming the Microsoft Bot Framework

A Multiplatform Approach to Building Chatbots

Joe Mayo

Programming the Microsoft Bot Framework: A Multiplatform Approach to Building Chatbots

Published with the authorization of Microsoft Corporation by:
Pearson Education, Inc.

Copyright © 2018 by Pearson Education Inc.

All rights reserved. Printed in the United States of America. This publication is protected by copyright, and permission must be obtained from the publisher prior to any prohibited reproduction, storage in a retrieval system, or transmission in any form or by any means, electronic, mechanical, photocopying, recording, or likewise. For information regarding permissions, request forms, and the appropriate contacts within the Pearson Education Global Rights & Permissions Department, please visit www.pearsoned.com/permissions/. No patent liability is assumed with respect to the use of the information contained herein. Although every precaution has been taken in the preparation of this book, the publisher and author assume no responsibility for errors or omissions. Nor is any liability assumed for damages resulting from the use of the information contained herein.

ISBN-13: 978-1-5093-0498-1
ISBN-10: 1-5093-0498-3

Library of Congress Control Number: 2017953261

1 17

Trademarks

Microsoft and the trademarks listed at https://www.microsoft.com on the "Trademarks" webpage are trademarks of the Microsoft group of companies. All other marks are property of their respective owners.

Warning and Disclaimer

Every effort has been made to make this book as complete and as accurate as possible, but no warranty or fitness is implied. The information provided is on an "as is" basis. The authors, the publisher, and Microsoft Corporation shall have neither liability nor responsibility to any person or entity with respect to any loss or damages arising from the information contained in this book or programs accompanying it.

Special Sales

For information about buying this title in bulk quantities, or for special sales opportunities (which may include electronic versions; custom cover designs; and content particular to your business, training goals, marketing focus, or branding interests), please contact our corporate sales department at corpsales@pearsoned.com or (800) 382-3419.

For government sales inquiries, please contact governmentsales@pearsoned.com.

For questions about sales outside the U.S., please contact intlcs@pearson.com.

Editor-in-Chief	Greg Wiegand
Acquisitions Editor	Trina MacDonald
Development Editor	Songlin Qiu
Managing Editor	Sandra Schroeder
Senior Project Editor	Tracey Croom
Editorial Production	Backstop Media
Copy Editor	Christina Rudloff
Indexer	Julie Grady
Proofreader	Christina Rudloff
Technical Editors	Ankit Sinha and Shahed Chowdhuri
Cover Designer	Twist Creative, Seattle
Cover image	Alexandr.v02/ ShutterStock

I would like to dedicate this book to Aaliyah Mayo.
—Joe Mayo

Contents at a Glance

Introduction   *xv*

PART I   GETTING STARTED

CHAPTER 1	Exploring Bot Framework Architecture	3
CHAPTER 2	Setting Up a Project	15
CHAPTER 3	Building Conversations: The Essentials	39
CHAPTER 4	Fine-Tuning Your Chatbot	65

PART II   BOT BUILDER

CHAPTER 5	Building Dialogs	91
CHAPTER 6	Using FormFlow	119
CHAPTER 7	Customizing FormFlow	153
CHAPTER 8	Using Natural Language Processing (NLP) with LUIS	183
CHAPTER 9	Managing Advanced Conversation	203

PART III   CHANNELS AND GUI

CHAPTER 10	Attaching Cards	237
CHAPTER 11	Configuring Channels	279
CHAPTER 12	Creating Email, SMS, and Web Chatbots	291

PART IV   APIS, INTEGRATIONS, AND VOICE

CHAPTER 13	Coding Custom Channels with the Direct Line API	303
CHAPTER 14	Integrating Cognitive Services	321
CHAPTER 15	Adding Voice Services	345

Index   *361*

Contents

Introduction . . . .*xv*
    Who should read this book. . . . .*xv*
        Assumptions. . . . . *xvi*
    This book might not be for you if…. . . . . *xvi*
    Organization of this book. . . . . *xvi*
    Conventions and features in this book . . . .*xvii*
    System requirements. . . . . *xviii*
    Downloads: Code samples. . . . . *xviii*
        Installing the code samples . . . . *xix*
        Using the code samples. . . . . *xix*
    Acknowledgments . . . . *xix*
    Errata, updates, & book support. . . . . *xx*
    Free ebooks from Microsoft Press . . . . *xx*
    We want to hear from you . . . . *xxi*
    Stay in touch. . . . . *xxi*

PART I GETTING STARTED

Chapter 1 Exploring Bot Framework Architecture 3

What Is a Chatbot? 3
    Defining a Chatbot 3
    Why Conversation? 4
    Chatbot Benefits. 5
    To Bot or Not 7
Bot Framework Architecture 8
    Visualizing Chatbots, Connector, and Channels. 8
    Overview of Channels. 9
    Bot Connector Services 10

　　　　　　　Characteristics of a Chatbot 12
　　　　　　　Chatbot Communications. 13
　　　　Summary 14

Chapter 2 Setting Up a Project　　　　　　　　　　　　　　　　　　15

　　　　Steps to Building a Chatbot 15
　　　　Creating a Bot Framework Project 16
　　　　　　　Installing the Project Template 16
　　　　　　　Starting a New Project 17
　　　　Examining the Default Code 18
　　　　　　　Assembly References 19
　　　　　　　Folder and File Layout 21
　　　　　　　The Default Chatbot 22
　　　　Initial Testing with the Emulator 26
　　　　　　　Installing the Bot Emulator 26
　　　　　　　Configuring the Chatbot 27
　　　　　　　Communicating with the Chatbot 29
　　　　Publishing and Registering a Chatbot. 31
　　　　　　　Publishing the Chatbot 31
　　　　　　　Registering a Chatbot. 35
　　　　Summary 38

Chapter 3 Building Conversations: The Essentials　　　　　　　　　39

　　　　The Rock, Paper, Scissors Game Bot. 39
　　　　　　　The PlayType Enum 39
　　　　　　　The Game Class. 40
　　　　　　　The MessagesController Class 44
　　　　Conversation State Management. 45
　　　　　　　Elements of a Conversation 45
　　　　　　　Saving and Retrieving State 47
　　　　Participating in Conversations. 59
　　　　　　　Responding to Conversations 59
　　　　　　　Building a Custom Message Activity 60

Using a Custom Message Activity 61

Summary 63

Chapter 4   Fine-Tuning Your Chatbot   65

Reviewing Bot Emulator Details 65

Handling Activities 67

    The *Activity Class* 67

    The *ActivityType Class* *68*

    Code Design Overview. 69

    Sending Activities with the Bot Emulator 72

    Relationship Changes 72

    Conversation Updates 73

    Deleting User Data 75

    Pinging 76

    Typing Indications 77

Advanced Conversation Messages. 77

    Sending *Typing Activities* *78*

    Sending Independent Messages. 80

Summary 87

PART II   BOT BUILDER

Chapter 5   Building Dialogs   91

Introducing WineBot 91

Using the Wine.com API 92

Implementing a Dialog 98

    Creating a Dialog Class. 102

    Dialog Initialization and Workflow. 103

    Examining *IDialogContext* *103*

    Dialog Conversation Flow 105

    Dialog Prompt Options 108

    Performing the Search 114

Calling a Dialog 115

Summary 116

Chapter 6 Using FormFlow — 119

- Out-of-the-Box Features 119
- *A Basic FormFlow Chatbot* *124*
  - The Wine API Interface 124
  - WineForm: A FormFlow Form 124
  - Using *WineForm as a Dialog* *129*
- Enhancing FormFlow Conversations 132
  - The *Describe Attribute* *133*
  - The *Numeric Attribute* *134*
  - The *Optional Attribute* *135*
  - The *Pattern Attribute* *135*
  - The *Prompt Attribute* *135*
  - The *Terms Attribute* *136*
- Advanced Templates and Patterns 137
  - Pattern Language 137
  - Basic Templates 139
  - Template Usage 139
  - Template Options 146
- Summary 151

Chapter 7 Customizing FormFlow — 153

- Understanding the FormFlow Fluent Interface 153
- The *Configuration* Property 154
  - Configuring Responses 156
  - Configuring Templates 157
  - Configuring Commands 160
- The *Message* Method and Common Parameters 162
  - The *condition Parameter* *164*
  - The *dependencies Parameter* *164*
  - The *prompt Parameter* *165*
  - The *generateMessage Parameter* *166*
- The *Confirm* Method 166
- Working with Fields 168

	Dynamic Field Definition	171
	Field Validation	173
	The *AddRemainingFields* Method	*175*
	The *HasField* Method	175
	The *OnCompletion* Method	176
	The *Build* Method	178
	Initializing FormFlow	178
	Summary	180
Chapter 8	**Using Natural Language Processing (NLP) with LUIS**	**183**
	Learning Essential LUIS Concepts	183
	Setting up LUIS and Training Models	185
	Creating Models	185
	Building Intents	186
	Specifying Entities	187
	Training and Deploying	189
	Using LUIS in Your Chatbots	193
	Introducing *WineBotLuis*	*193*
	Adding Intents	195
	Handling Entities	196
	Continuous LUIS Model Improvement	200
	Summary	201
Chapter 9	**Managing Advanced Conversation**	**203**
	Managing the Dialog Stack	203
	What is the Dialog Stack?	203
	Navigating to Other Dialogs	204
	Navigating via *Forward*	*207*
	Navigating via *Call*	*209*
	Finishing a Dialog	211
	Managing Conversations with Chaining	213
	The *WineBotChain* Program	*214*
	Chain.From	*219*

Contents **xi**

        Chain.Loop *220*

        Chain.Switch. *220*

        Chain.ContinueWith. *223*

        An Assortment of Posting and Waiting *Chain methods* *223*

        LINQ to Dialog *225*

    Handling Interruptions with *IScorable* 226

    Formatting Text Output. 230

    Summary 233

PART III   CHANNELS AND GUI

Chapter 10 Attaching Cards   237

    Music Chatbot Overview 237

        The Groove API. 238

        The Root Dialog 242

        The Profile Dialog 243

        The Browse Dialog. 243

        The Playlist Dialog 244

        The Search Dialog 245

    Building Blocks 246

        Presenting Suggested Actions. 246

        Working with Attachments. 250

    Displaying Cards 257

        Implementing *BrowseDialog* *258*

        Implementing *PlaylistDialog*. *263*

    Adaptive Cards 267

        Layout with Containers 271

        Using Controls. 273

        Handling Actions 274

    Summary 276

Chapter 11 Configuring Channels — 279

Channel Overview 279
    The Channel Configuration Page 280
    Chatbot Analytics...280
Configuring Teams 284
    Channel Setup..284
    Using the Chatbot 286
Configuring Bing 286
    Channel Setup..287
    Using in Search 288
The Channel Inspector 289
Summary 289

Chapter 12 Creating Email, SMS, and Web Chatbots — 291

Emailing Chatbot Conversations.......................................291
    Creating the Email Account 291
    Configuring the Email Channel 292
Texting a Chatbot 293
Embedding the Webchat Control......................................295
    Adding the Webchat IFrame Placeholder 295
    Client-Side Coding..296
    Handling the Server Request 297
Summary 300

PART IV APIS, INTEGRATIONS, AND VOICE

Chapter 13 Coding Custom Channels with the Direct Line API — 303

Overview of the Console Channel 303
    Console Channel Components 304
    Examining Console Channel Code 305
Starting a Conversation 308

Listening for New Activities 310

Keeping the Conversation Open. 315

Sending Activities 316

Ending Conversations. 318
    Examining *CancellationTokenSource* and *CancellationToken* *318*
    Handling User Exits 319

Summary 320

Chapter 14 Integrating Cognitive Services   321

Searching with Bing 321

Interpreting an Image 330

Translating Text 332

Building FAQ Chatbots with QnA Maker 339

Summary 344

Chapter 15 Adding Voice Services   345

Adding Speech to Activities 345

Adding Speech with *SayAsync*. *347*

Adding Speech to *PromptDialog*. *350*

Specifying Input Hints 353

Setting up Cortana 355

Summary 359

What do you think of this book? We want to hear from you!

Microsoft is interested in hearing your feedback so we can improve our books and learning resources for you. To participate in a brief survey, please visit:

https://aka.ms/tellpress

Introduction

Microsoft first introduced the Bot Framework on March 30th 2016, during their annual Build developer conference. Essentially, the Bot Framework lets developers build apps, called chatbots, that surface in common messaging apps. To get started, you can open your favorite messaging app, such as Messenger, Skype, or Slack and add chatbots to your friends list. You can then interact with the chatbot with buttons, cards, or text for nearly any type of application. The difference is that the chatbot excels at conversational communication, rather than graphical, which can be very efficient. Developers can also write one chatbot with the Bot Framework and target a growing list of platforms, called channels. I've already mentioned a few messaging channels and you can literally surface the same chatbot anywhere, including web pages, SMS text, and Cortana Skills. There's even a Direct Line API that lets you create your own channels. Combining the power of conversation, multiple-channels, and voice, developers have a powerful tool to bring their applications through chatbots to anyone anywhere.

Programming the Microsoft Bot Framework: A Multiplatform Approach to Building Chatbots tells the story of the Bot Framework from a high-level architecture to managing conversations with dialogs and on to natural language processing and voice communications. Each step of the way, you learn the core principles of why a subject is important and how it applies to the work you need to do. While chatbots exist for a myriad of enterprise applications and industries, this book focuses on three primary examples: a game, a storefront, and entertainment. The examples and explanations are designed to not only give you ideas for your own chatbots, but put you on solid footing to start building your own chatbots right away.

Who should read this book

This book is for C# developers who want to learn about what chatbots are and what they can do for you. If you can write code and are able to read C# syntax, you can probably keep up. Though it is developer focused, Architects should find this to be a useful reference for how the Bot Framework operates and the considerations affecting integrations and systems design.

Assumptions

At a minimum, you should be able to read C# code. Any experienced C# developer won't have a problem. Knowledge of ASP.NET MVC Web API is helpful because that's the project type for all of the examples. Even if you're new to ASP.NET, you shouldn't have a problem because the book explains many of the details of how the Bot Framework handles web related subject material.

This book might not be for you if...

This book might not be for you if you're a brand new programmer who doesn't know how to write code. Although the Bot Framwork has a Node.js SDK, and could potentially support other languages in the future, this book uses C# exclusively. When people hear the term "bot," they often think of physical robots and autonomous machines, and this book is neither of these things. It's a book on how to write software for conversational user interfaces, which is another reason I use the term "chatbot." There are other chatbot platforms available, but this book only covers the Microsoft Bot Framework. That said, if you're a C# developer who's excited about chatbots and other Microsoft AI technologies, then don't hesitate to buy this book right now.

Organization of this book

This book has 15 chapters, divided into four parts. Part I, *Getting Started*, sets up the fundamentals that you need for the rest of the book. The first chapter provides a bird's eye view of the Bot Framework architecture, which is important because many of the decisions you make depend on knowing how the pieces fit together. After learning how to set up a project, Chapter 2 and Chapter 3 introduce the core parts of a conversation, explaining messages and conversation state. In Chapter 4, you learn how to use the Bot Emulator for testing, different types of information a chatbot can handle, and ways a chatbot communicates with a user.

Part II covers different types of dialogs. In Chapter 5, you'll learn about tools for managing user interaction and how to transition from one question to another. Chapter 6 introduces FormFlow—a way to build question and answer forms with a small amount of code. FormFlow is simple, yet powerful, and Chapter 7 continues with more techniques in customizing FormFlow. Chatbots come alive with natural language processing (NLP) and Chapter 8 shows how Microsoft's Language Understanding Intelligence Service (LUIS) lets you easily add NLP to a chatbot with a special dialog type. You'll learn even more advanced conversation management in Chapter 9, which covers how Bot Framework manages communication with a dialog stack. Chapter 9 also covers chaining, including LINQ to Dialog, and shows how to handle situations where a user changes subjects in the middle of a conversation.

Part III discusses channels and the graphical user interface (GUI). Although chatbots are primarily conversational, Chapter 10 shows how the Bot Frameworks supports various types of cards, allowing users to interact with GUI controls. Multi-platform support is so important that there are two chapters dedicated to channels. Chapter 11 provides a channel overview, explains how to configure analytics, and shows how to configure Bing and Team channels. There's a growing list of channels and these are general examples of how you approach the task of configuring channels. Chapter 12 covers how to configure email, SMS, and web chatbots, showing how to add a chatbot to something other than a messaging app channel. The explanation also shows how to add the Webchat control to a page in a secure manner.

The last 3 chapters in Part IV show how to add custom channels, intelligence, and use voice services. The Bot Framework has a powerful Direct Line API that lets you create your own channel, discussed in Chapter 13. Essentially, this lets you surface a chatbot on any platform. You can write sophisticated chatbots without artificial intelligence (AI), but AI can make them better. That's why Chapter 14 shows you how to leverage Microsoft Cognitive Services to add AI to your chatbot. Finally, Chapter 15 introduces one of the most compelling and exciting techologies that we have to work forward to by surfacing a chatbot as a Cortana Skill. For all the discussion of what an intuitive user interface is, what is more intuitive than the ability to use your voice and talk to a computer?

Conventions and features in this book

This book presents information using conventions designed to make the information readable and easy to follow.

- Boxed elements with labels such as "Note" provide additional information or alternative methods for completing a step successfully.
- Text that you type (apart from code blocks) appears in bold.
- A plus sign (+) between two key names means that you must press those keys at the same time. For example, "Press Alt+Tab" means that you hold down the Alt key while you press the Tab key.
- For long lines of code in listings, the code line will break and align flush left.

- A vertical bar or greater than sign between two or more menu items (e.g. File | Close or File > Close), means that you should select the first menu or menu item, then the next, and so on.

System requirements

You will need the following hardware and software to complete the practice exercises in this book:

- One of Windows 10, Windows 8.1, Windows 8, Windows 7 SP 1, Windows Server 2012 R2, Windows Server 2012, or Windows Server 2008 R2 SP1
- Visual Studio 2015—any edition (works fine with Visual Studio 2017, but tested and verified to be compatible with Visual Studio 2015)
- Computer that has a 1.6 GHz or faster processor
- 1 GB of RAM (1.5 GB if running on a virtual machine)
- 4 GB of available hard disk space
- 5400 RPM hard disk drive
- DirectX 9 capable video card running at 1024 x 768 or higher-resolution display
- Internet connection to download software, refresh NuGet packages, or chapter examples

Depending on your Windows configuration, you might require Local Administrator rights to install or configure Visual Studio 2015.

Downloads: Code samples

Most of the chapters in this book include exercises that let you interactively try out new material learned in the main text. All sample projects, in both their pre-exercise and post-exercise formats, can be downloaded from the following page:

https://aka.ms/ProgBotFramework/downloads

Note In addition to the code samples, your system should have Visual Studio 2015 (or Visual Studio 2017) installed.

Installing the code samples

Follow these steps to install the code samples on your computer so that you can use them with the exercises in this book.

1. If you downloaded the code, unzip the MSBotFrameworkBook-master.zip file that you downloaded from the book's website (name a specific directory along with directions to create it, if necessary).

2. Locate the file with the chapter name you want to see and open the *.sln file for that chapter.

Using the code samples

The code is organized into 15 folders with the pattern ChapterXX, where XX holds the chapter numbers 01 through 15. Chapter 01 doesn't have code because that's an architectural overview chapter. Each folder contains a ChapterXX.sln file you can double click in File Explorer to open in Visual Studio. Other folder contents are the project and other files for the code in that chapter.

Each chapter folder has a packages folder for NuGet assemblies. If you have trouble building the solution, delete the packages folder and rebuild. This restores NuGet packages for a successful build.

Acknowledgments

The first person I'd like to thank is my wife, May. Writing a book takes a significant amount of time and I sincerely appreciate her love and support throughout the process. Thanks to the rest of my family for their patience in my absence.

Thanks to the editing team. Devon Musgrave brought me into the project. Trina Fletcher MacDonald was my Aquisitions Editor and I sincerely appreciate her advice, patience, and how she kept the project on track. This is the third project I've done with Songlin Qiu as Development Editor, who provided insight that continuously improved the organization and quality of the work-another excellent experience. Our copy editor, Christina Rudloff, smoothed out my choppy grammar and helped me stay within the guidelines. Troy Mott expertly managed the copyediting, layout and composition, proofreading, and indexing of chapters. There are many other people who contributed that I haven't worked with individually, but I am grateful for their contributions.

We had a couple of excellent technical editors: Ankit Sinha and Shahed Chowdhuri. I can't count how many times these guys identified areas for improvement that made significant differences in the quality of the work. Thanks a lot and I really appreciate everything you've done.

Errata, updates, & book support

We've made every effort to ensure the accuracy of this book and its companion content. You can access updates to this book—in the form of a list of submitted errata and their related corrections—at:

https://aka.ms/ProgBotFramework/errata

If you discover an error that is not already listed, please submit it to us at the same page.

If you need additional support, email Microsoft Press Book Support at *mspinput@microsoft.com*.

Please note that product support for Microsoft software and hardware is not offered through the previous addresses. For help with Microsoft software or hardware, go to *https://support.microsoft.com/*.

Free ebooks from Microsoft Press

From technical overviews to in-depth information on special topics, the free ebooks from Microsoft Press cover a wide range of topics. These ebooks are available in PDF, EPUB, and Mobi for Kindle formats, ready for you to download at:

https://aka.ms/mspressfree

Check back often to see what is new!

We want to hear from you

At Microsoft Press, your satisfaction is our top priority, and your feedback our most valuable asset. Please tell us what you think of this book at:

https://aka.ms/tellpress

We know you're busy, so we've kept it short with just a few questions. Your answers go directly to the editors at Microsoft Press. (No personal information will be requested.) Thanks in advance for your input!

Stay in touch

Let's keep the conversation going! We're on Twitter: *http://twitter.com/MicrosoftPress*

PART I

Getting Started

CHAPTER 1   Exploring Bot Framework Architecture 3

CHAPTER 2   Setting Up a Project 15

CHAPTER 3   Building Conversations: The Essentials 39

CHAPTER 4   Fine-Tuning Your Chatbot 65

It's easy to set up a first project with the Bot Framework and build a simple chatbot. For a more sophisticated chatbot, however, you'll need a foundation to help you know the platform you're working on and the essential features available to you. That's what this part of the book helps you with.

There are four chapters in the Getting Started part of this book: Chapter 1, *Exploring Bot Framework Architecture*, Chapter 2, *Setting up a Project*, Chapter 3, *Conversation Essentials*, and Chapter 4, *Fine-Tuning Your Chatbot*. In Exploring Bot Framework Architecture you'll get a birds eye view of all the major components of the Bot Framework. This is important because it could affect design, construction, and deployment decisions for your chatbot and will help throughout the book as you learn about how each concept fits together with others. In *Setting up a Project*, you'll learn how to create a Visual Studio project, the different artifacts in that project, and how the artifacts of a project work together. You'll also run and interact with your first chatbot. The *Conversation Essentials* chapter explains the code you'll need to work with to manage conversations and other types of communication with your chatbot. In *Fine-Tuning Your*

Chatbot, you'll take a deep dive into the Bot Emulator and see what options are available for testing your chatbot locally. You'll also learn about advanced activities and additional ways to communicate with users.

These essential, Getting Started chapters will set you up for Part II where you'll need this information before learning how to manage conversation flow with dialogs.

CHAPTER 1

Exploring Bot Framework Architecture

When you pick up a new technology, it's reasonable to look for what the benefit is. How does this toolset or framework help get the job done? Is it more productive than writing the same code from scratch? Why not create an app, website, or desktop program with existing (or familiar) technology? The goal of this chapter is to help answer some of these questions.

Beyond talking about the Bot Framework, this chapter starts with a discussion of what a chatbot is and why you might need to build one. This book takes you further into understanding and validating what you know. The Bot Framework has several major components, and this chapter shows you what they are, how they relate to each other, and their benefits.

What Is a Chatbot?

Though the Microsoft Bot Framework uses the term "bot," this book uses the term "chatbot" to refer to the type of programs that the Bot Framework helps you build. This section explores what a chatbots are, their benefits, and some pros and cons of whether a chatbot is appropriate for a given task.

Defining a Chatbot

There are many opinions, but no common agreed upon formal definition of a chatbot. A couple of dictionaries mention bots, but don't clearly capture the essence of what a modern day chatbot is. The following definition might not find its way into a dictionary, though it contains a description of what a chatbot is, what it can do, and the type of platform it resides upon:

A chatbot is an application, often available via messaging platforms and using some form of intelligence, that interacts with a user via a conversational user interface (CUI).

Examining this definition in greater detail shows that a chatbot is an application, meaning it's a program that a developer writes. The chatbot essentially takes input from a user, processes it, and responds to the user.

The next phrase on being available via messaging platforms is important because that's where users find a lot of moden chatbots today. These messaging platforms include Skype, Messenger, Slack, and others. The reason that's important is because messaging applications are some of the most popular places for people to interact with each other. Because of this, it makes sense that applications, in the form of chatbots, are available where people normally are.

The definition includes Intelligence because there are a whole set of Artificial Intelligence (AI) services available via Microsoft Cognitive Services and many other third parties. Though it isn't required, it's common to see chatbots leveraging some type of AI. Notice that we use the term "often" associated with messaging and intelligence to denote that it's common, but not absolute or required.

The last part of the definition, about interacting with a user via CUI, is a distinguishing factor of chatbots. People communicate with chatbots via text and possibly even voice. Conversation can be text and/or voice. Most of the chatbots in this book use text, but some use cards for quick interaction and others, as discussed in Chapter 15, Adding Voice Services, use Cortana Skills for a voice interface. The next section explains why conversation is so important.

> **Note** Although the community continues to enthusiastically debate the meaning of a chatbot, one thing most people seem to have coalesced around is to use the term "chatbot" rather than "bot." In the past, bots have represented automated server processes, spiders, and web crawlers. The word is even shorthand for robot. Looking at the definition in this chapter and considering the potential for ambiguity, many people consider the term chatbot to be more descriptive for programs they interact with via plain language.

Why Conversation?

Conversation is an important characteristic of chatbots. Let's dive deeper into what a conversation is and why it's important.

Chatbots use Conversational User Interfaces (CUI) rather than Graphical User Interfaces (GUI). A GUI is a program that appears on screen and has graphical elements and pictures for mouse or touch interaction. CUI is text that a user types into a messaging app. Many of the chatbots available today are primarily text, but a chatbot can also use voice to interact with users. As you'll learn, a chatbot can have some GUI elements, but a chatbot is still characterized primarity by its use of text to facilitate conversations.

Some software developers may look at this and say what's old is new again because CUI sounds like the same thing as a command-line or console application. However, there's a huge difference because CUI is conversational as opposed to command driven. Chatbots often use Natural Language Processing (NLP) to understand normal language text from users. On the other hand, a typical command-line application requires precise command syntax to understand what the user wants. You could certainly use NLP and converse with users in a command-line application, but then you've created a chatbot that happens to reside in a command-line application. In the past, command-line applications didn't normally use NLP or engage in conversations, which are unique characteristics of chatbots. We say "normally" because there are several AI projects in the past that interact with users via conversation and it would be reasonable to consider those to be chatbots by the definition in this chapter.

The previous paragraphs explain how conversation makes chatbots different from earlier application types, command-line, and GUI. To appreciate why, think of how humans communicate with each other. They use conversation. Typing terse commands into a command-line application or touching a GUI screen are learned behaviors. Much has been researched and written on how to create great user experiences through GUI interfaces. However, the core issue remains—people still have to learn how to use the interface. Not only do they have to learn the interface of one program, but they have to learn the interface of every other program they use. People familiar with computers often figure out how to use a GUI through common patterns across vendors or other conventions. However, non-technical people still encounter a learning curve, regardless of how technically intuitive a user interface might be. The alternative, through chatbots, is conversation.

Conversation is natural. It's what people have done before recorded time to communicate. People learn to communicate, through conversation, from the time they're born and lasting all their lives. Why can't people use normal conversation when trying to get their computers to work? Now that goal is a lot closer because the primary interface of a chatbot is conversation.

Because messaging apps are so popular, entire generations are accustomed to interacting via text. Think of Skype messaging, Facebook Messenger, Slack, and many more applications that people use every day. In fact, people have been using chat applications, like AOL Messenger, IRC, and SMS for years. The advent of conversations with chatbots, on messaging platforms, providing service via text is an advantageous and normal evolution of computing in the lives of many people.

Conversation is one of the benefits of chatbots and there are others, covered in the next section.

Chatbot Benefits

There are more reasons why chatbots are a good choice of platform for building applications, including: conversation, ease of deployment, device versatility, and platform independence.

Conversation

The previous section discussed conversation in-depth. Because conversation is natural, a chatbot is an ideal platform for creating an interface for anyone to use.

> **Note** In addition to conversation, messaging platforms have some GUI support. However, it's minimal and supplementary to the conversational aspects of the chatbot.

Ease of Deployment

One problem with desktop applications or apps is all of the additional work associated with deployment. With desktop applications, you have to download and install the program. For apps, you visit a store and install on your device. A problem with websites is that they sometimes don't work on a certain browser. The deployment experience for existing technologies has some subjective degree of friction that affects people in different ways.

In contrast, chatbots reside in messaging apps that people already use. The deployment process is as simple as inviting the bot into their workspace. There isn't any heavy deployment or installation. People just say "Hi" to the chatbot and it begins communicating.

Any Device

Today, people often think of chatbots as something that resides in a messaging application. However, note that we used the term "often" in the earlier definition of a chatbot. This is intentional because there isn't anything to say that a chatbot "must" reside in a messaging app. Further, there isn't a requirement for interacting with a chatbot via typed text. The interaction could be voice.

Think of Cortana from Microsoft or Siri from Apple, each communicating via voice. Other commercial devices, such as Amazon's Alexa or Google's Home communicate with voice also. Developers are even able to build chatbots that use voice, translated via text-to-speech/speech-to-text interfaces offered through APIs like the Cognitive Services offered by Microsoft.

Besides messaging, chatbots can reside in apps, websites, and desktop applications. Anyone could build their own hardware device that has audio/voice and communicates across the Internet to interact with a chatbot. A command-line or GUI app requires a screen, but a chatbot is more versatile because it isn't constrained to traditional computing devices.

Platform Independence

Apps tend to adopt the design and functionality conventions of the platform they're built for. For example, Universal Windows Programs (UWP) have their recommended design patters, Google has Material Design, and Apple has their own recommendations. If you wanted to build a single app with a cross-platform tool like Xamarin, you either throw away those conventions or tweak the UI to move it closer to the expectations of that platform. Furthermore, different versions of those platforms change over time and conventions and capabilities change, fragmenting the app base and confusing developers and users. The alternative to reach different platforms is to build separate apps with technologies specific to those platforms, which is even more work. With the Microsoft Bot Framework, one chatbot serves all platforms.

> **Note** Xamarin is an excellent platform for developing cross-platform apps. The comments here serve to help illuminate the difference between apps and chatbots.

With chatbots you have one convention: plain language with the user. The conversations with the user are driven by the purpose of the chatbot and the desires of the user. Instead of following a call tree or hierarchy for a user to arrive at the desired functionality, users just ask the chatbot directly what they want. Logically, there needs to be some context for many types of requests, but again, the chatbot doesn't have to climb back out of a hierarchy to answer the next question a user has. The chatbot isn't tied to a corporate driven idea of what everyone's user experience for a generation of a platform should be. The chatbot is free to converse based on how a company/developer builds it for intended users.

These are a few of the benefits of chatbots and there is much promise in their implementation. However, there might be times when a chatbot isn't the right solution and the next section discusses that in more detail.

To Bot or Not

Most software developers who have been in the business for a while have seen their share of programming language debates. Some people enjoy these interactions of endless exchanges of bit, nuance, and theory, and it's possible to learn a thing or two from the intellectual tidbits that occasionally surface in these threads. Often you'll see a seasoned developer chime into these conversations and remind people that not every programming language, platform, or tool is perfect for every scenario and that an informed approach is to consider what is the right tool for the job. In this spirit, there are valid reasons why a chatbot might not be the best solution to every problem. There are likely various reasons, but this section covers a few to give you an idea on times you might question using a chatbot when planning a project, including: need to change, appropriate UI, and criticality.

Need to Change

Consider that a certain app already provides a service that satisfies users. They're happy, are trained to use the app, and the job is getting done. If a change is going to cause those users discomfort with little or no apparent benefit, why re-write it as a chatbot? On the other hand, if the users like new tools and there's a clear benefit, a chatbot might be a possible solution.

Other times, you might have a lack of resources, such as no budget or enough people to build the chatbot. Sometimes the technical desire to re-write an app as a chatbot doesn't outweigh the business constraints to make what you have last a little longer. However, if you have the budget and the chatbot promises significant improvements, the decision could change.

Appropriate UI

A chatbot UI is conversational. Chapter 10, Attaching Cards, discusses some GUI elements you can use with a chatbot, but that's limited. In particular, imagine needing to use a map where users need to pan, zoom, and perform other manipulations. It might be possible to talk to the chatbot and get it to re-render images, but that would be cumbersome for the user.

Another area where a chatbot wouldn't be logical is as an augmented/virtual reality (AR/VR) application. It might sound obvious, but the point here is to mention a few items to facilitate a thinking process of whether a chatbot is appropriate for a scenario. In this case, a chatbot might not be the program, but maybe an AR program might have a character that is a chatbot.

Criticality

One of the things about working with chatbots are that conversations can be hard to design. There are so many different ways to say something, and so many different directions a conversation can take. By constraining the domain, it's possible to manage this complexity. However, there are real-life situations where it isn't practical to constrain a conversation, such as a 911 service, hospital emergency rooms, or first responder coordination. These are all situations that are unpredictable and conversations could go in any direction. Emergencies would also be catastrophic in a case where a chatbot was unable to understand what a person is saying. Given all the emotion, unpredictability, and complexity of these situations, a chatbot is probably not the correct solution.

That said, a chatbot might not be the right solution today, but as technology and tools advance, a chatbot might not only be capable, but also desired. If the chatbot had the ability to properly navigate a conversation, it might actually be safer by not allowing emotion and other human mistakes to make incorrect decisions. A similar discussion is being had for autonomous cars right now in that they might not be 100 percent reliable today, but are likely to be much safer than human drivers in the future.

Bot Framework Architecture

When working with the Bot Framework, it's important to know how the components fit together. e.g. What are the paths a message takes, what does the bot communicate with, or which services are available? Throughout this book, you'll read about these components and having a visual of everything can help with the technical details of a concept. This section takes a high-level view and gradually digs into each component. You'll learn about the relationships between each component and the services that each component offers.

Visualizing Chatbots, Connector, and Channels

A birds-eye view includes the major components of the Bot Framework, including channels, the Bot Connector, and chatbot. Figure 1-1 shows the communication flows between Bot Framework components.

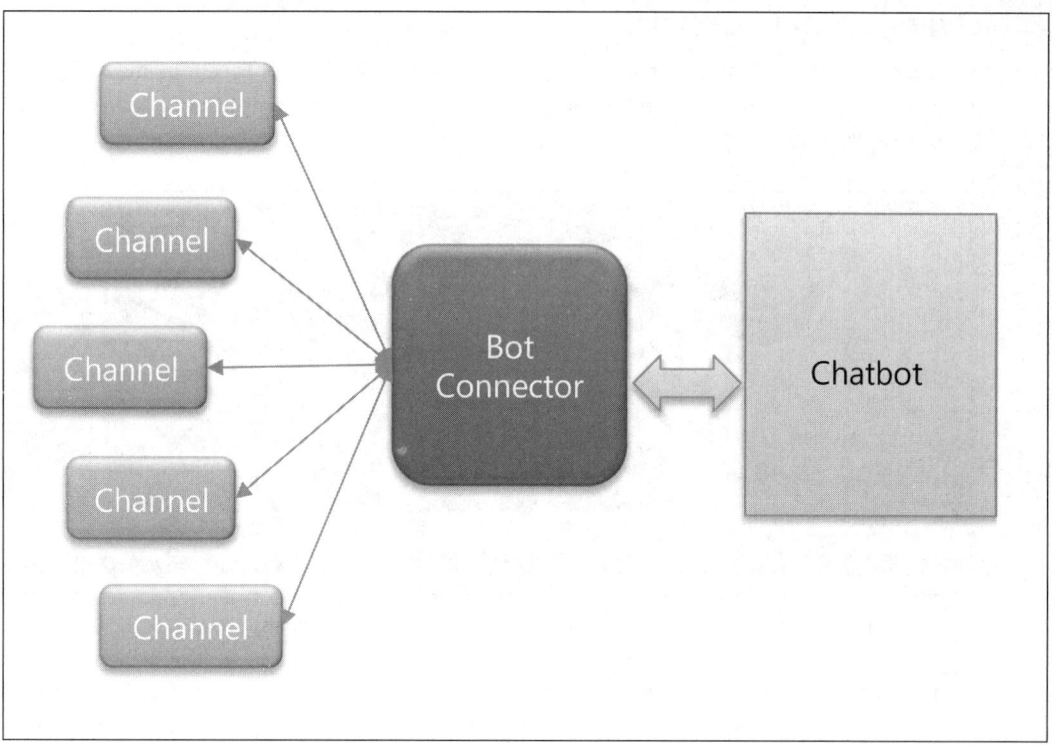

FIGURE 1-1 Channels communicate with the Bot Connector and the Bot Connector communicates with the chatbot.

Channels are apps, like Skype or Facebook Messenger, they're used to communicate with your chatbot. The Bot Connector is a Microsoft cloud component that channels send messages to and receive messages from. The Bot Connector then sends and receives messages with the chatbot. The chatbot component is something the developer builds and this book goes into detail on how to do that.

Each of these components: channel, Bot Connector, and chatbot; offers its own features and services, which you learn about in the following sections.

Overview of Channels

A channel is often associated with a messaging app, like Skype, Slack, or Facebook messenger. While this is true, a channel can also be any program that sends and receives messages to and from the Bot Connector, illustrated in Figure 1-2.

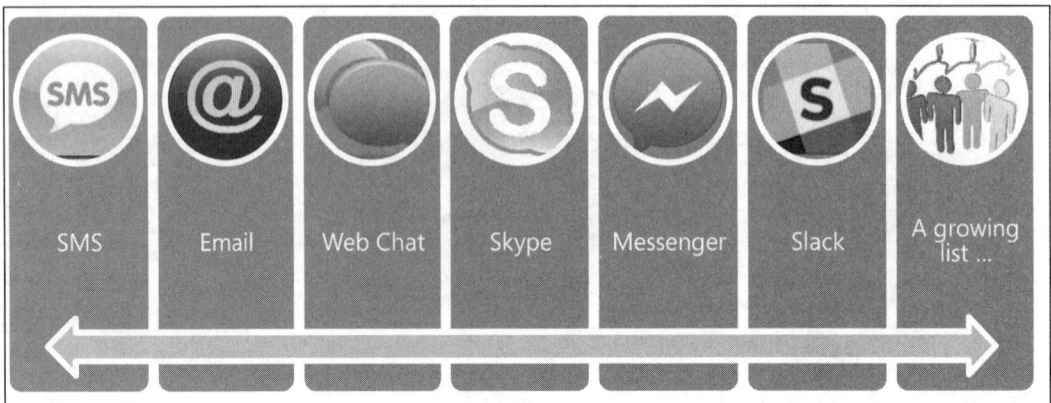

FIGURE 1-2 The Microsoft Bot Framework supports several third-party channels, email, SMS, and websites.

In addition to messaging applications, the Bot Framework supports email, SMS, and websites. The Bot Framework offers a webchat control that resides in a web page. Chapter 11, Configuring Channels, describes how to set up channels to work with a chatbot and Chapter 12, Creating Email, SMS, and Web Chatbots, shows how to set up and use email, SMS, and the webchat control.

Essentially, the channel represents a place where a user can interact with a chatbot. Microsoft integrates with several third-party apps and has its own channels, but chatbots aren't limited as to what type of channel they can offer a user. Imagine a company that needs to expose a chatbot through its own application. That's possible because the Bot Framework supports building custom channels and you can learn more about that in Chapter 14, Coding Custom Channels with the Direct Line API.

Bot Connector Services

Previous sections showed how the Bot Connector facilitates communication between channels and chatbots, which is called routing. However, there are more considerations for the Bot Connector's role, as shown in Figure 1-3.

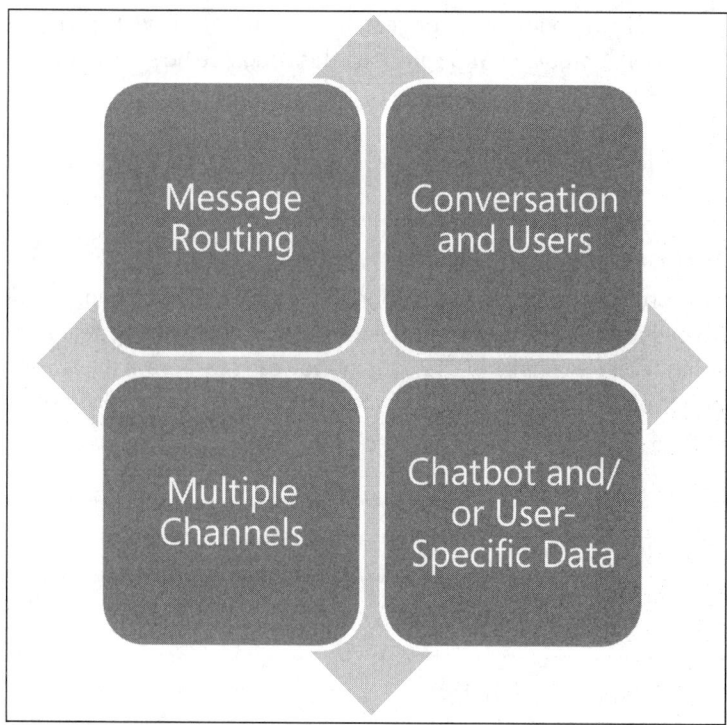

FIGURE 1-3 The Bot Connector offers several different services.

Besides routing, the Bot Connector stores *state*, which is custom information a chatbot can save. A chatbot is able to store custom information for a conversation, a user, or a user with conversation. Each of these types of state can store up to 32kb. Additionally, the Bot Framework will serialize the contents of Dialog types for storage in the Bot Connector state service.

Chatbots communicate with the Bot Connector via messages called *activities*. These activities can be text between the user and chatbot or other conversational events, like updating a conversation. These activities have various identifiers for a channel, conversation, and users. While the Bot Connector performs routing of these activities, it's important to note that the Bot Connector doesn't manage identifiers. Conversations and users, along with associated identities, are managed by channels. What the Bot Connector does manage is the wire format of messages between a channel and chatbot. Bot Framework chatbots always receive messages in the form of an Activity, deserialized from a JSON string object.

An interesting technical feature of the Bot Connector is that its interface is a REST API. The implications of this is that while the Bot Framework SDK is written in C# and resides on Windows Azure, the Bot Connector is platform agnostic. Anyone can build a custom channel that communicates via the Direct Line API. While the Bot Framework SDK supports C# (.NET programming languages) and node.js, the Bot Connector supports any programming language because it exposes a Connector REST API. There's also a State REST API that any programming language can access to manage chatbot state.

Tip The Bot Framework SDK is an open-source project, hosted on GitHub, which is located at *https://github.com/Microsoft/BotBuilder*. You can clone this code, see how it works, and interact with the Microsoft Bot Framework team. We've seen a couple related open source projects in Java and Python, and there are more.

Characteristics of a Chatbot

The chatbot serves whatever purpose its creator decides upon. There's currently a growing list of chatbots for nearly any imaginable domain. e.g. entertainment, information, retail, gaming, team management, and more. This book has several examples of chatbots and you can see more in the Skype bot directory. After forming an idea of what the bot should do, the next step is to start thinking about the design of the chatbot. Figure 1-4 shows a few characteristics of a chatbot that can help.

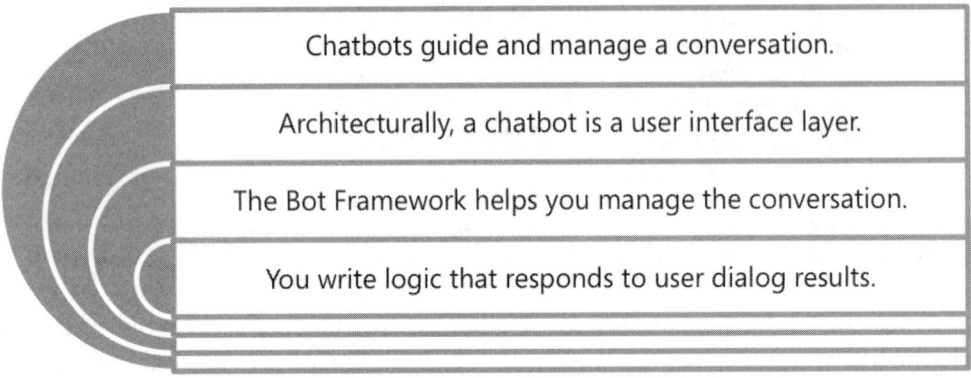

FIGURE 1-4 Chatbots have unique characteristics that guide their design and development.

From the time a user says 'Hi' until the conversation completes, the chatbot is responsible for the flow of the conversation. Certainly, the user drives the conversation and that conversation can go in a number of directions, many of which the chatbot was never designed for. For example, if a chatbot is built to sell phones and the user starts asking about desktop computers, the chatbot might not be able to talk about desktop computers if the developer didn't program it to do so. It's common for conversations to get off track and that means it takes significant thought to design a workflow that handles not only the path that the chatbot knows about, but to gracefully handle alternate paths. The definition of graceful depends on the developer and the nature of the application, though the need to map different conversation paths is a prime consideration when designing a chatbot.

Besides the unique requirements of managing conversations, a chatbot should be designed like any other computer program. In general terms, there are normally different layers of an application, which correspond to the design philosophy of the developer (or team) doing the work. In most applications, there is a user interface layer, which could be an HTML page for websites, a window for desktop applications, or a touch screen for phone apps. These are typically graphical user interfaces (GUI). However, a chatbot is primarily a conversational user interface (CUI). From an architectural perspective, the chatbot is the user interface. The code that manages the conversation is part of the user interface (or presentation layer). The developer builds code that the user interface calls to perform business logic, just like layered applications in other technologies. You'll see examples in this book where layered design isn't used to simplify an example and other examples that do separate chatbot CUI handling from business logic. Developers are free to design their chatbots any way they want and the concepts here are for ideas to help think about how a chatbot could be designed.

The final section of this chapter takes a step back and examines the inherent distributed architecture of a chatbot, providing information that affects chatbot design.

Chatbot Communications

A lot of applications are responsive because they reside on the same device or have tiers (e.g. database) on the same LAN. Even websites perform decently because they pass text between a server, which contains most of the application. Chatbots are a bit different because they're built on a distributed architecture. Figure 1-5 shows the communication paths between channel, Bot Connector, chatbot, and additional services.

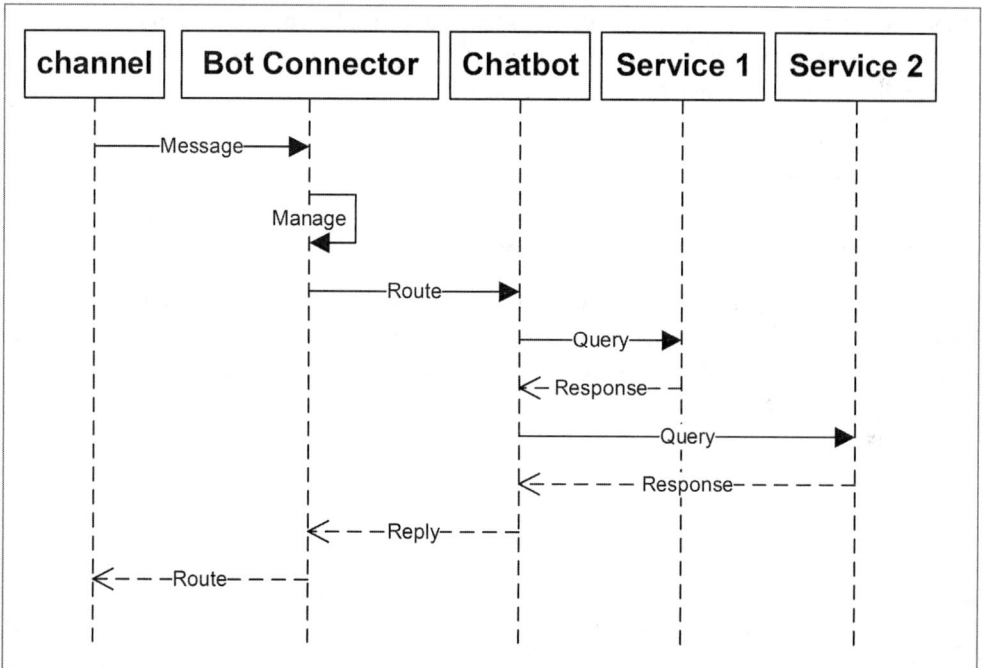

FIGURE 1-5 Each component in the Bot Framework communicates across the Internet.

There are distributed applications that communicate across the Internet and some of the concerns with this are around performance and scalability. Chatbots are distributed applications. Each of the

CHAPTER 1 Exploring Bot Framework Architecture   13

arrows between objects in Figure 1-5 represents a message being passed between components and, more specifically, each of those arrows represent communication across the Internet. The communication occurring between channels, Bot Connector, and chatbot are the normal part of the Bot Framework architecture. If a user has a slow Internet connection, their experience suffers. Alternatively, if the Internet connection is good, the user experience is better.

Figure 1-5 also contains an interesting set of features that developers must be aware of with services (labeled Service 1 and Service 2) that the chatbot uses. It is normal to call external services, like Language Understanding Intelligence Service (LUIS) for natural language processing (NLP). There might be a couple other services of interest, such as those offered by Microsoft Cognitive Services. If those services are necessary, by all means use them. However, be aware of their nature and their potential to affect a chatbot's performance and scalability.

Summary

The definition of a chatbot involves many features, including being another type of application, potential messaging platforms and intelligence services, and an inherent conversational nature. There are several benefits of using a chatbot over other application types like ease of deployment, multiple devices, and platform independence.

The Bot Framework architecture has components for channels, Bot Connector, and chatbots. The channels are the interfaces, such as messaging apps, websites, email, or SMS where a user interacts with a chatbot. The Bot Connector offers a set of routing and state management services. A chatbot is the application that a developer writes and contains the logic for interacting with the user.

Now you have the core knowledge needed to understand all the major components of the Bot Framework and where the chatbot fits in. The next chapter, Chapter 2, Setting Up a Project, shows how to build a simple chatbot and how to use the basic services of the Bot Framework SDK.

CHAPTER 2

Setting Up a Project

The Microsoft Bot Framework team created a starter template for building a chatbot. This starter template helps set up a project with folders, files, configuration, and assembly references that represent the basic configuration for a chatbot. This chapter tours a Bot Framework project and examines the minimal amount of code necessary for a chatbot.

While the ultimate goal is publishing a chatbot to one or more platforms, testing is a required activity. That's why the Bot Framework includes an emulator for communicating with a bot and testing its functionality. After testing, you can publish a chatbot and then register with the Bot Framework and request placement in the Bot Framework directory.

Steps to Building a Chatbot

Getting started with building a chatbot with the Microsoft Bot Framework is similar to building any software application that you build with Visual Studio: Create a new project, replace default code with your own, test, and deploy. Some developers have a specialized process to automating and performing a couple of these steps, but the end result is that these are the basic steps, shown in Figure 2-1, for creating and deploying a chatbot.

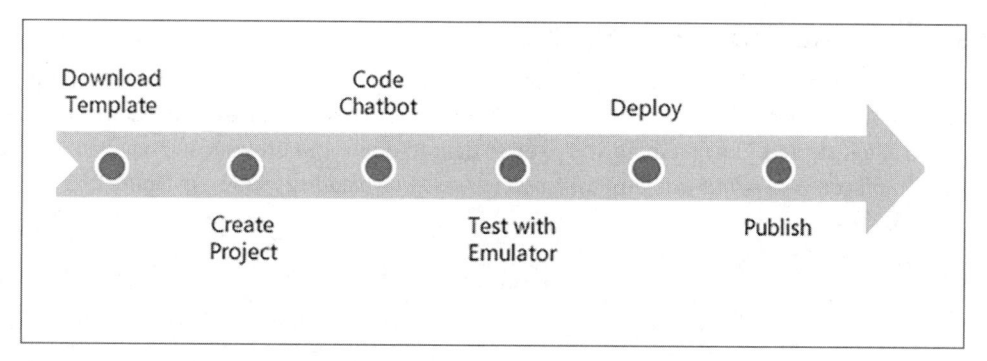

FIGURE 2-1 Steps to building a chatbot with the Microsoft Bot Framework.

Here's a quick overview of the steps, explained in greater detail in the following sections.

1. Download Visual Studio Template from the Microsoft Bot Framework site.
2. Create a Bot Framework Project in Visual Studio.
3. Write chatbot code.
4. Download the Bot Emulator from the Microsoft Bot Framework site and test it.
5. Deploy to a web server, such as Microsoft Azure.
6. Publish to channels.

This doesn't imply any type of process or guidance, but is a learning tool to understand how to go from 0 to chatbot in a short amount of time. The following sections take you through these steps to help get you started with your first chatbot.

> **Note** This chapter is a quick introduction with an abstract view of chatbot creation steps. Later chapters in this book go into greater depth on coding, testing, and deploying chatbots.

Creating a Bot Framework Project

Before getting started with a Bot Framework project, it's useful to download the Microsoft Bot Framework project template so you don't have to do the work yourself. This section explains how to do that and how to create a new project.

Installing the Project Template

If you've never created a Bot Framework chatbot before, the first task should be to download the Microsoft Bot Framework template. The template contains all the folders, files, and assembly references you need to get started. The first thing you need to do is to open your browser and navigate to *https://docs.microsoft.com/en-us/bot-framework/resources-tools-downloads*, as shown in Figure 2-2.

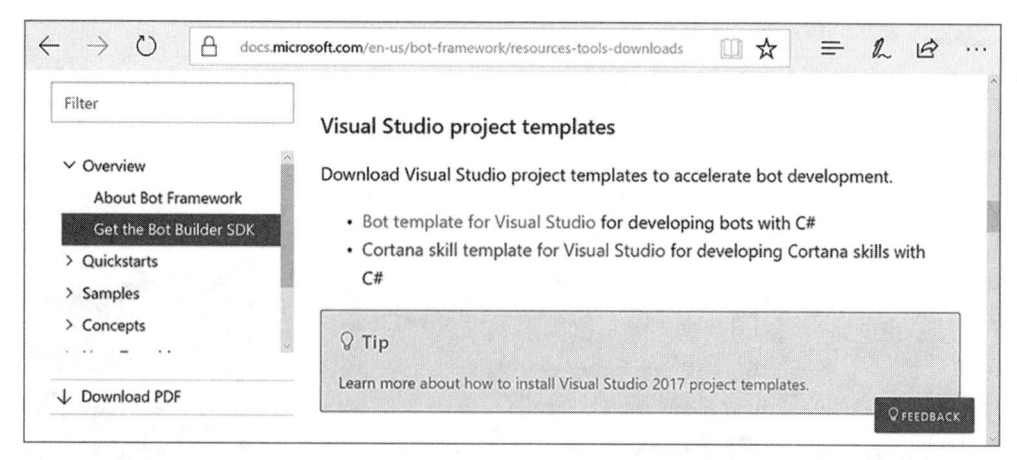

FIGURE 2-2 To get the template, visit https://docs.microsoft.com/en-us/bot-framework/resources-tools-downloads and click the Bot template for Visual Studio link.

Clicking the *Bot template for Visual Studio* link will download the template file named *Bot Application.zip* and you can find that file in your downloads folder. The Microsoft Edge and IE browsers ask if you want to open the folder and other browsers have similar features.

> **Warning** The inherent nature of the web is that pages on an active site change. Therefore, the exact steps for downloading could change between the time this is written and when you download the Bot Framework template. That means you might have to search for the template in case it's on another page and the exact steps in this book are different from the changed site.

Copy *Bot Application.zip* into the *%USERPROFILE%\Documents\Visual Studio 2015\Templates\ProjectTemplates\Visual C#* folder. This effectively installs the template so you can create a new project in Visual Studio, and the next section explains how to do that.

> **Tip** If Visual Studio is running when you installed the Bot Application.zip template, you might want to restart Visual Studio to see the Bot Framework template in the new applications list, discussed in the section titled Starting a New Project.

Starting a New Project

To create a new Bot Framework project, select the menu options **File | New | Project**, which displays the *New Project* dialog, shown in Figure 2-3. Click the *Visual C#* branch and scroll to *Bot Application*.

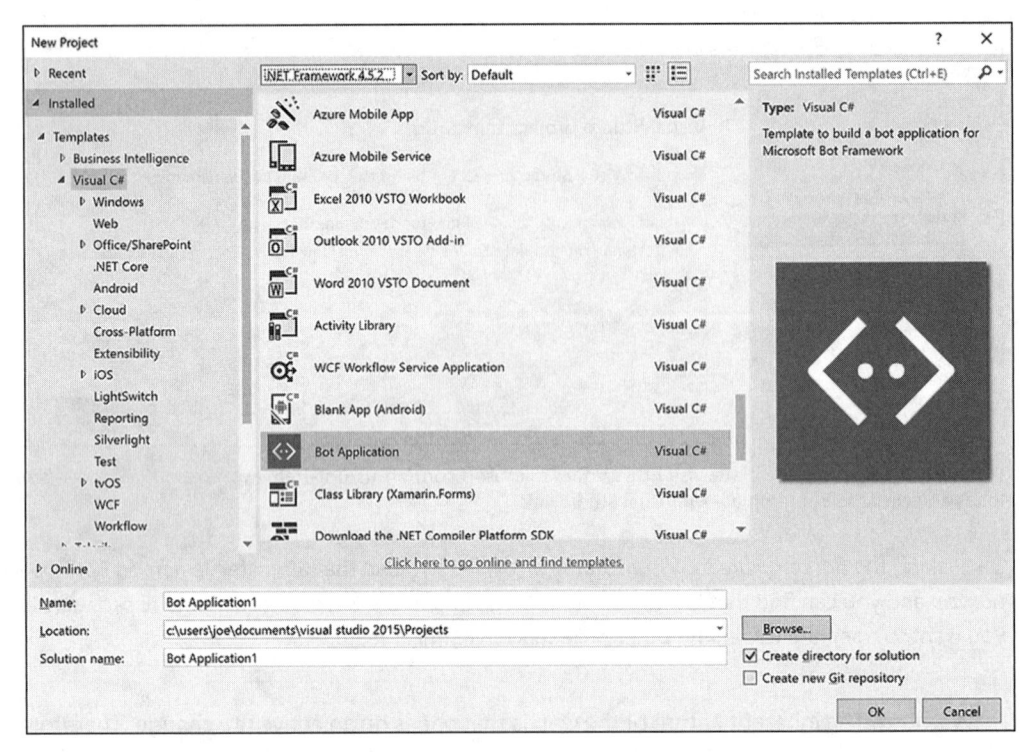

FIGURE 2-3 Select Bot Application as a new template to start a Bot Framework project.

When you create a new project, notice that it's very similar to creating most other projects in that it asks for project *Name*, *Location*, and *Solution name*. You're welcome to change any of the settings or just take the defaults to see what the template produces. Click *OK* to generate the project. The next section explains the project that this template creates.

Examining the Default Code

When creating a new project with the Bot Framework template, the resulting project is an ASP.NET MVC Web API project. It contains folders laid out with ASP.NET MVC conventions and the appropriate file references. In addition to ASP.NET MVC assembly references, it includes Bot Framework assembly references. The following sections cover the details of a new Bot Framework project.

Assembly References

The Bot Framework template includes all of the assemblies required for a new chatbot. This includes all of the ASP.NET MVC and Bot Framework assemblies, shown in Figure 2-4.

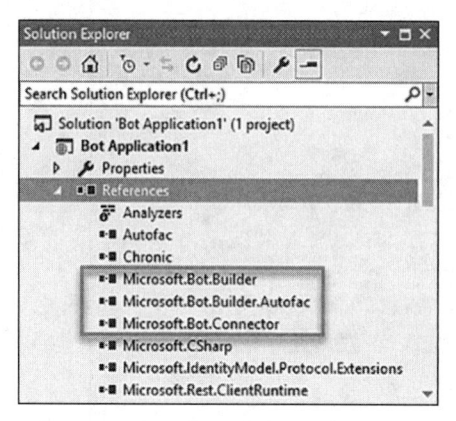

FIGURE 2-4 The Microsoft.Bot.Builder, Microsoft.Bot.Builder.Autofac, and Microsoft.Bot.Connector install via NuGet.

The *Microsoft.Bot.Builder* assembly installs via NuGet and depends on *Microsoft.Bot.Connector*. *Microsoft.Bot.Builder* (Bot Builder) has types for managing a chatbot conversation while *Microsoft.Bot.Connector* has types for communicating with the Bot Connector. Bot Builder also uses Autofac as its Inversion of Control (IoC) container and has additional types in *Microsoft.Bot.Builder.Autofac* to help work with Autofac. You'll see a few examples throughout this book that rely on Autofac and there is more info on how to use Autofac at *https://autofac.org/*. So, all you need to do is keep *Microsoft.Bot.Builder* up-to-date via NuGet, as shown in Figure 2-5.

> **Note** NuGet is the Microsoft package manager. It makes it easy to install, update, and remove assembly references. Library writers, including Microsoft and third-parties, configure these libraries to automatically update dependencies, install necessary project files, and update configuration to help the library work with a project.

CHAPTER 2 Setting Up a Project **19**

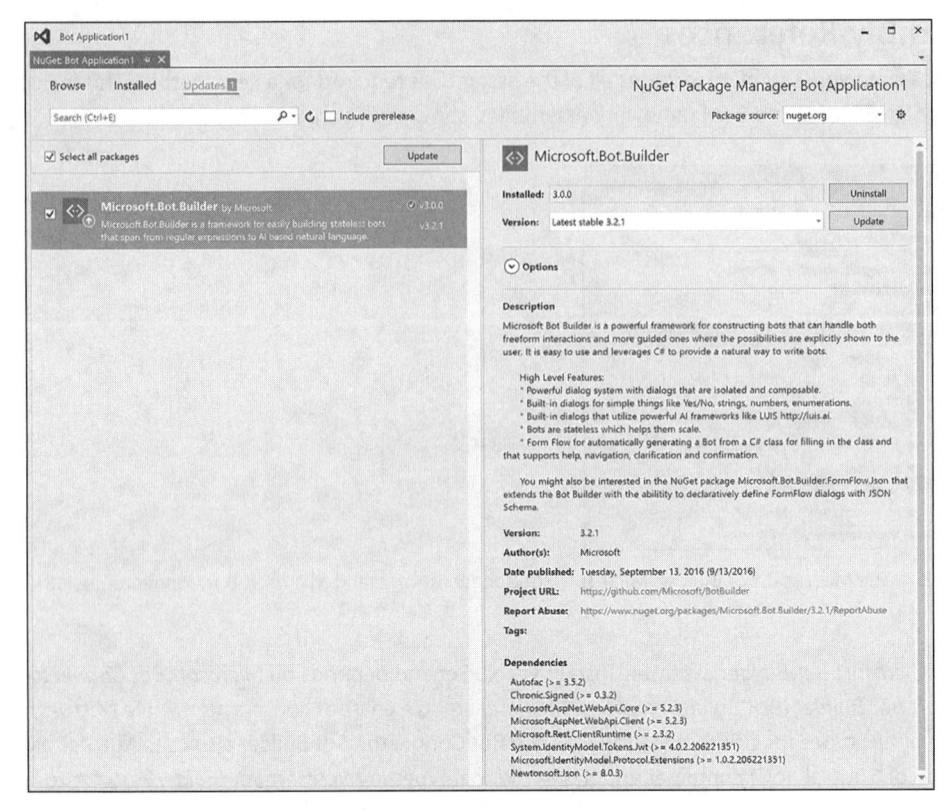

FIGURE 2-5 NuGet helps keep Bot Framework assemblies up-to-date.

Often, library authors upate NuGet assemblies to versions that are newer than the assemblies referenced by a new project template. These updates can include new features and/or security updates so it could be useful to periodically review updates. You can do this via NuGet by right-clicking the *References* folder (in the Visual Studio project) and selecting *Manage NuGet Packages*, showing a dialog similar to Figure 2-5.

Figure 2-5 shows (at the time this is written) that the current version of *Microsoft.Bot.Builder* is v3.0.0 and can be upgraded to v3.2.1. Unless you have an exceptional circumstance that makes you not want to install the upgrade (e.g. existing code relying on a deprecated feature), you should check NuGet to make sure you start a project with the most recent version. Select *Microsoft.Bot.Builder*, click *Update*, and accept the confirmation and licensing dialogs.

Another way to update the Microsoft.Bot.Builder is via the Package Manager Console, which you can find in Visual Studio by selecting the **Tools | NuGet Package Manager | Package Manager Console** menu. Then you can update with fhe following command:

>**Update-Package Microsoft.Bot.Builder**

With assemblies updated, it's time to look at the rest of the project structure, which is next.

Folder and File Layout

Because the project is for a web API, the folder and file structure follows the conventions of ASP.NET MVC. Figure 2-6 shows the *Controllers* folder with a *MessagesController.cs* and other web-related files.

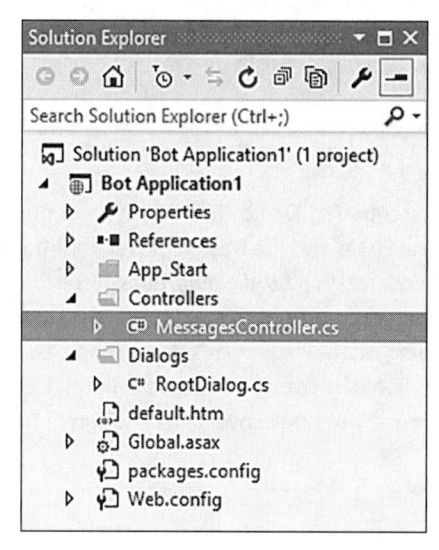

FIGURE 2-6 The folder structure of a Bot Framework project follows ASP.NET MVC conventions.

The *default.htm* file is an initial web page that has basic information about the chatbot. It's your choice whether to use this page or not, but it's useful when testing a chatbot. One potential use of *default.htm* is to host a Web Chat control, discussed in Chapter 12, and including links to various channels that the the chatbot is configured for. *Global.asax* and *Web.config* are the standard ASP.NET MVC application and configuration files. The *App_Start* folder contains web API initialization code that this book doesn't need to cover and generally won't need to change.

Figure 2-6 shows a *Dialogs* folder with a *RootDialog.cs* file. Right now, we're going to ignore that file because it contains a type derived from *IDialog<T>*, which we'll discuss in-depth in Chapter 5.

> **Note** ASP.NET is the Microsoft web application development framework. ASP is an acronym for Active Server Pages and it works on the .NET platform. MVC is an acronym for Model View Controller, which is an interface separation pattern with an implementation geared toward the Microsoft ASP.NET framework. This book doesn't require deep knowledge of ASP. NET MVC and will explain the parts that are fundamental to building chatbots.

What is most interesting in this project is the *MessagesController.cs* file. Pay particular attention to the name of this file because the ASP.NET MVC convention matches the prefix of this controller, *Messages*, as part of the addressable URL that the Microsoft Bot Connector uses to locate a chatbot. The next section examines the contents of *MessagesController.cs*.

The Default Chatbot

The *MessagesController.cs* file is the entry point to the chatbot. It's a *Representational State Transfer (REST)* interface with an HTTP POST endpoint. As shown in Listing 2-1, the ASP.NET web API models the HTTP POST endpoint with a method named *Post* inside of a controller class, *MessagesController* in the default case. As mentioned earlier, we're ignoring the default *RootDialog.cs* implementation for now. Replace the code in *MessagesController.cs* with the code from Listing 2-1.

> **What is REST**
>
> REST is an architectural style of communication, first described by Dr. Roy Fielding in his 2000 University of California, Irvine, Ph.D. thesis, Architectural Styles and the Design of Network-based Software Architectures, which can be found at *http://www.ics.uci.edu/~fielding/pubs/dissertation/top.htm*. The ASP.NET MVC implementation relies on web communcations and HTTP verbs, with a capability for more formal implementations, depending on the developer's design choices. REST has taken over as a dominant form of communication over the web in modern times. Deep understanding of REST isn't required to build a chatbot and this book covers just enough details on REST to help build a chatbot.

LISTING 2-1 The entry point of a chatbot is the Post method inside a controller.

```csharp
using System;
using System.Linq;
using System.Net;
using System.Net.Http;
using System.Threading.Tasks;
using System.Web.Http;
using System.Web.Http.Description;
using Microsoft.Bot.Connector;
using Newtonsoft.Json;

namespace Bot_Application1
{
    [BotAuthentication]
    public class MessagesController : ApiController
    {
        /// <summary>
        /// POST: api/Messages
        /// Receive a message from a user and reply to it
        /// </summary>
        public async Task<HttpResponseMessage> Post([FromBody]Activity activity)
        {
            if (activity.Type == ActivityTypes.Message)
            {
                ConnectorClient connector =
                    new ConnectorClient(new Uri(activity.ServiceUrl));
                // calculate something for us to return
                int length = (activity.Text ?? string.Empty).Length;
```

```csharp
                // return our reply to the user
                Activity reply = activity.CreateReply(
                    $"You sent {activity.Text} which was {length} characters");
                await connector.Conversations.ReplyToActivityAsync(reply);
            }
            else
            {
                HandleSystemMessage(activity);
            }
            var response = Request.CreateResponse(HttpStatusCode.OK);
            return response;
        }

        private Activity HandleSystemMessage(Activity message)
        {
            if (message.Type == ActivityTypes.DeleteUserData)
            {
                // Implement user deletion here
                // If we handle user deletion, return a real message
            }
            else if (message.Type == ActivityTypes.ConversationUpdate)
            {
                // Handle conversation state changes, like members
                // being added and removed Use Activity.MembersAdded
                // and Activity.MembersRemoved and Activity.Action
                // for info Not available in all channels
            }
            else if (message.Type == ActivityTypes.ContactRelationUpdate)
            {
                // Handle add/remove from contact lists
                // Activity.From + Activity.Action represent what happened
            }
            else if (message.Type == ActivityTypes.Typing)
            {
                // Handle knowing tha the user is typing
            }
            else if (message.Type == ActivityTypes.Ping)
            {
            }
            return null;
        }
    }
}
```

Listing 2-1 shows a *MessagesController* class, containing a *Post* method, and decorated with a *BotAuthentication* attribute. The *BotAuthentication* attribute is a Bot Framework type for securely authenticating that only the Microsoft Bot Connector can communicate with the chatbot. It relies on the following *Web.config appSettings* element keys:

```xml
<appSettings>
  <!-- update these with your BotId, Microsoft App Id and your Microsoft App Password-->
  <add key="BotId" value="YourBotId" />
  <add key="MicrosoftAppId" value="" />
  <add key="MicrosoftAppPassword" value="" />
</appSettings>
```

The default *appSettings* elements have empty values and the developer fills these in based on values obtained during bot registration, discussed in a later section of this chapter.

The *Post* method accepts an *Activity* parameter and returns a *Task<HttpResponseMessage>*, shown below:

```csharp
public async Task<HttpResponseMessage> Post([FromBody]Activity activity)
{
}
```

The *Activity* is a Bot Framework type, representing a communication between the Bot Connector and a chatbot. The available *Activity* types are for user messages, conversation status changes, and various others. Chapter 4 discusses the activity types in more depth, while this chapter gives a quick view into the *Message* Activity type.

Post is a normal C# async method, so the return type is *Task<T>* with a payload type of *HttpResponseMessage*. The *[FromBody]* parameter attribute indicates that the parameter comes from the body of the HTTP message, which is normal for an HTTP POST.

The *Post* method logic, shown below, includes an *if* statement to filter for a *Message* Activity type, branches appropriately, and responds to the Bot Connector with an HTTP message.

```csharp
if (activity.Type == ActivityTypes.Message)
{
}
else
{
    HandleSystemMessage(activity);
}
var response = Request.CreateResponse(HttpStatusCode.OK);
return response;
```

When the *Activity* type isn't *Message*, the default code calls a helper method, *HandleSystemMessage*, to handle the message type. After handling the message, the code uses the ASP.NET MVC *Request.CreateResponse* method to return *HttpStatusCode.OK*, which is status code 200 for the HTTP protocol. You'll read more about how to work with the *Activity* types in *HandleSystemMessages* in Chapter 4, though this chapter goes into more depth about the *Message* type because it's foundational in it's contribution to building a chatbot.

The main part of a chatbot will start when the *Activity* type is a *Message*. This example takes the message the user passed in and replies with the text and the length of that text. The code will instantiate a type to communicate with the Bot Connector, process the user's message, create a reply activity, and send that reply activity to the Bot Connector as shown below:

```
ConnectorClient connector =
    new ConnectorClient(new Uri(activity.ServiceUrl));
// calculate something for us to return
int length = (activity.Text ?? string.Empty).Length;

// return our reply to the user
Activity reply = activity.CreateReply(
    $"You sent {activity.Text} which was {length} characters");
await connector.Conversations.ReplyToActivityAsync(reply);
```

The *ConnectorClient* class allows the code to communicate with the Bot Connector. The *Activity* parameter has a *Text* property, containing the message that the user entered. The code calculates the length of that text.

The *Activity* type also has a *CreateReply* method for creating a new *Activity* instance that is sent back to the connector. In this example, *CreateReply* builds a string with the original user text and the length of that text, and returns an *Activity* instance, assigned to *reply*.

In addition to formatting the text, the *CreateReply* message takes communications information from the original *Activity*, with the parameter name *activity*, to populate values of the new *Activity* instance. In particular, the original *Activity* contains information about the conversation and user and performs actions such as swapping the *From* and *Recipient* user properties. This convenience method helps avoid instantiating a new response *Activity* and populationg property values manually.

Using the *ConnectorClient* instance, connector, the code sends the new *Activity*, *reply*, to the Bot Connector. The Bot Connector knows how to translate and routes the *Activity* to the original channel that the user is communicating with.

> **Tip** Whether going through old Bot Framework documentation or watching HTTP traffic, you'll see URLs that refer to the Bot Connector, where the communication came from. Don't be tempted to hard-code these URLs. Because of versioning and other system changes, the Bot Connector URL can change. It's better to always rely on the ServiceUrl property of the incoming Activity parameter, as shown in the ConnectorClient instantiation from Listing 2-1.

This section demonstrated how to create the default chatbot with the Bot Application template and modify that template to simplify the discussion. Now it's time to test the user experience and see the chatbot in action.

Initial Testing with the Emulator

The Bot Framework comes with an emulator that lets you test the user experience for a chatbot. Chapter 4 covers emulator details in more depth and this section is a quick introduction to testing essentials such as installing the emulator, performing basic configuration, and communicating with the chatbot.

Installing the Bot Emulator

Your first task in performing user testing of a chatbot is to download the Bot Framework Emulator. You can find it by browsing to the Bot Framework site at *https://docs.microsoft.com/en-us/bot-framework/debug-bots-emulator*, shown in Figure 2-7.

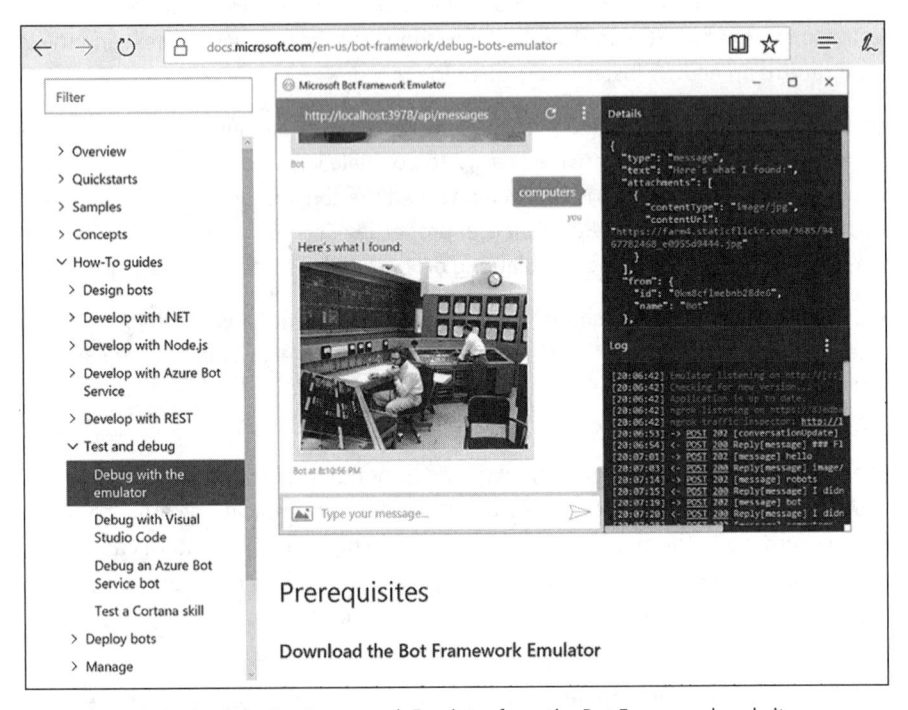

FIGURE 2-7 Download the Bot Framework Emulator from the Bot Framework website.

To perform the installation, click *emulator download page*. This downloads a setup file that you should *Run* and click the *Install* button on the dialog that appears.

> **Warning** The inherent nature of the web is that pages on an active site change. Therefore, the exact steps for downloading could change between the time this is written and when you download the Bot Framework Emulator. That means you might have to search for the emulator in case it's on another page and the exact steps in this book are different from the changed site.

26   PART I   Getting Started

The Bot Framework Emulator is an open source project and the code runs on Linux, Mac, and Windows. You can find the source code by clicking the *GitHub releases page* link, shown in Figure 2-7.

With the Bot Framework Emulator installed, it's time to configure the chatbot.

Configuring the Chatbot

Bot Framework Emulator configuration consists of properly addressing the chatbot and correctly setting credentials. The emulator has many features, and Chapter 4 discusses them in-depth. This section looks at the minimal work required to perform basic user testing.

After installation, the Bot Framework Emulator opens. You should manually start the program if it isn't already open. You'll see a window similar to Figure 2-8.

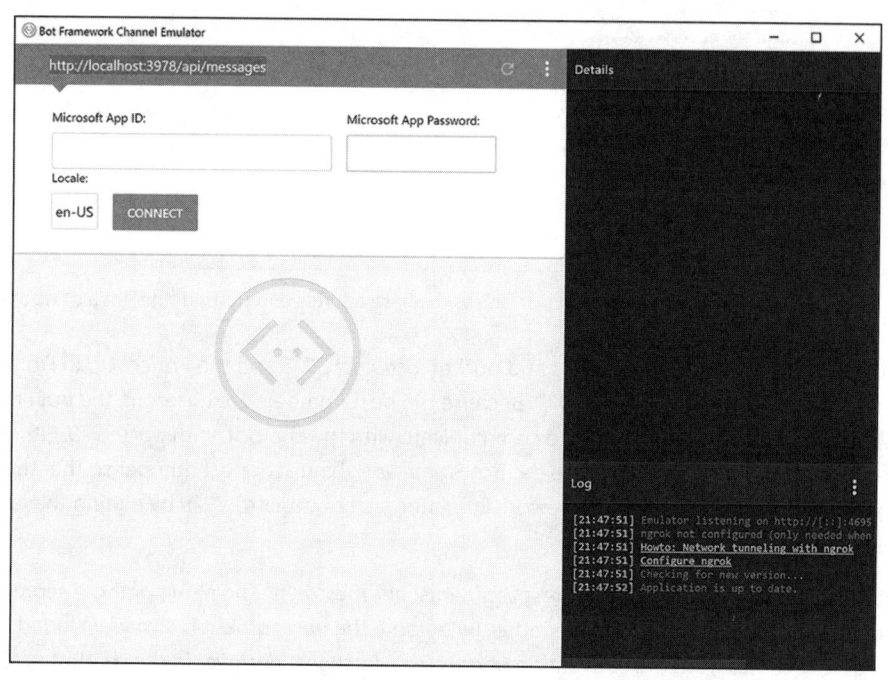

FIGURE 2-8 The Bot Framework Emulator allows user testing of a chatbot.

The parts of the Bot Framework Emulator of current interest are the URL, *Microsoft App ID, and Microsoft App Password*. When you first open the emulator, the URL box at the top of the emulator is empty with *Enter your endpoint URL* placeholder text, but we added one here to illustrate points that follow.

The URL consists of base address, port number, and segments addressing the chatbot. The chatbot runs on the current computer, launched by Visual Studio via IIS Express by default, which is addressed as *localhost*. The port number must match the port number of the chatbot being tested. To see how this base address and port are configured in the Bot Application project template, open the default chatbot from previous sections of this chapter, double-click the *Properties* folder, and click the web tab to see a screen similar to Figure 2-9.

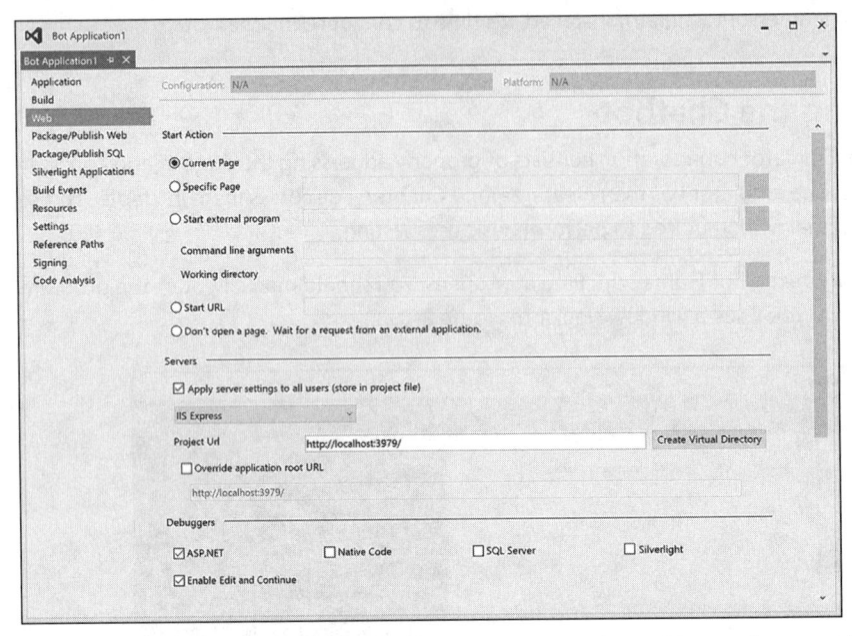

FIGURE 2-9 The Properties, Web, Servers, Project Url shows the base address and port number for a chatbot.

The URL in Figure 2-8 has a base address and port of *localhost:3978* and the *Project Url* in Figure 2-9 has a base address and port of *localhost:3979*. Because the port numbers are different, the Bot Framework Emulator, with port *3978*, is not able to communicate with the chatbot, with port *3979*. Either change the chatbot port number or change the Bot Framework Emulator port number so that they match. This example changes the Bot Framework Emulator port number to *3979* by editing the emulator URL field.

The URL field in Figure 2-8 also has address segments, *api/messages*. The name of these segments are based on ASP.NET MVC Web API conventions. By default, the *api* segment is always included. The *messages* segment comes from the convention of how the *MessagesController* in the chatbot project is named. The convention removes the *Controller* suffix and uses the prefix as the address, resulting in *api/messages*. Hypothetically, if you changed the controller name to *DefaultBotController*, the new address would be *api/defaultbot*. This book leaves the chatbot controller name as the default, *MessagesController*.

> **Tip** If you have multiple chatbots in the same project, perhaps testing or doing demos, you might want to give each chatbot a separate port number. This will help avoid confusion caused by accidentally communicating with the wrong chatbot.

28 PART I Getting Started

Based on the information and values from this section, set your URL, in the Bot Framework Channel Emulator, to *http://localhost:3979/api/messages* so it matches the *Project Url*.

The next configuration task is to set *Microsoft App Id* and *Microsoft App Password* correctly. In Figure 2-8 the values are empty, as they should be. The reason they should be empty is because their values, in *Web.config*, are not set yet, as shown below:

```
<appSettings>
  <!-- update these with your BotId, Microsoft App Id and your Microsoft App Password-->
  <add key="BotId" value="YourBotId" />
  <add key="MicrosoftAppId" value="" />
  <add key="MicrosoftAppPassword" value="" />
</appSettings>
```

The *appSettings* section of *Web.config* holds the *MicrosoftAppId* and *MicrosoftAppPassword* keys and those should match the *Microsoft App Id* and *Microsoft App Password* fields in the Bot Framework Channel Emulator, which are corresponding credentials. These *appSettings* keys are required for the *BotAuthentication* attribute. When communicating with a chatbot, the Bot Framework Channel Emulator sends the *Microsoft App Id* and *Microsoft App Password*. The *BotAuthentication* attribute reads the values sent from the Bot Framework Channel Emulator, and compares them with the matching *appSettings* keys. In the "Publishing and Registering a Chatbot" section later in this chapter, you'll see how the registration process produces these keys and that you need to populate *appSettings* with the new values when publishing the chatbot to communicate securely with the Bot Connector. For the purposes of user testing with the Bot Framework Channel Emulator, leave the credentials blank in both *appSettings* and the Bot Framwork Channel Emulator.

With the Bot Framework Channel Emulator installed and configured, you're ready to communicate with the chatbot.

Communicating with the Chatbot

To communicate, both the chatbot and the Bot Framework Channel Emulator must be running. Start by opening the chatbot project in Visual Studio and pressing *F5* to run. You'll see a test web page similar to the one shown in Figure 2-10.

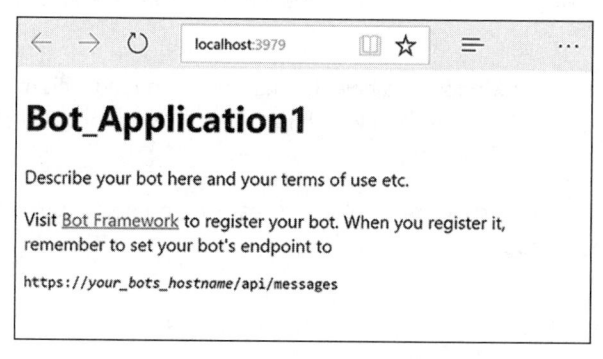

FIGURE 2-10 Run the chatbot before communicating with it.

This *default.htm* page is made of HTML and is the home page when the chatbot is running. Notice that the port number, in the address bar, is *3979*. Next, open the Bot Framework Channel Emulator and you'll see a screen similar to Figure 2-11, except the chat text will be empty the first time. You might need to click the URL and click the *Connect* button, as shown in Figure 2.8.

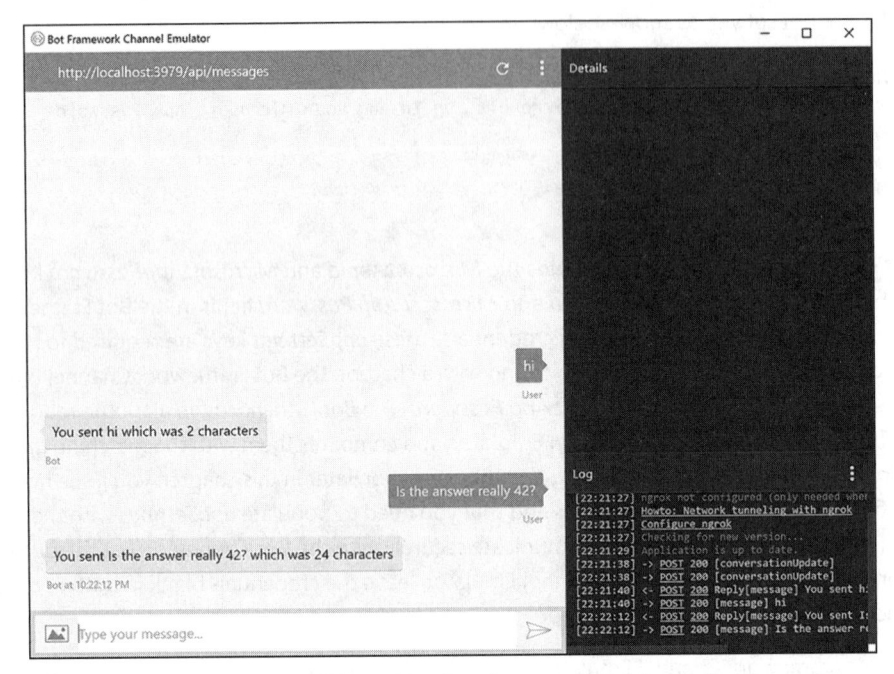

FIGURE 2-11 The Bot Framework Channel Emulator lets you chat with a chatbot.

Two parts of Figure 2-11 pertain to the current discussion: The chat window and the text entry box on the left side of the screen. This is where an entire conversation appears. It scrolls vertically as the conversation grows. The lighter colored statements on the left of the chat window is from the chatbot and the darker boxes on the right are from the user.

The text box at the bottom of the window, with the arrow on the right side, lets the user type text to the chatbot. You can type in this box and either press Enter, or click the arrow button to send text to the chatbot.

You can see a couple of messages in the chat window and how the chatbot responded. This matches the chatbot functionality from Listing 2-1. The chatbot responds with the text the user types, as well as the number of characters in that text.

At this point, you've successfully built and tested a chatbot. Now, it's ready for others to use and you do that by publishing and registering the chatbot.

Publishing and Registering a Chatbot

For other people to use a chatbot, it has to be published somewhere they can interact with. You also need to register with the Bot Framework website so the Bot Connector can find the chatbot. The following sections describe how to accomplish these tasks.

Publishing the Chatbot

Publishing a chatbot means that it will be deployed to a server. This book uses Windows Azure and you're free to select any hosting provider of your choice. The requirements are that the host must be publicly accessible over the Internet, so the Bot Connector can find it, and the hosting provider must support ASP.NET MVC.

> **Note** Technically, your chatbot can be built with any technology stack and hosted anywhere with a public endpoint that accepts HTTP Post requests. As long as it conforms to the Bot Connector interface, it should work. Specifics aside, the Microsoft .NET and Azure technology stack with the Bot Framework SDK are compelling technologies, which is why we are excited to write this book.

If you don't already have an Azure account, you can visit *https://azure.microsoft.com/* and sign up for a free trial. The following steps explain how to deploy to Azure:

1. Open the chatbot project in Visual Studio.

2. Right-click the project and select *Publish*.

3. When you see the *Publish* dialog in Figure 2-12, click the *Microsoft Azure App Service* button.

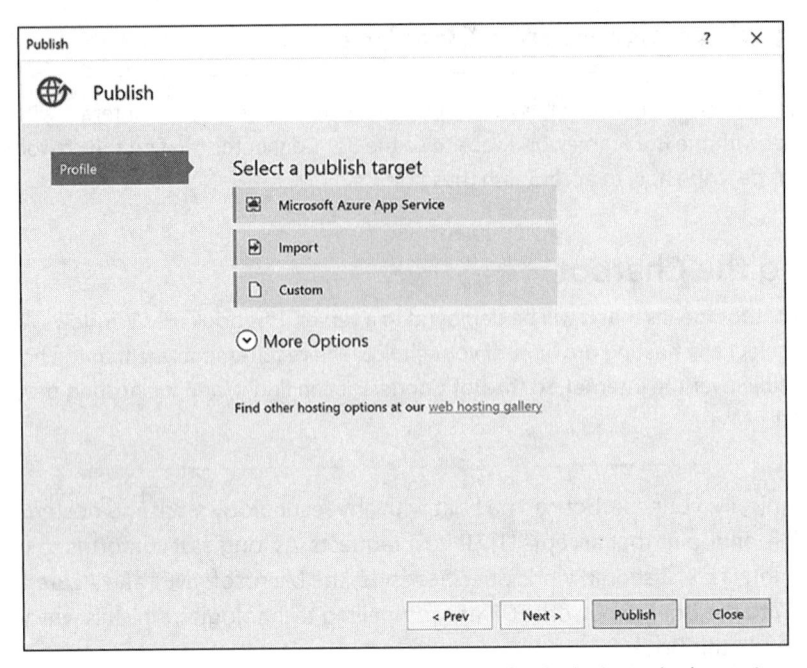

FIGURE 2-12 The Publish dialog starts off the process for deploying a chatbot to Azure.

4. When you see the *App Service* dialog in Figure 2-13, click the New button (on the middle-right side).

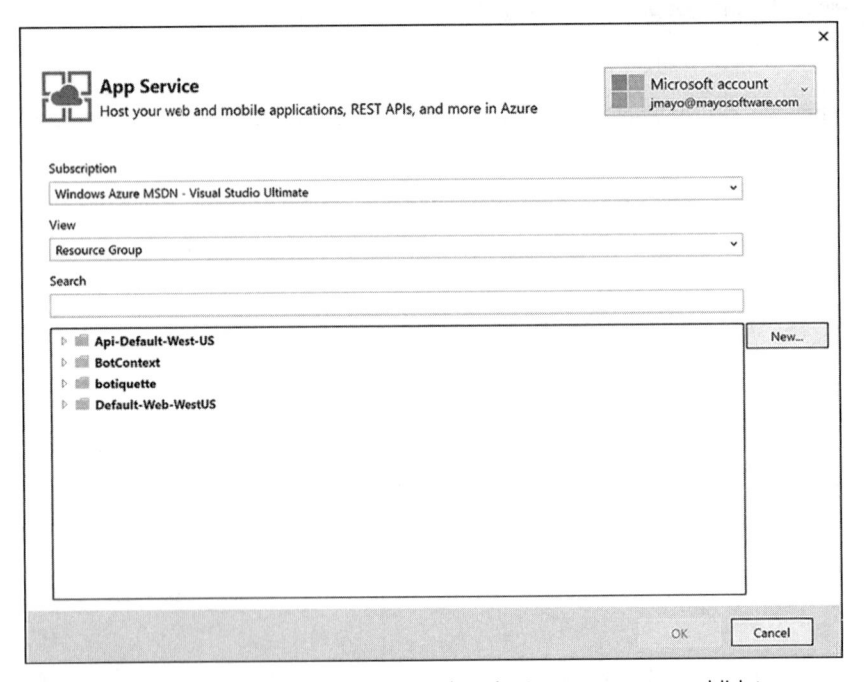

FIGURE 2-13 The App Service dialog lets you select the Azure account to publish to.

5. When you see the *Create App Service* dialog in Figure 2-14, select the *Change Type* drop-down list to *Web App* and give the bot a unique *Web App Name*. We used *JoeMayoTestBot* as the name and you'll need to use something different. The dialog shows an error icon on the right side of *Web App Name* if the name isn't unique.

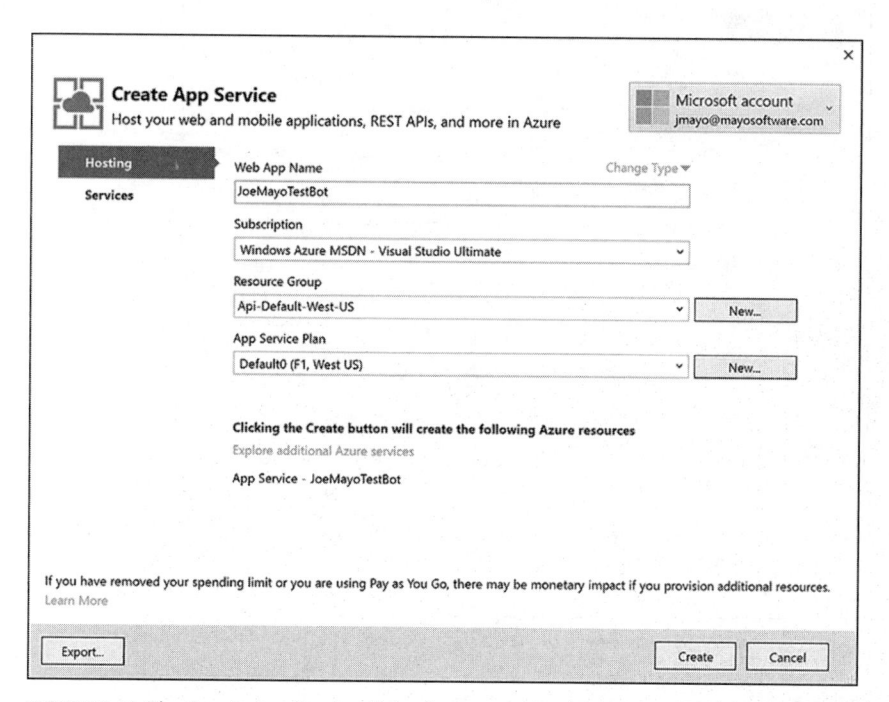

FIGURE 2-14 The Create App Service dialog lets you select the App type and change the Web App Name.

6. Click the *Create* button in the *Create App Service* dialog.

7. When you see the *Publish* dialog in Figure 2-15, click the *Publish* button.

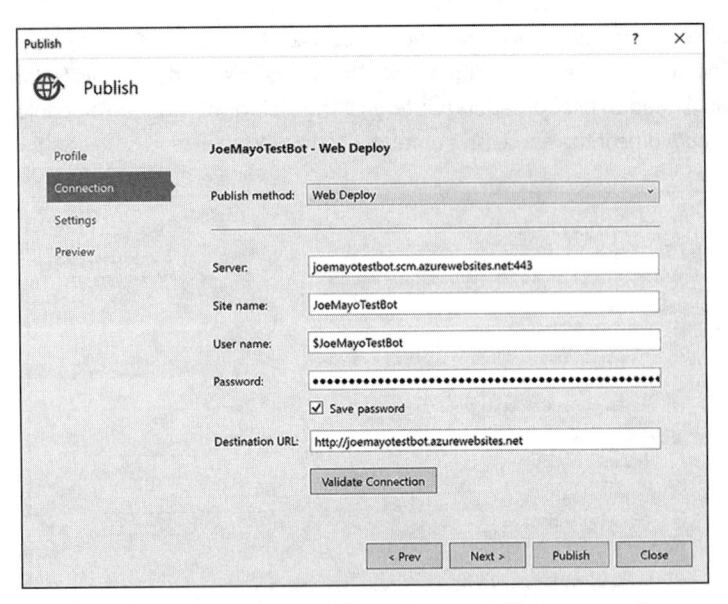

FIGURE 2-15 The Publish dialog performs the actual deployment to Azure.

The publishing wizard deploys the chatbot and opens a new browser window, shown in Figure 2-16, addressing the chatbot on Windows Azure.

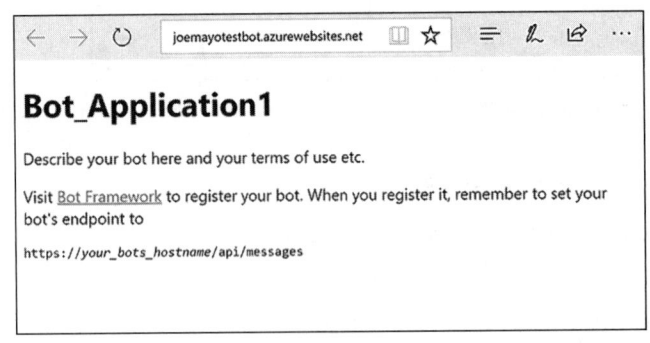

FIGURE 2-16 When publishing a chatbot to Azure, the wizard concludes by opening a browser addressing the chatbot.

At this point, you can't test the chatbot yet because the Bot Connector doesn't know it exists. Take note of the address in the browser window because you'll use that in the next section to register the chatbot.

> **Tip** Azure has paid plans that allow you to assign your own domain name to an App Service. They also have free plans and you can use the address a chatbot is deployed to as a link to the web page for a chatbot – like *http://joemayotestbot.azurewebsites.net* shown in Figure 2-16.

Registering a Chatbot

You need to register with the Bot Framework website so that the Bot Connector can find your chatbot. During this process, you'll also receive credentials to enable secure communications between your chatbot and the Bot Connector.

To register a chatbot, visit *https://dev.botframework.com/bots/new*, sign in, and you'll see a page similar to Figure 2-17.

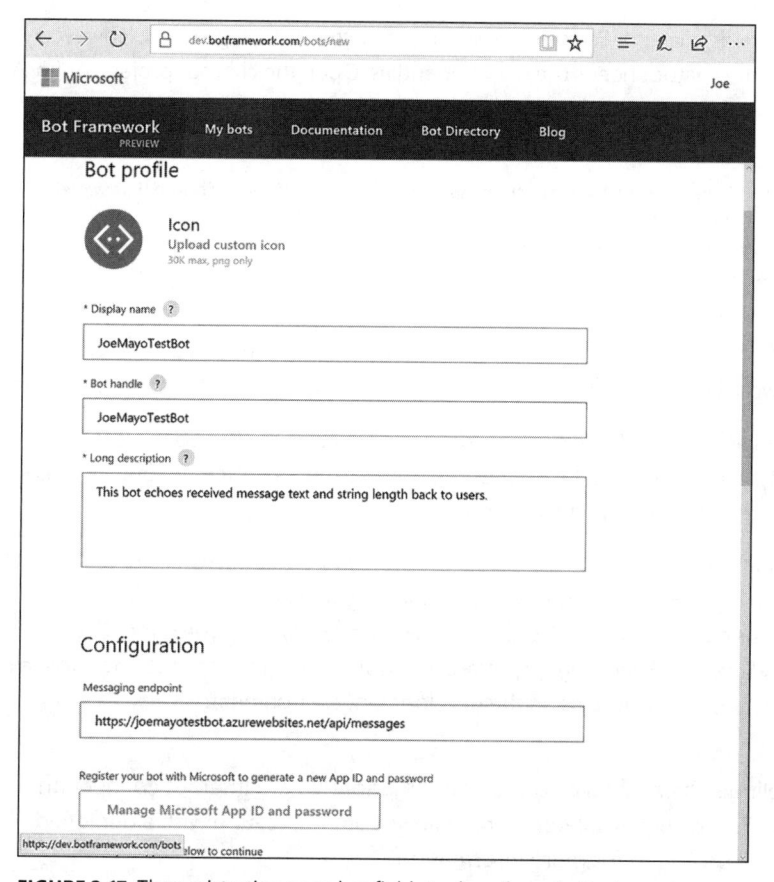

FIGURE 2-17 The registration page has fields to describe a chatbot.

At the Bot Profile page, fill out the form and click any question marks for guidance on fields. A couple of interesting fields are the *Messaging endpont* and the button under *Register your bot with Microsoft to generate a new App ID and password*.

CHAPTER 2 Setting Up a Project   **35**

The *Messaging endpoint* is the base address of your chatbot on Azure, as shown in the address bar in Figure 2-16, appended with *api/messages*. You must change the address protocol to *https*. Earlier sections of this chapter explained why *api/messages* is part of the address. For the example chatbot, the *Messaging Endpoint* is *https://joemayotestbot.azurewebsites.net/api/messages*.

Click the *Create Microsoft App ID and password* button and follow the instructions to generate a new App ID and password. Make sure you copy the App ID and password to add to the chatbot configuration.

> **Warning** When Microsoft recommends you copy the password, they're serious. After you complete generating the password and click OK, the password is then obfuscated so you can't read it. There is no way to change it or read it. In case you didn't copy your password and don't remember it, your only option is to re-generate a new set of credentials.

After filling in all the fields, click the *Register* button and your chatbot is registered. However, you aren't done yet because the chatbot needs the new credentials. Open the chatbot project, open *Web.config*, and add the credentials to their corresponding *appSettings* keys as follows:

```
<appSettings>
  <!-- update these with your BotId, Microsoft App Id and your Microsoft App Password-->
  <add key="BotId" value="JoeMayoTestBot" />
  <add key="MicrosoftAppId" value="<Your App ID>" />
  <add key="MicrosoftAppPassword" value="<Your App Password>" />
</appSettings>
```

BotId should be the value used in *Bot Handle* when registering the chatbot and the *MicrosoftAppId* and *MicrosoftAppPassword* keys hold the corresponding credentials from the chatbot registration.

After you set credentials, re-publish the chatbot. You can verify that the Bot Connector can communicate with the chatbot by re-visiting the Bot Framework site https://dev.botframework.com, clicking *My Bots*, clicking your chatbot, and using the *Test* tab on your chatbot page.

Before testing the chatbot deployed to Azure, you need one more step to install ngrok. Visit *https://ngrok.com/*, download, unzip, and copy ngrok.exe to a folder of your choice. I put ngrok in the *C:\ngrok* folder. Following the steps demonstrated in Figure 2-18, click **More | App Settings** menu, enter the path to ngrok, and click *Save*. The *Bypass ngrok for local addresses* causes the emulator to avoid extra network traffic when the address is *localhost* in the development environment.

> **Tip** ngrok is tunneling software, which is useful in situations where Internet connection ports are restricted. For the Bot Emulator, ngrok allows communication with a deployed chatbot. Visit https://ngrok.com for more information.

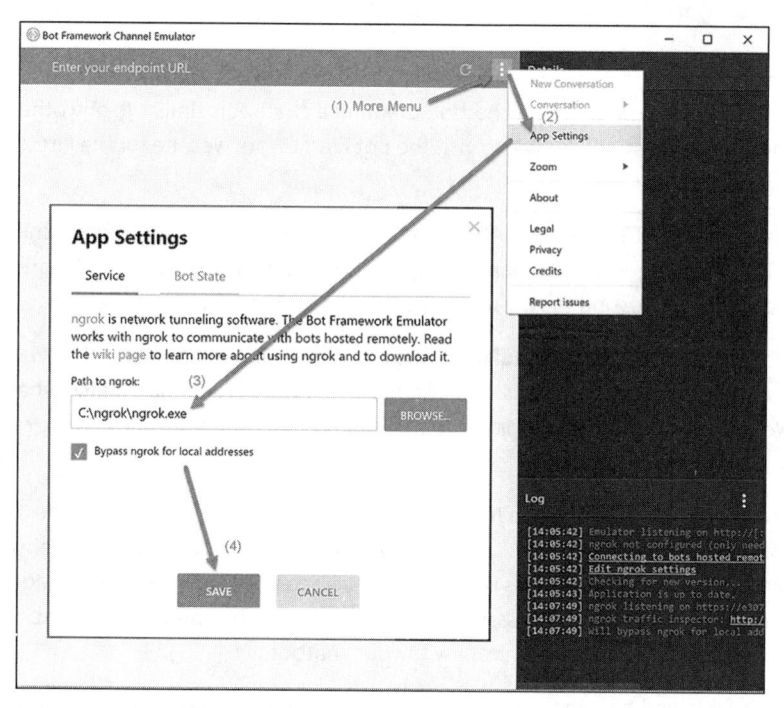

FIGURE 2-18 Installing ngrok for remote Azure testing.

To test your chatbot on Azure, open the Bot Framework Channel Emulator. Replace the *Bot Url* with the Azure chatbot address, remembering the *https* protocol prefix, base address, and append the *api/messages* segment. Instead of leaving the credentials blank, fill in the credentials received from registering the chatbot, which you can probably locate in the *appSettings* section in *Web.config*. You should be able to communicate with your chatbot now.

If you type the address incorrectly, you'll see a red error message with an HTTP code 405, saying "Method Not Allowed". Besides looking for typing mistakes, verify the protocol is *https* and that you've appended *api/messages*.

> **Tip** Generally, it's good to publish a new chatbot, before writing any code. This validates the communication and credentials and eliminates bugs in code as a cause of problems. It also helps you reserve a meaningful handle early in the process.

CHAPTER 2 Setting Up a Project 37

Summary

You now know how to create, test, and deploy a chatbot. Creating a chatbot is similar to any other type of new .NET application project creation. Be sure to visit the Bot Framework website for the latest Bot Application template.

A chatbot is essentially an ASP.NET MVC Web API project, where the Bot Connector communicates with the chatbot via HTTP POST requests. You learned that the Bot Connector sends activities and the chatbot receives the activities, processing the message activities as requests from the user.

To test a chatbot, you must configure the emulator, which relies on addressing the chatbot via base address, port, and web API segment conventions. You also need to ensure credentials match what is in the chatbot project Web.config file. The emulator has a chat interface, allowing messages to be sent and received from the chatbot.

Before configuring channels for others to use the chatbot, you must publish and register the chatbot. This chapter showed how to publish to Azure. You need to visit the Bot Framework website to register the chatbot. Remember that the address is to the deployed chatbot with the api/messages segments appended. During the registration process, you obtain credentials to configure the chatbot with. These steps allow the Bot Connector to communicate with your chatbot.

Although this chapter explained how to publish a chatbot, we won't set up any channels until a later chapter. For now, there are additional skills you need and the next chapter starts that off with a real chatbot and explanation of Bot Framework communications essentials.

CHAPTER 3

Building Conversations: The Essentials

At its core, a chatbot is a conversational user interface (CUI). This is different from traditional graphical user interface (GUI) development in that the primary interface for a chatbot is text. The elements of the interface aren't a page layout and styles as much as they are about users and the conversations that users and chatbots engage in. This approach shifts the concept of an interface from graphical to conversational and this chapter focuses on CUI essentials.

While covering conversation essentials, this chapter explains how to track users and conversations. You'll learn how to manage conversation state with the Bot State Service and how to read and respond to user messages. You'll also learn about activities, which are any type of information that flows from and to your chatbot.

The Rock, Paper, Scissors Game Bot

The demo chatbot for this chapter is a game called Rock, Paper, Scissors. The game is based on a two person interaction where both players count one, two, three and, on the count of three, show their hand in a gesture of a fist (Rock), flat (Paper), or two fingers (Scissors). The winner is either a rock smashes scissors, paper covers rock, or scissors cut paper. A tie occurs when both players display the same gesture. The Rock, Paper, Scissors chatbot (refered to as just chatbot in following paragraphs) plays with the user via text. Listing 3-1 though 3-3 shows the entire program. This is the RockPaperScissors1 project in the accompanying source code.

The PlayType Enum

The chatbot has a *PlayType* enum, a *Game* class, and logic in the *MessagesController* class that manages a game and responds to the user. Listing 3-1 though 3-3 shows each of these types, starting with Listing 3-1, showing the *PlayType* enum.

LISTING 3-1 The Rock, Paper, Scissors Game - *PlayType* Enum

```
namespace RockPaperScissors1.Models
{
    public enum PlayType
    {
        Rock, Paper, Scissors
    }
}
```

PlayType has a value for each of the choices a user or chatbot can make. The program uses *PlayType* to represent a user's play, a chatbot's play, and as part of the logic to determine who won.

The Game Class

The *Game* class, shown in Listing 3-2, has game result state and methods for managing a single play.

LISTING 3-2 The Rock, Paper, Scissors Game – *Game* Class

```
using System;
using System.Collections.Generic;

namespace RockPaperScissors1.Models
{
    public class Game
    {
        readonly Dictionary<PlayType, string> rockPlays =
            new Dictionary<PlayType, string>
            {
                [PlayType.Paper] = "Paper covers rock - You lose!",
                [PlayType.Scissors] = "Rock crushes scissors - You win!"
            };
        readonly Dictionary<PlayType, string> paperPlays =
            new Dictionary<PlayType, string>
            {
                [PlayType.Rock] = "Paper covers rock - You win!",
                [PlayType.Scissors] = "Scissors cuts paper - You lose!"
            };
        readonly Dictionary<PlayType, string> scissorsPlays =
            new Dictionary<PlayType, string>
            {
                [PlayType.Rock] = "Rock crushes scissors - You lose!",
                [PlayType.Paper] = "Scissors cut paper - You win!"
            };

        public string Play(string userText)
        {
            string message = "";

            PlayType userPlay;
            bool isValidPlay = Enum.TryParse(
```

```csharp
                userText, ignoreCase: true, result: out userPlay);

    if (isValidPlay)
    {
        PlayType botPlay = GetBotPlay();
        message = Compare(userPlay, botPlay);
    }
    else
    {
        message = "Type \"Rock\", \"Paper\", or \"Scissors\" to play.";
    }

    return message;
}

public PlayType GetBotPlay()
{
    long seed = DateTime.Now.Ticks;
    var rnd = new Random(unchecked((int) seed) );
    int position = rnd.Next(maxValue: 3);

    return (PlayType) position;
}

public string Compare(PlayType userPlay, PlayType botPlay)
{
    string plays = $"You: {userPlay}, Bot: {botPlay}";
    string result = "";

    if (userPlay == botPlay)
        result = "Tie.";
    else
        switch (userPlay)
        {
            case PlayType.Rock:
                result = rockPlays[botPlay];
                break;
            case PlayType.Paper:
                result = paperPlays[botPlay];
                break;
            case PlayType.Scissors:
                result = scissorsPlays[botPlay];
                break;
        }

    return $"{plays}. {result}";
    }
  }
}
```

In the *Game* class, the *Dictionary<PlayType, string>* dictionaries hold the rules for each play. The *Play* accepts a *PlayType* enum for both the user and the chatbot and compares them to see who won. *GetBotPlay* implements the chatbot's choice, and *Compare* matches the user's choice with the chatbot's choice to indicate who the winner is. Here's the implementation of *GetBotPlay*:

```csharp
public PlayType GetBotPlay()
{
    long seed = DateTime.Now.Ticks;
    var rnd = new Random(unchecked((int) seed) );
    int position = rnd.Next(maxValue: 3);

    return (PlayType) position;
}
```

GetBotPlay shows how the chatbot determines its next play. The .NET *Random* class generates a pseudo-random number, within the range of the underlying values of the *PlayType* enum, then converts that number to a *PlayType* enum to return to the caller.

The *Compare* method, repeated below, uses the *Game* class dictionaries to determine who won:

```csharp
public string Compare(PlayType userPlay, PlayType botPlay)
{
    string plays = $"You: {userPlay}, Bot: {botPlay}";
    string result = "";

    if (userPlay == botPlay)
        result = "Tie.";
    else
        switch (userPlay)
        {
            case PlayType.Rock:
                result = rockPlays[botPlay];
                break;
            case PlayType.Paper:
                result = paperPlays[botPlay];
                break;
            case PlayType.Scissors:
                result = scissorsPlays[botPlay];
                break;
        }

    return $"{plays}. {result}";
}
```

The first task of *Compare* is to determine if both the user and chatbot played the same choices. If so, the game is a tie. Otherwise, it uses the *switch* statement to see who won. In this case, the *userPlay* determines which dictionary to use and the *botPlay* represents the member of that dictionary. The chosen dictionary represents the user's choice and the index into the dictionary represents the chatbot's choice. The result is the indexed value from the chosen dictionary.

The *Play* method, repeated below, pulls all of this together by obtaining the user's *PlayType*, then the chatbot's *PlayType*, and calling *Compare* to learn who won, shown below:

```
public string Play(string userText)
{
    string message = "";

    PlayType userPlay;
    bool isValidPlay = Enum.TryParse(
        userText, ignoreCase: true, result: out userPlay);

    if (isValidPlay)
    {
        PlayType botPlay = GetBotPlay();
        message = Compare(userPlay, botPlay);
    }
    else
    {
        message = "Type \"Rock\", \"Paper\", or \"Scissors\" to play.";
    }

    return message;
}
```

The *Play* method uses *TryParse* because the user could potentially type in anything, possibly not matching a value in the *PlayType* enum. If the user input is valid, *Play* calls *GetBotPlay* to obtain a random *PlayType* for the chatbot, as discussed previously. Then *Play* calls *Compare* with both plays and returns the resulting message to the caller.

When the *userText* doesn't match a value in the *PlayList* enum, that means the chatbot can't understand what the user wants. In that case, the *else* clause prints out a help message to the user to let them know what the proper plays are.

Note Designing conversations is sometimes like designing algorithms. If you write your code to be correct for the happy path, then there won't be a problem. However, when user input differs from what you expect, you'll encounter errors and exceptions. Just like any other input, conversation text can be something you didn't expect. In the PaperRockScissors program, the code expects a response that matches the *PlayType* enum. The code is also written to handle the situation where user input is something that isn't expected and responds with a simple help message. While the PaperRockScissors program is simple, this situation highlights the type of thinking that you need to do when designing your own chatbot conversations. e.g. Would it be acceptable to attempt to interpret misspellings or abbreviations? Throughout this book, you learn different techniques for managing conversation flow. Because the conversation is the user interface, you should keep the concept of stray input on your mind while designing the chatbot conversation.

The MessagesController Class

This program delegates all of its logic into a single class, *Game*. The *Post* method, from the *MessagesController* class, uses *Game* to make a play and return the result to the user, shown in Listing 3-3.

LISTING 3-3 The Rock, Paper, Scissors Game – *MessagesController* Class

```
using System;
using System.Linq;
using System.Net;
using System.Net.Http;
using System.Threading.Tasks;
using System.Web.Http;
using Microsoft.Bot.Connector;
using RockPaperScissors1.Models;

namespace RockPaperScissors1
{
    [BotAuthentication]
    public class MessagesController : ApiController
    {
        public async Task<HttpResponseMessage> Post([FromBody]Activity activity)
        {
            if (activity.Type == ActivityTypes.Message)
            {
                var connector = new ConnectorClient(new Uri(activity.ServiceUrl));

                var game = new Game();

                string message = game.Play(activity.Text);

                Activity reply = activity.CreateReply(message);
                await connector.Conversations.ReplyToActivityAsync(reply);
            }

            HttpResponseMessage response = Request.CreateResponse(HttpStatusCode.OK);
            return response;
        }
    }
}
```

The *MessagesController* class has the *Post* method, described in Chapter 2, which is the HTTP endpoint that the Bot Connector calls when the user sends a message from the chat window. After instantiating the *ConnectorClient*, the code instantiates a *Game* instance, calls *Play* with the user's message, *activity.Text*, and responds to the user with the string returned from *Play*.

In practice, you don't want *Post* to contain any business logic, but rather act as a controller for accepting user response, passing parameters to the business logic for processing, and responding to the user. This helps manage separation of concerns between the various layers of your application.

If you run the RockPaperScissors application and run the emulator, addressing the chatbot endpoint, as described in Chapter 2, you should see a similar experience to Figure 3-1. Remember that the port number specified in the emulator must match the port number from the address of the chatbot, which you can find in the browser address bar, such as how. the port number in Figure 3.1 is 3979. You also need the *api/messages* suffix to properly identify the Web API endpoint where the chatbot resides.

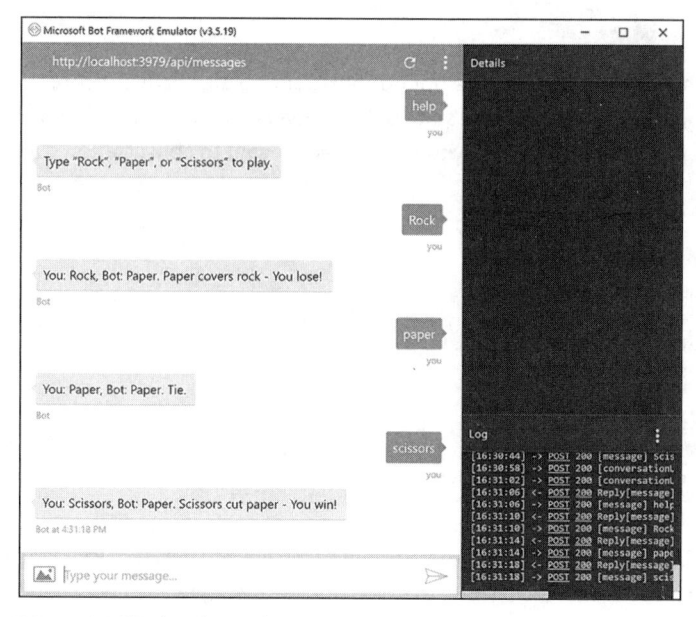

FIGURE 3-1 Playing the Rock, Paper, Scissors Game.

Figure 3-1 shows that typing a case insensitive member of the *PlayType* enum results in a play and anything else results in a help message. In the next section, you learn more about the participants in the conversation and ways to work with conversation state.

Conversation State Management

In the Bot Framework, you can track both users and conversations. This section explains how to read user and conversation data. There's also a Bot Connector service that allows you to save user and conversation state, which is covered here too.

Elements of a Conversation

Conversations have identifiers for users, chatbots, and the conversation between the user and chatbot. The following sections discuss this and show how to access these values with code.

The Conversation

A conversation is a set of request and response message activities between a user and a chatbot. You can track a conversation by a conversation identifier associated with a message. The first message in a conversation is always initiated by a user, when they invite the chatbot into their channel and send their

CHAPTER 3   Building Conversations: The Essentials   **45**

first message. That message arrives with a conversation identifier. In Chapter 4, you' learn how a chatbot can initiate a conversation, but that can only happen if the chatbot knows which user it wants to communicate with, which can only happen if the user has previously communicated with the chatbot.

Activity State

The basic element of a conversation is an *Activity*. The Bot Connector populates *Activity* properties to make it convenient for a chatbot to reason about the state of the conversation on a single interaction with the user. Here's an excerpt of the *Activity* class, with properties relevant to conversation state:

```
public partial class Activity
{
    public string Type { get; set; }
    public string Id { get; set; }
    public DateTime? Timestamp { get; set; }
    public DateTimeOffset? LocalTimestamp { get; set; }
    public string ServiceUrl { get; set; }
    public string ChannelId { get; set; }
    public ChannelAccount From { get; set; }
    public ConversationAccount Conversation { get; set; }
    public ChannelAccount Recipient { get; set; }
}
```

As shown in Listing 3-3, the *Type* property indicates the purpose of the *Activity*, which checked for *ActivityType.Message* in that case. Later sections of this chapter examine other *ActivityType* members. *ServiceUrl* is useful because you don't need to remember the URL of the Bot Connector, which might not always be the same. The *Id* property is useful for logging the identity of that particular *Activity* or help with debugging and you'll learn how important it is in building replies to a message. *Timestamp* is the UTC time of the message, but *LocalTimestamp* is the local time of the user. These might help approximate the user's timezone in addition to other logic that might be useful for a chatbot. The *ChannelId* tells you which channel the user is communicating from. The next section explains *From* and *Recipient* properties.

User and Chatbot Identities

To personalize chatbots, you need to identify the user, which is what the *From* property allows. Similarly the *Recipient* property identifies the chatbot. There might be future scenarios, like support for group conversations, but this chapter assumes a single conversation between chatbot and user, meaning that the *Recipient* is the chatbot. Both the *From* and *Recipient* properties are type *ChannelAccount*, shown here:

```
public partial class ChannelAccount
{
    public string Id { get; set; }
    public string Name { get; set; }
}
```

> **Note** This section describes *Recipient* as the chatbot and *From* as the user. However, in the Building a Custom Message Activity section of this chapter, you'll see how the *From* becomes the chatbot and the *Recipient* becomes the user when the chatbot needs to send a response to the user.

The channel that the user is communicating on assigns the *Id* property, which uniquely identifies the user. The channel also populates the *Name* property, which you might want to use to address the user by their name. Many messaging clients allow users to change their name, so the *Name* property can change, while the *Id* property is constant for that channel.

> **Note** The *From.Id* property is unique to a channel, identified by *Activity.ChannelId*. If you publish to different channels, it's possible that the same user communicates with a chatbot over different channels. In that case, it's the same user, but *ActivityChannelId* and *From.Id* are different.

Now that you can find state associated with users and chatbots, the next section discusses the *Conversation* property.

Conversation Identity

All communication between users and chatbots occur in the scope of a conversation. The *Conversation* property, shown below, contains properties to represent details of the conversation:

```
public partial class ConversationAccount
{
    public bool? IsGroup { get; set; }
    public string Id { get; set; }
    public string Name { get; set; }
}
```

Each conversation has a unique *Id* and *Name*. The *IsGroup* property indicates that a channel supports groups. Only a limited number of channels support groups, so this doesn't always apply.

> **Note** The Bot Connector doesn't define the lifetime of a conversation. Rather, channels define when conversations start and end. A new conversation starts when a user first communicates with your chatbot, but when that conversation ends and a new conversation begins is undefined. Unless a channel specifically states what the rules are for a conversation beginning and endings, you shouldn't make any assumptions about conversation lifetime.

The user, chatbot, and conversation identification just discussed help you track information and state associated with conversations, and the next section discusses how the Bot Framework can help with this.

Saving and Retrieving State

One of the services of the Bot Connector is state management, called the Bot Framwork State Service. For Bot Framework purposes, state is any data or information you need to save to support the operation of your chatbot. You can save state for a user, a conversation, or a user in a conversation (aka private).

The Bot Connector state management service capacity is 32kb for either user, conversation, or private states. The largest amount of data you can have in user state is 32kb. Additionally, this state is scoped by channel. For example, you can have 32kb for user state on channel A, 32kb for user state on channel B, 32kb for conversation state on channel A. The following sections describe how to use the Bot Connector state management service.

> **Tip** The Bot Connector state management service isn't meant to be a generalized store of all user and conversation data. If you need more space for tracking user and conversation data, it's best to use your own database solution.

Caching with Multiple Chatbots in the Same Project

There are times when you might have solutions with more than one chatbot project. This can confound testing because the default setup with the Visual Studio development environment and IIS Express uses aggressive web page caching. The result is that once a chatbot is running, it's cached by the server. The problem occurs when running the second chatbot because the first is cached and you never see the second. There might be multiple fixes to this situation such as disabling browser link, closing IIS Express applications, and even clearing browser cache. However, a quick fix is to change the port number for the chatbot's URL by double-clicking the project's *Properties* folder, going to the *Web* tab, and changing the port number in the *Project URL* box, as shown in Figure 3-2.

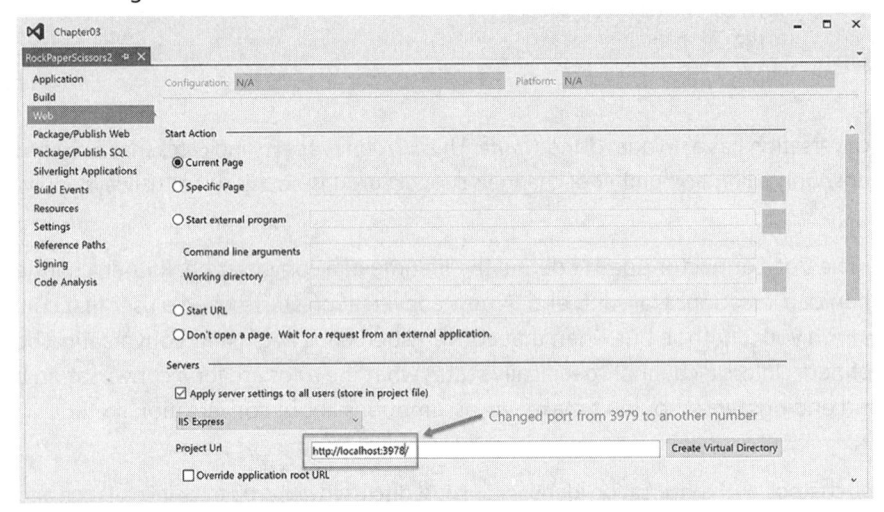

FIGURE 3-2 Change the port number of the *Project URL* to avoid page caching problems when testing multiple chatbots in the same project.

In Figure 3-2, you can see that the port for the *Project URL* is changed from the default *3979* to *3978*. The reason this works is because the web page caching strategy affecting this is based on the page URL. Changing the port number changes the URL and ensures the browser displays the proper page.

Bot State Service Capabilities and Organization

As mentioned earlier, the Bot State Service allows you to manage state for either a single user, a conversation, or private (user in a conversation). The difference between user and private is that, for a single user, the user state is maintained across all conversations, whereas private is user data in only that one conversation, but not in other conversations. Conversation state holds data for only one conversation and is available for all users in that conversation.

You'll see code for how to use this service in more detail in the Using the Bot State Service section of this chapter, but first you'll see an overview of the types required to make this happen and their relationships. The first type you'll need is an instance of the Bot Framework's *StateClient*, which implements *IStateClient*, shown below:

```
public partial interface IStateClient : IDisposable
{
    Uri BaseUri { get; set; }
    JsonSerializerSettings SerializationSettings { get; }
    JsonSerializerSettings DeserializationSettings { get; }
    ServiceClientCredentials Credentials { get; }
    IBotState BotState { get; }
}
```

The *BaseUri* and *Credentials* are the address and username/password, respectively, for communicating with the Bot State Service. Fortunately, you won't have to set up these details because the *Activity* type has a convenience method that you'll learn about in Responding to Conversations section of this chapter. The Bot Framework uses the popular open-source Json.NET package for managing data payloads, so the *SerializationSettings* and *DeserializationSettings* are types belonging to Json.NET, which is where you can find guidance on how to configure those. Generally, using default serialization as you'll see in this chapter, is sufficient for serialization and you won't need to work with the *StateClient* serialization settings yourself.

When working with state, you always use the *BotState*, which allows you to choose between user, conversation, and private state. *BotState* implements *IBotState*, shown below:

```
public partial interface IBotState
{
    public static async Task<BotData> GetUserDataAsync(
        this IBotState operations, string channelId, string userId,
        CancellationToken cancellationToken = default(CancellationToken));

    public static async Task<BotData> SetUserDataAsync(
        this IBotState operations, string channelId, string userId, BotData botData,
        CancellationToken cancellationToken = default(CancellationToken));

    public static async Task<string[]> DeleteStateForUserAsync(
        this IBotState operations, string channelId, string userId,
        CancellationToken cancellationToken = default(CancellationToken));

    public static async Task<BotData> GetConversationDataAsync(
        this IBotState operations, string channelId, string conversationId,
        CancellationToken cancellationToken = default(CancellationToken));
```

```
public static async Task<BotData> SetConversationDataAsync(
    this IBotState operations, string channelId,
    string conversationId, BotData botData,
    CancellationToken cancellationToken = default(CancellationToken));

public static async Task<BotData> GetPrivateConversationDataAsync(
    this IBotState operations, string channelId,
    string conversationId, string userId,
    CancellationToken cancellationToken = default(CancellationToken));

public static async Task<BotData> SetPrivateConversationDataAsync(
    this IBotState operations, string channelId,
    string conversationId, string userId, BotData botData,
    CancellationToken cancellationToken = default(CancellationToken));
}
```

The *IBotState* shows set and get methods for user, conversation, and private data. The *DeleteStateForUserAsync* method deletes all user data across all conversations for the specified channel. You might have noticed that each of the methods shown in the *IBotState* interface are async.

There are also non-async methods that mirror each of the async methods. The non-async methods use the same implementation as async methods, so either is technically correct. Our preference of the two is async because we believe it hints at the fact that there is an out-of-process Internet call happening, which might help with design and maintenance for performance and scalability.

> **Note** The *IBotState* code in this chapter shows methods that you won't see on the *IBotState* in the Bot Framework source code. The physical implementation is that these methods are really extension methods of the *BotStateExtensions* class that extend *IBotState*. In this chapter, we depict the abstraction, rather than the more complex implementation, which we believe is more practical.

The response type from *DeleteStateForUserAsync* is *string[]*, which would hold any messages that the server responded with. It the server doesn't return response messages, the return values is *null*. For all *IBotState* methods, if the server returns an error, the method throws an *HttpOperationException* with *Request*, *Response*, and *Body* properties for code to examine and handle as it sees fit.

Besides *DeleteStateForUserAsync*, all other *IBotState* methods return a *BotData*, shown below.

```
public partial class BotData
{
    public string ETag { get; set; }
    public object Data { get; set; }
}
```

The Bot State Service uses *ETag* for optimistic concurrency. If a first chatbot reads, changes, and writes data back to the state service, while a second chatbot reads that same data after the first, but writes its changes before the first chatbot, the first chatbot's write throws an *HttpOperationException*. This indicates to the first chatbot that it's operating on stale data.

Remembering that the *BotData* instance came from a method call to get state in either user, conversation, or private context, the *Data* property holds the JSON-formatted data for the requested context. *BotData* also has *GetProperty*, *SetProperty*, and *RemoveProperty* methods, which do exactly what their names imply. The next section shows you how to use the types and methods just presented to manage chatbot state.

Using the Bot State Service

The examples in this section, showing how to work with the Bot Framework State Service, build on the RockPaperScissors program from Listings 3-1 to 3-3. In this case, you'll see a new class that sets and reads state in addition to modifications to the *Post* method in Listing 3-3. This program is in the RockPaperScissors2 project of the accompanying source code.

The goal is to use the Bot State Service to hold scores. The scores show the tallies of the 10 most recent plays with how many user wins, how many chatbot wins, and how many ties. Listing 3-4 shows the *GameState* class for managing state and Listing 3-5 (shown later in this chapter) presents the new *Post* method for handling score requests and keeping the score updated.

LISTING 3-4 The Rock, Paper, Scissors Game – *GameState* Class

```csharp
using System;
using System.Collections.Generic;
using System.Linq;
using System.Threading.Tasks;
using Microsoft.Bot.Connector;

namespace RockPaperScissors2.Models
{
    public class GameState
    {
        [Serializable]
        class PlayScore
        {
            public DateTime Date { get; set; } = DateTime.Now;
            public bool UserWin { get; set; }
        }

        public async Task<string> GetScoresAsync(Activity activity)
        {
            using (StateClient stateClient = activity.GetStateClient())
            {
                IBotState chatbotState = stateClient.BotState;
                BotData chatbotData = await chatbotState.GetUserDataAsync(
                    activity.ChannelId, activity.From.Id);

                Queue<PlayScore> scoreQueue =
                    chatbotData.GetProperty<Queue<PlayScore>>(property: "scores");

                if (scoreQueue == null)
                    return "Try typing Rock, Paper, or Scissors to play first.";
```

```csharp
            int plays = scoreQueue.Count;
            int userWins = scoreQueue.Where(q => q.UserWin).Count();
            int chatbotWins = scoreQueue.Where(q => !q.UserWin).Count();

            int ties = chatbotData.GetProperty<int>(property: "ties");

            return $"Out of the last {plays} contests, " +
                   $"you scored {userWins} and " +
                   $"Chatbot scored {chatbotWins}. " +
                   $"You've also had {ties} ties since playing.";
        }
    }

    public async Task UpdateScoresAsync(Activity activity, bool userWin)
    {
        using (StateClient stateClient = activity.GetStateClient())
        {
            IBotState chatbotState = stateClient.BotState;
            BotData chatbotData = await chatbotState.GetUserDataAsync(
                activity.ChannelId, activity.From.Id);

            Queue<PlayScore> scoreQueue =
                chatbotData.GetProperty<Queue<PlayScore>>(property: "scores");

            if (scoreQueue == null)
                scoreQueue = new Queue<PlayScore>();

            if (scoreQueue.Count >= 10)
                scoreQueue.Dequeue();

            scoreQueue.Enqueue(new PlayScore { UserWin = userWin });

            chatbotData.SetProperty<Queue<PlayScore>>(property: "scores", data: scoreQueue);
            await chatbotState.SetUserDataAsync(activity.ChannelId, activity.From.Id, chatbotData);
        }
    }

    public async Task<string> DeleteScoresAsync(Activity activity)
    {
        using (StateClient stateClient = activity.GetStateClient())
        {
            IBotState chatbotState = stateClient.BotState;

            await chatbotState.DeleteStateForUserAsync(activity.ChannelId, activity.From.Id);

            return "All scores deleted.";
        }
    }

    public async Task AddTieAsync(Activity activity)
    {
        using (StateClient stateClient = activity.GetStateClient())
        {
```

```
            IBotState chatbotState = stateClient.BotState;
            BotData chatbotData = await chatbotState.GetUserDataAsync(
                activity.ChannelId, activity.From.Id);

            int ties = chatbotData.GetProperty<int>(property: "ties");

            chatbotData.SetProperty<int>(property: "ties", data: ++ties);

            await chatbotState.SetUserDataAsync(activity.ChannelId, activity.From.Id, chatbotData);
        }
    }
}
```

The *GameState* class has a nested *PlayScore* class that holds the result of a play, where either the user or chatbot won, shown here:

```
[Serializable]
class PlayScore
{
    public DateTime Date { get; set; } = DateTime.Now;
    public bool UserWin { get; set; }
}
```

The *Date* tracks when the play occurred and a bool *UserWin* property, indicating if the user won that game. Notice the *Serializable* attribute decorating *PlayScore*. Because the chatbot communicates over the Internet, all types must be serializable, allowing them to be formatted properly for transmission.

The *GetScoresAsync* method, shown below, reads scores from the Bot State Service. It uses the types and methods you learned about in the previous section.

```
public async Task<string> GetScoresAsync(Activity activity)
{
    using (StateClient stateClient = activity.GetStateClient())
    {
        IBotState chatbotState = stateClient.BotState;
        BotData chatbotData = await chatbotState.GetUserDataAsync(
            activity.ChannelId, activity.From.Id);

        Queue<PlayScore> scoreQueue =
            chatbotData.GetProperty<Queue<PlayScore>>(property: "scores");

        if (scoreQueue == null)
            return "Try typing Rock, Paper, or Scissors to play first.";

        int plays = scoreQueue.Count;
        int userWins = scoreQueue.Where(q => q.UserWin).Count();
        int chatbotWins = scoreQueue.Where(q => !q.UserWin).Count();

        int ties = chatbotData.GetProperty<int>(property: "ties");
```

```
            return $"Out of the last {plays} contests, " +
                $"you scored {userWins} and " +
                $"Chatbot scored {chatbotWins}. " +
                $"You've also had {ties} ties since playing.";
        }
    }
```

If you recall from the previous section, the *StateClient* implements *IStateClient*. The *using* statement in the code above shows how to create a *StateClient* instance, using the *CreateStateClient* factory method of the *Activity* instance parameter, *activity*.

The *chatbotState* variable holds the *IBotState* from *stateClient*. From *chatbotState*, the code can get a *BotData* instance for either the user, conversation, or private context, but calls *GetUserDataAsync* for user context. Also, observe that the arguments indicate which channel this state is associated, *activity.ChannelId*, with and the user's id, *activity.From.Id*.

At this point, the code has an instance of *BotData*, *chatbotData*. This is what you will always do in preparing to get, set, or remove data. You'll see the exact same pattern in all of the *GameState* methods.

The *Queue* of *PlayScore* helps manage the top 10 scores. The code calls *chatbotData.GeProperty* to get a reference to the *scores* property. The first time a user plays, the *Data* property of *BotData* is *null* because there isn't any data. The code could check that too, but because *Data* can have more than one property, and it does in this program, that wouldn't indicate whether the *Queue* was missing. The *Queue* collection is serializable, so it works fine as a state service property.

> **Tip** By design, the example in this chapter only keeps track of the last 10 plays. While this might not be the exact strategy everyone would use, it demonstrates one way to manage resources. Remember, there are resource considerations, like storage space and bandwidth, leading to the intentional choices made in the Rock, Paper, Scissors program.

The next statements collect number of plays, userWins, and chatbotWins, where a *chatbotWin* is defined as the user not winning. One thing you might be wondering is how the program keeps track of ties. That's handled in the *chatbotData.GetProperty* call for the *ties* property, which is the second property, besides *scores*, in *chatbotData*. After that, we return a message to the user. Before examining the code that calls *GetScoresAsync*, let's look at the other *GameState* methods.

You've seen how to read state and the *UpdateScoresAsync* shows how to write state. Essentially, it gets the scores *Queue*, updates the *Queue* with a new score, and writes the *Queue* back to the Bot State Service, as shown below.

```
        public async Task UpdateScoresAsync(Activity activity, bool userWin)
        {
            using (StateClient stateClient = activity.GetStateClient())
            {
                IBotState chatbotState = stateClient.BotState;
                BotData chatbotData = await chatbotState.GetUserDataAsync(
```

```
                    activity.ChannelId, activity.From.Id);

            Queue<PlayScore> scoreQueue =
                chatbotData.GetProperty<Queue<PlayScore>>(property: "scores");

            if (scoreQueue == null)
                scoreQueue = new Queue<PlayScore>();

            if (scoreQueue.Count >= 10)
                scoreQueue.Dequeue();

            scoreQueue.Enqueue(new PlayScore { UserWin = userWin });

            chatbotData.SetProperty<Queue<PlayScore>>(property: "scores", data: scoreQueue);
            await chatbotState.SetUserDataAsync(activity.ChannelId, activity.From.Id,
chatbotData);
        }
    }
```

As mentioned in the *GetScoresAsync* walkthrough, the code to get *BotData* follows the same pattern. Similarly, this code calls *chatbotData.GetProperty* on the *scores* property. If the user hasn't ever played this game or deleted their data, as explained in a later part of this section, then the *scoreQueue* will be *null*, indicating that the code must instantiate a new *Queue* of *PlayScore*. Since the program only manages the 10 most recent scores, it calls *scoreQueue.Dequeue* to remove the oldest. Then it calls *scoreQueue.Enqueue* to add the new, most recent, score.

The program uses *chatbotData.SetProperty* to assign *scoreQueue* to the *scores* property in *BotData*. At this point, you could set multiple properties if it made sense for a chatbot. Calling *chatbotState.SetUserDataAsync* posts the property changes back to the Bot State Service for the user on the channel they're communicating on.

In addition to reading and writing, chatbots can delete state. The following *DeleteScoresAsync* Method, from Listing 3-4, shows how.

```
        public async Task<string> DeleteScoresAsync(Activity activity)
        {
            using (StateClient stateClient = activity.GetStateClient())
            {
                IBotState chatbotState = stateClient.BotState;

                await chatbotState.DeleteStateForUserAsync(activity.ChannelId, activity.From.
Id);

                return "All scores deleted.";
            }
        }
```

The *DeleteScoresAsync* method doesn't need *BotData* because it uses *chatbotState*, to call the *IBotState.DeleteStateForUserAsync*. This deletes all of the user's state in user, conversation, and private contexts.

> **Tip** The *IBotState.DeleteStateForUserAsync* method removes all properties from a user's data. If you needed a more surgical technique for removing only one property, get a reference to *BotData*, using the typical pattern shown in multiple code listings in this section, and call *chatbotData.RemoveProperty()*.

Most of the examples so far show how to work with complex objects, which is the *Queue* of *PlayScore* in this case. The only exception was the *GetScoresAsync* method, reading the ties property, which is a primitive type object. The *AddTieAsync* method, below, shows how to write a primitive type property:

```
public async Task AddTieAsync(Activity activity)
{
    using (StateClient stateClient = activity.GetStateClient())
    {
        IBotState chatbotState = stateClient.BotState;
        BotData chatbotData = await chatbotState.GetUserDataAsync(
            activity.ChannelId, activity.From.Id);

        int ties = chatbotData.GetProperty<int>(property: "ties");

        chatbotData.SetProperty<int>(property: "ties", data: ++ties);

        await chatbotState.SetUserDataAsync(activity.ChannelId, activity.From.Id, chatbotData);
    }
}
```

After the standard pattern for getting *BotData*, the method calls *chatbotData.GetProperty* to read the *ties* property. Then it calls *chatbotData.SetProperty*, also using the pre-increment operator on *ties* to increase the number of *ties* to its new value. Just as in the *UpdateScoresAsync* method, the *AddTieAsync* method calls *chatbotState.SetUserDataAsync* to post the new ties property value back to the Bot State Service.

Now you know how to use the Bot State Service, so the only thing left to do is tie it together to see how the user can interact with the chatbot to interact with scores. Figure 3-3 shows a user playing and using new **score** and **delete** commands. After playing a while, the user types **score** and receives a message indicating what the score is. Then the user types **delete** and receives a message that user information has been deleted. Finally, the user types **score** again and since the system doesn't have a score, it suggests that the user play to create a score first.

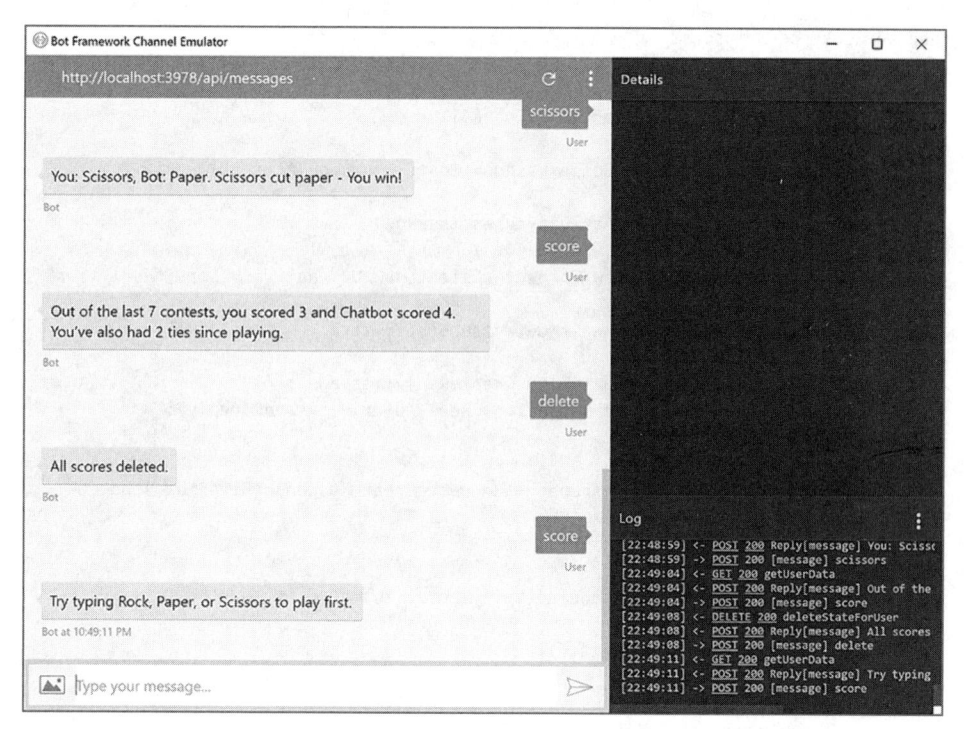

[Screenshot of Bot Framework Channel Emulator showing a Rock, Paper, Scissors conversation]

FIGURE 3-3 The score and delete commands show scores or removes scores, respectivly.

> **Warning** The Bot Emulator only stores state for the current session. If you shut down and restart the emulator, state is cleared. It will be as if you've starting from scratch, which you are. To work around this for testing state operations while working with code, leave the emulator running and only close the chatbot application (browser window). Then you can re-run your chatbot, which starts a new browser instance, and then begin using the emulator again, with the same state intact.

To make these commands work, the chatbot must read the user's message and execute the proper logic. Listing 3-5 shows the modifications to the *Post* method to make score and delete commands work.

LISTING 3-5 The Rock, Paper, Scissors Game – *MessagesController* Class With Score and Delete Commands

```
using System;
using System.Net;
using System.Net.Http;
using System.Threading.Tasks;
using System.Web.Http;
using Microsoft.Bot.Connector;
using RockPaperScissors2.Models;
```

CHAPTER 3 Building Conversations: The Essentials   57

```csharp
namespace RockPaperScissors2
{
    [BotAuthentication]
    public class MessagesController : ApiController
    {
        public async Task<HttpResponseMessage> Post([FromBody]Activity activity)
        {
            if (activity.Type == ActivityTypes.Message)
            {
                var connector = new ConnectorClient(new Uri(activity.ServiceUrl));

                string message = await GetMessage(activity);

                Activity reply = activity.CreateReply(message);
                await connector.Conversations.ReplyToActivityAsync(reply);
            }

            HttpResponseMessage response = Request.CreateResponse(HttpStatusCode.OK);
            return response;
        }

        async Task<string> GetMessage(Activity activity)
        {
            var state = new GameState();

            string userText = activity.Text.ToLower();
            string message = string.Empty;

            if (userText.Contains(value: "score"))
            {
                message = await state.GetScoresAsync(activity);
            }
            else if (userText.Contains(value: "delete"))
            {
                message = await state.DeleteScoresAsync(activity);
            }
            else
            {
                var game = new Game();
                message = game.Play(userText);

                bool isValidInput = !message.StartsWith("Type");
                if (isValidInput)
                {
                    if (message.Contains(value: "Tie"))
                    {
                        await state.AddTieAsync(activity);
                    }
                    else
                    {
                        bool userWin = message.Contains(value: "win");
                        await state.UpdateScoresAsync(activity, userWin);
                    }
                }
            }
```

```
            return message;
        }
    }
}
```

The first change in Listing 3-5 is that the logic is moved from *Post* to a *GetMessage* method. The *GameState* instance, state, is the same as the GameState class in Listing 3-4. The code first looks to see if the user typed score or delete and calls state.*GetScoresAsync* or *DeleteScoresAsync*, respectively. Otherwise, the program handles the user message as a game play, using the same *Game* class from Listing 3-2.

When the response starts with *Type*, that means the user entered something the chatbot doesn't recognize and the message is "Type \"Rock\", \"Paper\", or \"Scissors\" to play." If the game was a tie, the code calls state.*AddTieAsync* to increment the tie count. Otherwise, it determines if the user won and calls state.*UpdateScoresAsync* to record the new score.

That covers the Bot State Service. The next section goes deeper into the *Activity* type to help understand *Activity* internals.

Participating in Conversations

Previous sections of this chapter discussed ways of listening and interpreting the meaning of an *Activity*, but only briefly addressed the response, which was a *CreatReply* method that did all the work of building a response object. This section discusses the other side of the conversation, when a chatbot sends a message to the user. Here, we'll look deeper into a message *Activity*.

Responding to Conversations

When an *Activity* arrives at a chatbot, as the parameter to the *Post* method, the Bot Connector already constructed the details. Similarly, the Bot Framework supports the common case of preparing an *Activity* for a reply via convenience methods. To help you understand the nature of an *Activity*, this section breaks an *Activity* down and shows how it's manually created in case you need to perform your own customization of responses in the future.

Previous listings use the *CreateReply* convenience method, shown below, to prepare a reply message:

```
Activity reply = activity.CreateReply(message);
```

The *CreateReply* method also offers an optional *locale* parameter to let the user's client software know which language the chatbot is communicating in. The default locale is *en-US*:

```
Activity reply = activity.CreateReply(message , locale: "en-US");
```

Building a Custom Message Activity

While most of your work can use *CreateReply* to save time, you might have a need to build your own *Activity*. Listing 3-6 shows how it's done. This is from the RockPaperScissors3 project in the source code for this chapter.

LISTING 3-6 The Rock, Paper, Scissors Game – *BuildMessageActivity* Extension Method

```csharp
using Microsoft.Bot.Connector;

namespace RockPaperScissors3.Models
{
    public static class ActivityExtensions
    {
        public static Activity BuildMessageActivity(
            this Activity userActivity, string message, string locale = "en-US")
        {
            IMessageActivity replyActivity = new Activity(ActivityTypes.Message)
            {
                From = new ChannelAccount
                {
                    Id = userActivity.Recipient.Id,
                    Name = userActivity.Recipient.Name
                },
                Recipient = new ChannelAccount
                {
                    Id = userActivity.From.Id,
                    Name = userActivity.From.Name
                },
                Conversation = new ConversationAccount
                {
                    Id = userActivity.Conversation.Id,
                    Name = userActivity.Conversation.Name,
                    IsGroup = userActivity.Conversation.IsGroup
                },
                ReplyToId = userActivity.Id,
                Text = message,
                Locale = locale
            };

            return (Activity)replyActivity;
        }
    }
}
```

The *ActivityExtensions* class, in Listing 3-6 holds the *BuildMessageActivity* extension method for the *Activity* class. The *BuildMessageActivity* has a *message* and optional *locale* parameter. The *userActivity* is the *Activity* that this message is being built to reply to.

This example uses object initialization syntax to create a new *Activity* instance. The *Activity* class also has *CreateXxx* factory methods that produce activities of various types, where *Xxx* is the type, which would have been *Activity.CreateMessageActivity()* in this case. This is another way to do the same thing.

Because the message is built to reply to the original message, the *From ChannelAccount* values populate from the *userActivity.Recipient* values, which is the chatbot. Similarly, the *Recipient ChannelAccount* values populate from the *userActivity.From* values, which is the user who sent the original message. The code builds the *Activity* to come from the chatbot to the user.

The *Conversation* is the same conversation as the user's. *Text* is the message parameter and *Locale* is the locale parameter.

Notice the *ReplyToId* parameter, being set with the *userActivity.Id*. This indicates to the Bot Connector that this *Activity*, from the chatbot, is a reply to the user's *Activity*.

> **Warning** Forgetting to set the *ReplyToId* results in a *ValidationException* thrown for *activityId cannot be null*. That means the code must set the *ReplyToId* of the reply *Activity* with the value of the Id property of the user's *Activity*.

Using a Custom Message Activity

Now that you know how to create a custom *Activity* from scratch, you can see how it works. Listing 3-7 shows modifications to the *Post* method that calls *BuildMessageActivity*.

LISTING 3-7 The Rock, Paper, Scissors Game – *Post* Method Changes for *BuildMessageActivity*

```
using System;
using System.Net;
using System.Net.Http;
using System.Threading.Tasks;
using System.Web.Http;
using Microsoft.Bot.Connector;
using RockPaperScissors3.Models;

namespace RockPaperScissors3
{
    [BotAuthentication]
    public class MessagesController : ApiController
    {
        public async Task<HttpResponseMessage> Post([FromBody]Activity activity)
        {
            if (activity.Type == ActivityTypes.Message)
            {
                var connector = new ConnectorClient(new Uri(activity.ServiceUrl));

                string message = await GetMessage(connector, activity);

                Activity reply = activity.BuildMessageActivity(message);

                await connector.Conversations.ReplyToActivityAsync(reply);
            }

            HttpResponseMessage response = Request.CreateResponse(HttpStatusCode.OK);
```

```csharp
        return response;
    }

    async Task<string> GetMessage(ConnectorClient connector, Activity activity)
    {
        var state = new GameState();

        string userText = activity.Text.ToLower();
        string message = "";

        if (userText.Contains(value: "score"))
        {
            message = await state.GetScoresAsync(activity);
        }
        else if (userText.Contains(value: "delete"))
        {
            message = await state.DeleteScoresAsync(activity);
        }
        else
        {
            var game = new Game();
            message = game.Play(userText);

            bool isValidInput = !message.StartsWith("Type");
            if (isValidInput)
            {
                if (message.Contains(value: "Tie"))
                {
                    await state.AddTieAsync(activity);
                }
                else
                {
                    bool userWin = message.Contains(value: "win");
                    await state.UpdateScoresAsync(activity, userWin);
                }
            }
        }

        return message;
    }
}
```

The code in Listing 3-7 is equivalent to Listing 3-5 with one exception. In the *Post* method, the call to *activity.CreateReply(message)* is replaced with *activity.BuildMessageActivity(message)*.

Summary

This chapter introduced the Rock, Paper, Scissors chatbot. Different sections of this chapter modified that program to show various concepts and ways to manage conversations. The *Conversation*, *Activity*, and identity types are core elements of conversations and you saw their properties and relationships between them. You learned how to use the Bot State Service and how it supports user, conversation, and private state. The explanation covered the essential types and their relationships and the code showed how to work with complex and primitive state. Finally, you built a custom message activity, showing some of the internals of the *Activity* type.

The next chapter builds on this one by discussing Activities and other ways to participate in conversations. The difference will be in illustrating how the Bot Emulator supports debugging and testing chatbots.

CHAPTER 4

Fine-Tuning Your Chatbot

Most software development projects have common tasks for design, coding, and deployment. There are also a common set of tasks surrounding maintenance, quality, security, and user experience that are often not documented, but remain vital considerations for a healthy project. These are continuous tasks, so it's important to learn about ways to approach them early with the Bot Framework and carry the skillset through the rest of this book and beyond. We are pulling these additional tasks together under a general category of fine-tuning.

You've used the Bot Emulator in previous chapters to interact with chatbots, but that was a minimal introduction. This chapter does a deep dive of Bot Emulator features and shows how to perform testing from user to chatbot. This chapter is also a continuation of Chapter 3, "Building Conversations: The Essentials," where there's a detailed explanation of *Activities*, communications between a chatbot and the Bot Connector, and how the Bot Emulator facilitates testing Activities.

Reviewing Bot Emulator Details

In addition to having conversations, the Bot Emulator offers several other features for testing and verifying the operation of a chatbot. At first glance, you can see that the Bot Emulator has sections for *Connection Details*, *Conversation Display*, *Message Input*, *Details*, and *Log*. Figure 4-1 shows each section of the Bot Emulator.

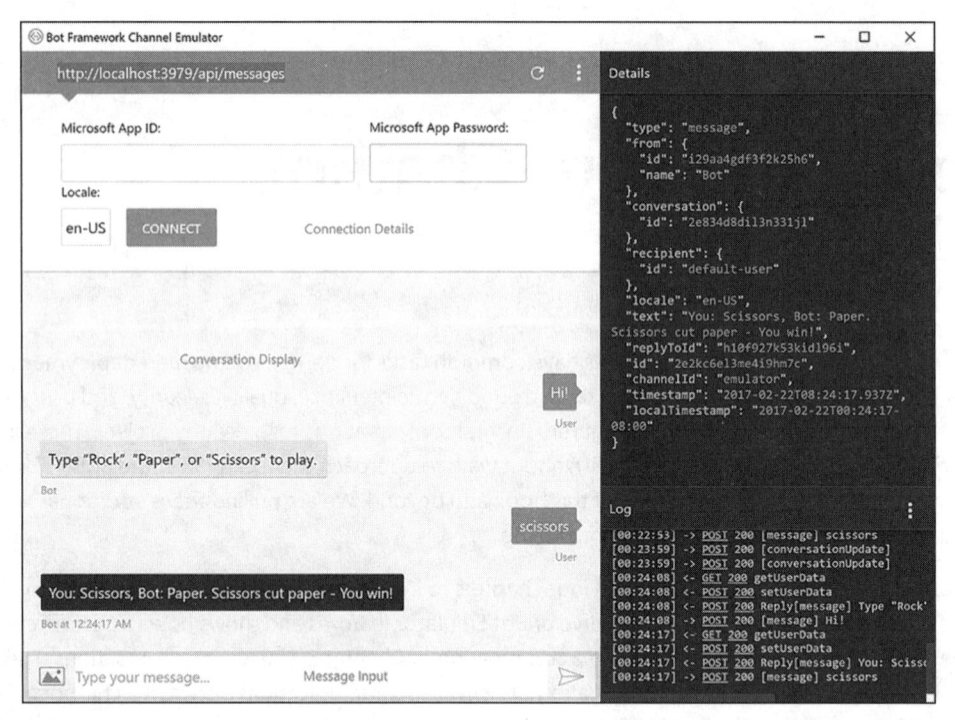

FIGURE 4-1 Sections of the Bot Emulator.

Clicking the address bar at the top left causes the Bot Emulator to display the *Connection Details* area. Chapter 2 explained how to obtain *Microsoft App ID* and *Microsoft App Password* from the Registration page on the Microsoft Bot Framework page. The *Locale* field, containing *en-US* lets you populate a custom locale for testing localization. In the code, you can read that value via the *Activity.Locale* field. Clicking *Connect* hides the *Connection Details* window and enables the *Message Input* box.

As shown several times in previous chapters, type any text into the *Message Input* field and either press **Enter** or click the arrow button on the right to send a message to the chatbot. Clicking the button on the left of the *Message Input*, with the picture icon, opens the File dialog so you can attach a file and sent it to the chatbot, which is covered in more detail in Chapter 10.

Sending a message to the chatbot also shows what you typed on the right side of the *Conversation Display*. Messages from the chatbot show on the left side of the *Conversation Display*. Notice the dark highlight on the final message in the Figure 4-1 *Message Display*, ending with *You win*. This displays the message sent to the chatbot to appear in the *Details* section.

The *Details* section shows messages between Bot Emulator and chatbot. *Details* shows JSON formatted messages. Read the message in *Details* any time you want to know the exact details of the data transferred between Bot Emulator and chatbot.

Use the *Log* section to see the type of messages between Bot Emulator and chatbot. This shows time, HTTP status, and a quick description of what transferred. Figure 4-1 shows POST and GET messages for the current conversation. The timings give you some idea of when a message transferred and how long it took between messages and the arrows show the direction of the messages, where pointing right is a message to the chatbot, and pointing left is a message to the Bot Emulator. There is a hyperlink on each status that you click to make the message display in the Details section.

The Bot Emulator also lets you test Activities, which you'll learn about in the next section.

Handling Activities

An *Activity* class is one of a set of notification types that the Bot Connector can send to your chatbot. So far, the only Activity you've seen in this book is a *Message* Activity and there are several more that notify your chatbot of events related to conversations. The following sections explain what activities are available, what they mean, and examples of how to write code for them.

The *Activity Class*

The *Activity* class contains members to represent all of the different types of activities there are. Some *Activity* members are only used in the context of the type of *Activity* sent to a chatbot. e.g. if the *Acitivity* is *Type Message*, you would expect the *Text* property to contain the user input but an *Activity* of *Type Typing* doesn't use any properties because the *Type* represents the semantics of the *Activity*. Here's an abbreviated *Activity* class definition, with all members removed, except for *Type*.

```
public partial class Activity :
    IActivity,
    IConversationUpdateActivity,
    IContactRelationUpdateActivity,
    IMessageActivity,
    ITypingActivity
    // interfaces for additional activities
{
    public string Type { get; set; }
}
```

> **Note** Version 1 of the Bot Framework passed a *Message* class instance as the single parameter to the *Post* method. In version 3, the Bot Framework replaced *Message* with an *Activity* class instance instead. This makes sense because *Message* is only a single type of information that the Bot Framework supports and the general semantics of the name, *Activity*, encompases other notification types.

As you can see, *Activity* implements several interfaces. Each of these interfaces specify members to support derived activity types. *Activity* also has a *Type* property, indicating the purpose of an *Activity* instance. Table 4-1 shows common *Activity* types, matching interfaces, and a quick description (explained in more detail in following sections).

TABLE 4-1 Activity Types

Activity Type	Interface	Description
ConversationUpdate	IConversationUpdateActivity	User(s) joined or left a conversation.
ContactRelationshipUpdate	IContactRelationshipUpdateActivity	User added or removed your chatbot from their list.
DeleteUserData	None	User wants to remove all Personally Identifiable Information (PII).
Message	IMessageActivity	Chatbot receives or sends a communication.
Ping	None	Sent to determine if bot URL is available.
Typing	ITypingActivity	Busy indicator by either chatbot or user.

When handling *Activity* instances, you might care about the interface so you can convert the activity to that interface to access relevant *Activity* members. You might also notice in Table 4-1 that some *Activity Types*, like *DeleteUserData* and *Ping*, don't have a matching interface on *Activity*. In those cases, there aren't any *Activity* members to access because the purpose of the *Activity Type* is for you to take an action, regardless of Activity instance state. The next section discusses the *ActivityType* class that contains members that define the type of an *Activity* instance.

The *ActivityType* Class

In the Bot Framework, an *Activity* has a purpose that is represented by an *ActivityType* class, with properties for each activity type, shown below:

```
public static class ActivityTypes
{
    public const string ContactRelationUpdate = "contactRelationUpdate";
    public const string ConversationUpdate = "conversationUpdate";
    public const string DeleteUserData = "deleteUserData";
    public const string Message = "message";
    public const string Ping = "ping";
    public const string Typing = "typing"
```

> **Note** As the Bot Framework matures, you might see new activity types introduced. Some members might be added to *ActivityTypes* to reflect a future possibility, yet still be undocumented until full implementation.

ActivityTypes is a convenient class to help avoid typing strings and you might recall from earlier examples where the code checked for *ActivityTypes.Message* to ensure the activity it was working with was a *Message*, like this:

```
if (activity.Type == ActivityTypes.Message)
{
    // handle message...
}
```

While the names of each *ActivityTypes* member suggests their purpose, the following sections offer more details and code examples.

Code Design Overview

The code in this chapter continues the Rock, Paper, Scissors game from Chapter 3. Chapter 4 code includes a project named *RockPaperScissors4*, with a new class, shown in Listing 4-1, for handling *Activities*.

LISTING 4-1 The Rock, Paper, Scissors Game - *SystemMessages* Class

```csharp
using Microsoft.Bot.Connector;
using System;
using System.Collections.Generic;
using System.Linq;
using System.Net;
using System.Threading.Tasks;
using System.Web;

namespace RockPaperScissors4.Models
{
    public class SystemMessages
    {

        public async Task Handle(ConnectorClient connector, Activity message)
        {
            switch (message.Type)
            {
                case ActivityTypes.ContactRelationUpdate:
                    HandleContactRelation(message);
                    break;
                case ActivityTypes.ConversationUpdate:
                    await HandleConversationUpdateAsync(connector, message);
                    break;
                case ActivityTypes.DeleteUserData:
                    await HandleDeleteUserDataAsync(message);
                    break;
                case ActivityTypes.Ping:
                    HandlePing(message);
                    break;
                case ActivityTypes.Typing:
                    HandleTyping(message);
                    break;
                default:
                    break;
            }
        }

        void HandleContactRelation(IContactRelationUpdateActivity activity)
        {
            if (activity.Action == "add")
            {
                // user added chatbot to contact list
            }
```

```csharp
            else // activity.Action == "remove"
            {
                // user removed chatbot from contact list
            }
        }

        async Task HandleConversationUpdateAsync(
            ConnectorClient connector, IConversationUpdateActivity activity)
        {
            const string WelcomeMessage =
                "Welcome to the Rock, Paper, Scissors game! " +
                "To begin, type \"rock\", \"paper\", or \"scissors\". " +
                "Also, \"score\" will show scores and " +
                "delete will \"remove\" all your info.";

            Func<ChannelAccount, bool> isChatbot =
                channelAcct => channelAcct.Id == activity.Recipient.Id;

            if (activity.MembersAdded.Any(isChatbot))
            {
                Activity reply = (activity as Activity).CreateReply(WelcomeMessage);
                await connector.Conversations.ReplyToActivityAsync(reply);
            }

            if (activity.MembersRemoved.Any(isChatbot))
            {
                // to be determined
            }
        }

        async Task HandleDeleteUserDataAsync(Activity activity)
        {
            await new GameState().DeleteScoresAsync(activity);
        }

        // random methods to test different ping responses
        bool IsAuthorized(IActivity activity) => DateTime.Now.Ticks % 3 != 0;
        bool IsForbidden(IActivity activity) => DateTime.Now.Ticks % 7 == 0;

        void HandlePing(IActivity activity)
        {
            if (!IsAuthorized(activity))
                throw new HttpException(
                    httpCode: (int)HttpStatusCode.Unauthorized,
                    message: "Unauthorized");
            if (IsForbidden(activity))
                throw new HttpException(
                    httpCode: (int) HttpStatusCode.Forbidden,
                    message: "Forbidden");
        }

        void HandleTyping(ITypingActivity activity)
        {
            // user has started typing, but hasn't submitted message yet
        }
    }
}
```

The *Handle* method in the Listing 4-1 *SystemMessages* class receives an *Activity* and passes it to a method, based on the matching type in the *switch* statement. Later sections of this chapter explain the *Activity* handling methods. Listing 4-2 has a modified *Post* method in *MessagesController*, showing how to call *SystemMessages.Handle*.

LISTING 4-2 The Rock, Paper, Scissors Game – *MessageController.Post* Method

```csharp
using System;
using System.Net;
using System.Net.Http;
using System.Threading.Tasks;
using System.Web.Http;
using Microsoft.Bot.Connector;
using RockPaperScissors4.Models;
using System.Web;

namespace RockPaperScissors4
{
    [BotAuthentication]
    public class MessagesController : ApiController
    {
        /// <summary>
        /// POST: api/Messages
        /// Receive a message from a user and reply to it
        /// </summary>
        public async Task<HttpResponseMessage> Post([FromBody]Activity activity)
        {
            HttpStatusCode statusCode = HttpStatusCode.OK;

            var connector = new ConnectorClient(new Uri(activity.ServiceUrl));

            if (activity.Type == ActivityTypes.Message)
            {
                // code that handles Message Activity type
            }
            else
            {
                try
                {
                    await new SystemMessages().Handle(connector, activity);
                }
                catch (HttpException ex)
                {
                    statusCode = (HttpStatusCode) ex.GetHttpCode();
                }
            }

            HttpResponseMessage response = Request.CreateResponse(statusCode);
            return response;
        }
    }
}
```

The *Post* method in Listing 4-2 checks to see if the incoming *Activity* is a *Message* and handles the *Activity*, as described in Chapter 3. For other *Activity* types, the code calls *Handle* on a new *SystemMessages* instance, passing the *ConnectorClient* instance and *Activity*.

The *try/catch* block sets *statusCode* if *Handle* throws an *HttpException*. Otherwise, *statusCode* is a *200 OK*. Stay tuned for the upcoming discussion of *Ping* Activities to see how this fits in.

Those are the essential changes to the Rock, Paper, Scissors game. The following sections describe how *SystemMessages* methods handle *Activity* types.

Sending Activities with the Bot Emulator

Clicking the three vertical dots on the Bot Emulator address bar reveals a menu with additional options for testing a chatbot. Figure 4-2 shows this menu, including the options for sending additional *Activity* types to the chatbot.

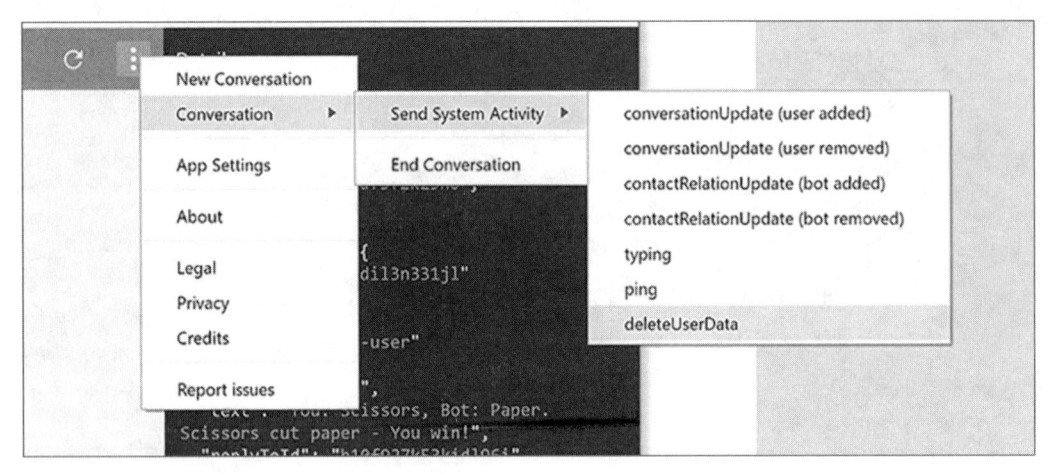

FIGURE 4-2 Bot Emulator Send System Activity Menu Options.

If you click on the menu **Conversation | Send System Activity | <Activity Type>**, the Bot Emulator sends the selected *Activity* type to your chatbot. These *Activities* match what you see in Table 4-1 with more granularity for added and removed options on *conversationUpdate* and *contactRelationUpdate*. The following sections walk through concepts of handing these *Activity* types, showing interfaces and handler method code from Listing 4-1.

Relationship Changes

When a user wants to interact with a chatbot, they add that chatbot to their channel. Similarly, when the user no longer wants to interact with a chatbot, they remove that chatbot from their channel. Whenever these channel add and remove events occur, the Bot Connector sends an *Activity* with Type *ContactRelationUpdate* to your chatbot. The *IContactRelationUpdateActivity* interface, below, shows properties relevant to a change in relationship.

```
public interface IContactRelationUpdateActivity : IActivity
{
    string Action { get; set; }
}
```

The *Action* property takes on string values of either *add* or *remove*. Here's the code that handles the *IContactRelationUpdateActivity*:

```
void HandleContactRelation(IContactRelationUpdateActivity activity)
{
    if (activity.Action == "add")
    {
        // user added chatbot to contact list
    }
    else // activity.Action == "remove"
    {
        // user removed chatbot from contact list
    }
}
```

The value of *Action* is either *add* or *remove*, making the logic to figure out which relatively easy. A possible use of this is for a chatbot that sends notifications to either *add* to a notification list or *remove* it. Another possibilitiy is to clean out any cached, transient, or un-used data associated with that user. You might even send an email to yourself as an alert when a user removes the chatbot from their user list, giving you the opportunity to review logs, learn why, and improve the chatbot.

As shown in Figure 4-2, you can open the menu and select the *contactRelationUpdate* item for testing. This sends a *ContactRelationUpdate Activity* to your chatbot.

Conversation Updates

Whenever a user first communicates with a chatbot, they join a conversation. When the user joins a conversation, the Bot Connector sends an *Activity* with *Type ConversationUpdate*, implementing the following *IConversationUpdateActivity* interface:

```
public interface IConversationUpdateActivity : IActivity
{
    IList<ChannelAccount> MembersAdded { get; set; }
    IList<ChannelAccount> MembersRemoved { get; set; }
}
```

> **Note** There's Bot Framework support for when a user leaves a conversation, represented by an *Activity* implementing *IConversationUpdateActivity* with the *MembersRemoved* property. However, the Bot Framework doesn't define when this will happen. While the leaving a conversation scenario might be defined in the future, you might not want to consider the undefined nature of this feature in your design.

Whenever the Bot Connector sends a *ConversationUpdate Activity*, *MembersAdded* contains the *ChannelAccount* information for the added user(s). The Bot Emulator sends a separate *ConversationUpdate Activity* for each user added – one for the user and another for the chatbot. The following handler code from Listing 4-1 shows one way to handle a *ConversationUpdate Activity*:

```
async Task HandleConversationUpdateAsync(
    ConnectorClient connector, IConversationUpdateActivity activity)
{
    const string WelcomeMessage =
        "Welcome to the Rock, Paper, Scissors game! " +
        "To begin, type \"rock\", \"paper\", or \"scissors\". " +
        "Also, \"score\" will show scores and " +
        "delete will \"remove\" all your info.";

    Func<ChannelAccount, bool> isChatbot =
        channelAcct => channelAcct.Id == activity.Recipient.Id;

    if (activity.MembersAdded.Any(isChatbot))
    {
        Activity reply = (activity as Activity).CreateReply(WelcomeMessage);
        await connector.Conversations.ReplyToActivityAsync(reply);
    }

    if (activity.MembersRemoved.Any(isChatbot))
    {
        // to be determined
    }
}
```

The *isChatbot* lambda detects whether added/removed user is the chatbot. The *Recipient* is the chatbot, so comparing its *Id* to the *Id* of the current *ChannelAccount* returns *true* if the *ChannelAccount* is also the chatbot.

Because the code works with the *IConversationUpdateActivity*, it needs to convert *activity* back to the base *Activity* type to call *CreateReply* when the chatbot has been added to the conversation. Then it uses the *connector* parameter to send that reply back to the user. Figure 4-3 shows the Bot Emulator interaction that results in the *ConversationUpdate Activity* arriving at your chatbot.

> **Tip** Sending an initial Hello message to a user is regarded as a best practice. The message should contain a greeting with information on how the user can use the chatbot. This lets the user know that the chatbot is alive and helps the user acclimate faster. Handling the *ConversationUpdate Activity* is the perfect place to do this.

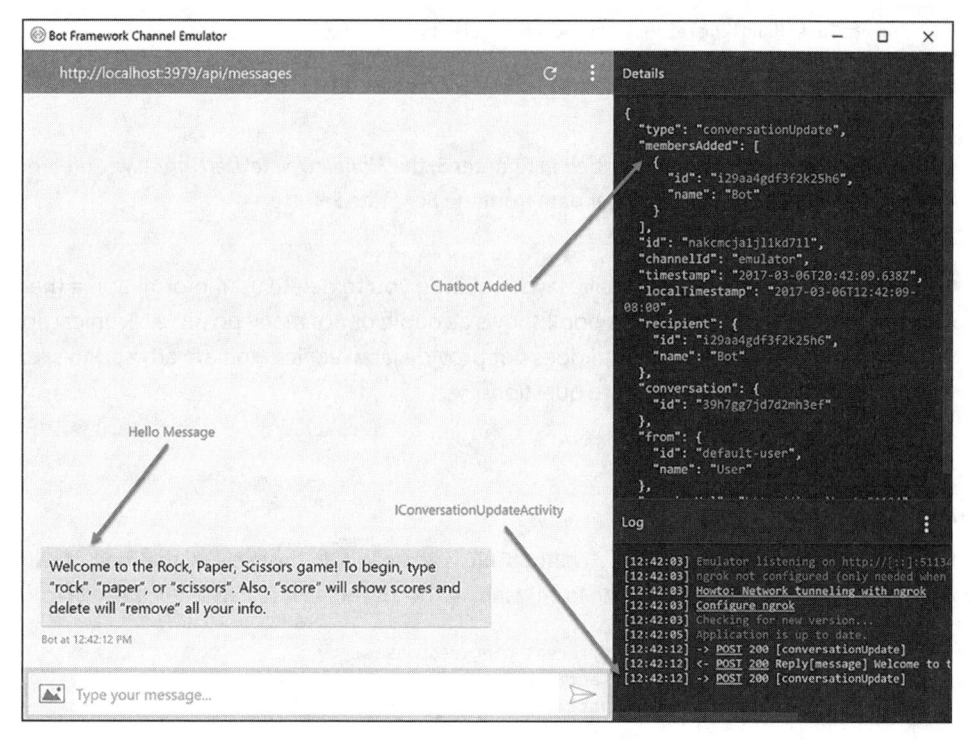

FIGURE 4-3 The Bot Emulator Sending ConversationUpdate Activity.

The three callouts in Figure 4-3 show how to examine the *ConversationUpdate* Activity. Start at the lower right and observe that at *12:42:12*, there are three *POST* messages. Two *POSTs* go to the chatbot and one to the Bot Emulator, labeled *IConversationUpdateActivity* to indicate the *Activity* implementation sent to the chatbot. The first *POST* sends a *ConversationUpdate Activity* with *membersAdded* set to the user. The second *POST* occurs when the chatbot receives the *ConversationUpdate Activity* and replies with the *Hello Message*. The third post is another *ConversationUpdate Activity* – this time adding the chatbot to the conversation. Clicking the *POST* link for the third *POST*, observe that *Details* contains the JSON formatted message sent to the chatbot. The type says *conversationUpdated* to indicate *Activity* type and *membersAdded* indicates that the chatbot is added to the conversation.

Any time you click the *Connect* button in the *Connection Details* panel, the Bot Emulator sends these *ConversationUpdate Activities* to your chatbot. You can also do this manually, as shown in Figure 4-2, you can open the menu and select the *conversationUpdate* item for testing. This sends a *ConversationUpdate Activity* to your chatbot.

Deleting User Data

A user can request that you delete their data. This request comes to you as an *Activity* with *Type DeleteUserData*. This *Activity* type doesn't have an interface and the *Activity* is treated as a command. The following code, from Listing 4-1, shows how you could handle a *DeleteUserData Activity*:

CHAPTER 4 **Fine-Tuning Your Chatbot** 75

```
async Task HandleDeleteUserDataAsync(Activity activity)
{
    await new GameState().DeleteScoresAsync(activity);
}
```

Reusing the *GameState* class, described in Chapter 3, the *HandleDeleteUserDataAsync* method calls *DeleteScoresAsync*. This deletes the user data from the Bot State Service.

> ⚠️ **Warning** Various governments have laws requiring you to delete user information if the user requests that you do so. This book shows a couple out of many possible technical approaches to deleting user data and does not provide legal advice. You are advised to seek the council of an attorney for more questions.

Pinging

A channel might want to test whether a chatbot URL is accessible. In those cases, the chatbot receives an *Activity* of *Type Ping*. Here's the code from Listing 4-1, showing how to handle *Ping Activities*:

```
// random methods to test different ping responses
bool IsAuthorized(IActivity activity) => DateTime.Now.Ticks % 3 != 0;
bool IsForbidden(IActivity activity) => DateTime.Now.Ticks % 7 == 0;

void HandlePing(IActivity activity)
{
    if (!IsAuthorized(activity))
        throw new HttpException(
            httpCode: (int)HttpStatusCode.Unauthorized,
            message: "Unauthorized");
    if (IsForbidden(activity))
        throw new HttpException(
            httpCode: (int) HttpStatusCode.Forbidden,
            message: "Forbidden");
}
```

With a *Ping Activity*, you have three ways to respond, where the number is the HTTP status code and the text is a short description of that code's meaning:

- 200 OK
- 401 Unauthorized
- 403 Forbidden

In Listing 4-2, the *Post* method sets *statusCode* to *OK* by default. An *HttpException* indicates that *Ping* handling resulted in something other than *OK*. The *IsAuthorized* and *IsForbidden* methods above are demo code to pseudo-randomly support *Unauthorized* and *Forbidden* responses. A typical application could implement those by examining the *Activity* instance and determining whether the user was authorized or allowed to use this chatbot – if those semantics make sense. Currently, there isn't any guidance from Microsoft or channels for associated Ping response protocols and it's fine to just return a *200 OK* response.

Typing Indications

Sometimes you might not want to send a message to the user if they are typing. e.g. What if they provided incomplete information and the next message was an *Activity* of *Type Typing*? You might want to wait a small period of time to receive another message that might have more details. Another scenario might be that a chatbot wants to be polite and avoid sending messages while a user is typing. Here's the *ITypingActivity* interface that the Bot Connector sends:

```
public interface ITypingActivity : IActivity
{
}
```

ITypingActivity doesn't have properties, so you simply need to handle it as a notification that the user, indicated by the *Activities From* property is preparing to send a message. Here's the handler, from Listing 4-1, for the *TypingActivity*:

```
void HandleTyping(ITypingActivity activity)
{
    // user has started typing, but hasn't submitted message yet
}
```

The demo code doesn't do anything with the *Typing Activity*. While the previous paragraph speculated that you might use this to decide whether to send the next response to the user, the reality might not be so simple. Consider the scenario where you have a Web API REST service scaled out across servers in Azure. The first user *Message Activity* arrives as one instance of the chatbot and the *Typing* instance arrives at a second instance of the chatbot because the user is rapidly communicating with the chatbot via multiple messages. That means if you want to react to a *Typing* activity, you also need to coordinate between running instances of your chatbot. How easy is that? It depends because in a world of distributed cloud computing you have to consider designs that might affect performance, scalability, and time to implement.

A more appropriate use of a *Typing Activity* might not be in receiving and handling the *Activity* from the user, but to send a *Typing Activity* to the user. The next section covers sending *Typing Activity* responses and other communication scenarios.

Advanced Conversation Messages

Chapter 3 went into depth on how to construct a reply to a user's *Message Activity*. That was for a text response, but this section covers a couple of other scenarios for communicating with the user: Sending *Typing Activities* to the user and sending a notification or alert to the user.

Sending *Typing Activities*

Occasionally, a chatbot needs to perform some action that might take longer than normal. Rather than make the user wait and wonder, it's polite to let the user know that the chatbot is busy preparing their response. You could send a text message to let the user know, but a common way to do this is by sending the user a *Typing Activity*. Listing 4-3 shows shows how to build a new *Typing Activity*.

LISTING 4-3 The Rock, Paper, Scissors Game – *BuildTypingActivity* Method

```csharp
using Microsoft.Bot.Connector;

namespace RockPaperScissors4.Models
{
    public static class ActivityExtensions
    {
        public static Activity BuildTypingActivity(this Activity userActivity)
        {
            ITypingActivity replyActivity = Activity.CreateTypingActivity();

            replyActivity.ReplyToId = userActivity.Id;
            replyActivity.From = new ChannelAccount
            {
                Id = userActivity.Recipient.Id,
                Name = userActivity.Recipient.Name
            };
            replyActivity.Recipient = new ChannelAccount
            {
                Id = userActivity.From.Id,
                Name = userActivity.From.Name
            };
            replyActivity.Conversation = new ConversationAccount
            {
                Id = userActivity.Conversation.Id,
                Name = userActivity.Conversation.Name,
                IsGroup = userActivity.Conversation.IsGroup
            };

            return (Activity) replyActivity;
        }
    }
}
```

Each of the *Activity* types has a factory method, prefixed with *Create* to create a new instance of that *Activity* type. The *BuildTypingActivity* takes advantage of that and calls the *CreateTypingActivity* factory method on the *Activity* class. For the resulting instance that implements *ITypingActivity*, you must populate *ReplyToId*, *From*, *Recipient*, and *Conversation*.

ReplyToId is the *Id* of the *Activity* instance being replied to. Notice that the *From* and *Recipient* populate from the *Recipient* and *From userActivity* properties, saying that the *Activity* is now *From* the chatbot and being sent to the user *Recipient*. You also need to populate the *Conversation* so the Bot Connector knows which conversation the *Activity* belongs to.

In this example, the return type is cast to *Activity*, rather than returning *ITypingActivity* for the convenience of the caller, which is the *GetScoresAsync* method in Listing 4-4.

LISTING 4-4 The Rock, Paper, Scissors Game – *GetScoresAsync* Method

```csharp
using System;
using System.Collections.Generic;
using System.Linq;
using System.Threading.Tasks;
using Microsoft.Bot.Connector;

namespace RockPaperScissors4.Models
{
    public class GameState
    {
        [Serializable]
        class PlayScore
        {
            public DateTime Date { get; set; } = DateTime.Now;
            public bool UserWin { get; set; }
        }

        public async Task<string> GetScoresAsync(ConnectorClient connector, Activity activity)
        {
            Activity typingActivity = activity.BuildTypingActivity();
            await connector.Conversations.ReplyToActivityAsync(typingActivity);
            await Task.Delay(millisecondsDelay: 10000);

            using (StateClient stateClient = activity.GetStateClient())
            {
                IBotState chatbotState = stateClient.BotState;
                BotData chatbotData = await chatbotState.GetUserDataAsync(
                    activity.ChannelId, activity.From.Id);

                Queue<PlayScore> scoreQueue =
                    chatbotData.GetProperty<Queue<PlayScore>>(property: "scores");

                if (scoreQueue == null)
                    return "Try typing Rock, Paper, or Scissors to play first.";

                int plays = scoreQueue.Count;
                int userWins = scoreQueue.Where(q => q.UserWin).Count();
                int chatbotWins = scoreQueue.Where(q => !q.UserWin).Count();

                int ties = chatbotData.GetProperty<int>(property: "ties");

                return $"Out of the last {plays} contests, " +
                        $"you scored {userWins} and " +
                        $"Chatbot scored {chatbotWins}. " +
                        $"You've also had {ties} ties since playing.";
            }
        }
    }
}
```

You can read about the *GetScoresAsync* method in Chapter 3 and the changes are the first three lines in the method, repeated below for convenience:

```
Activity typingActivity = activity.BuildTypingActivity();
await connector.Conversations.ReplyToActivityAsync(typingActivity);
await Task.Delay(millisecondsDelay: 10000);
```

This method calls the *BuildTypingActivity* extension method on the activity parameter. The code also passes in the *ConnectorClient* instance, connector, that the *Post* method of *MessagesController* instantiated. *ReplyToActivityAsync* sends the response back to the user, exactly like all the previous replies. *Task.Delay* simulates a longer running process because *GetScoresAsync* is too quick to see the *Typing Activity* in the Bot Emulator and that gives you a chance to see the *Typing* message before it goes away. After *Task.Delay* completes, the rest of the *GetScoresAsync* method runs as normal. Figure 4-4 shows how the Bot Emulator displays a *Typing Acivity*.

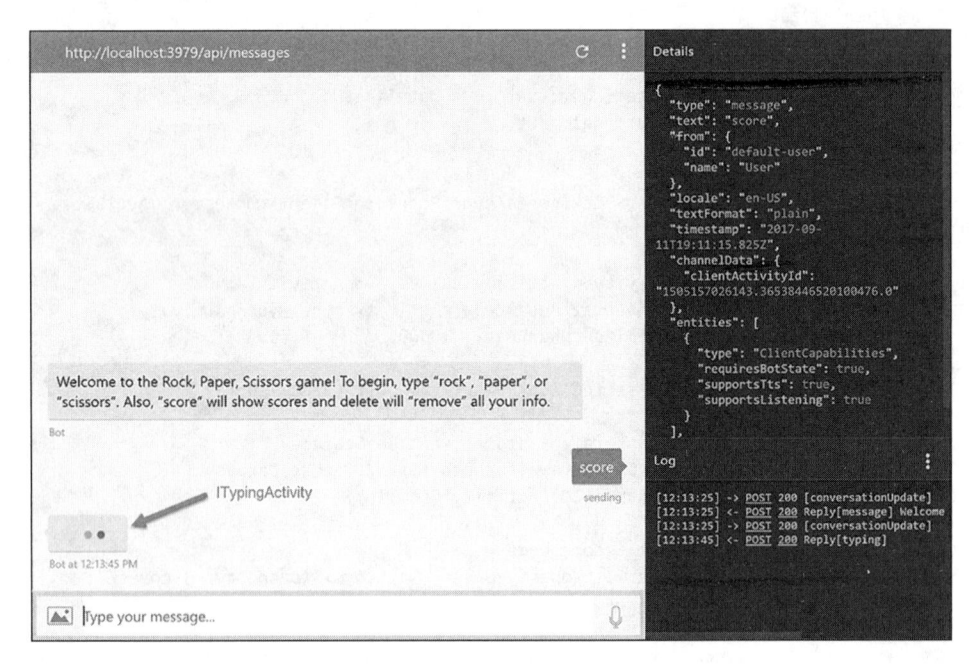

FIGURE 4-4 The Bot Emulator Receiving *Typing Activity*.

The *Typing* activity appears in Figure 4-4 as an ellipses animation, pointed to by the *ITypingActivity* label. In the emulator, the animation continues for a few seconds and goes away. The appearance and duration of the animation is dependent upon the channel the chatbot appears in.

Sending Independent Messages

Most of the time, a chatbot waits passively for the user to talk to it and then only replies. The following sections describe two scenarios where a chatbot might want to proactively send messages to the user on its own.

For the demos in this section, we created a new *Console* application that you can see in Listing 4-5. If you do this yourself, remember to add a reference to the *Microsoft.Bot.Builder* NuGet package. You'll also need a reference to the *System.Configuration* assembly. The project name for this demo is *RockPaperScissorsNotifier1*.

LISTING 4-5 Proactive Communication from Chatbot to User – Program.cs

```csharp
using Newtonsoft.Json;
using System;
using System.Configuration;
using System.IO;

namespace RockPaperScissorsNotifier1
{
    class Program
    {
        public static string MicrosoftAppId { get; set; }
            = ConfigurationManager.AppSettings["MicrosoftAppId"];
        public static string MicrosoftAppPassword { get; set; }
            = ConfigurationManager.AppSettings["MicrosoftAppPassword"];

        static void Main()
        {
            ConversationReference convRef = GetConversationReference();

            var serviceUrl = new Uri(convRef.ServiceUrl);

            var connector = new ConnectorClient(serviceUrl, MicrosoftAppId, MicrosoftAppPassword);

            Console.Write(value: "Choose 1 for existing conversation or 2 for new conversation: ");
            ConsoleKeyInfo response = Console.ReadKey();

            if (response.KeyChar == '1')
                SendToExistingConversation(convRef, connector.Conversations);
            else
                StartNewConversation(convRef, connector.Conversations);
        }

        static void SendToExistingConversation(ConversationReference convRef, IConversations conversations)
        {
            var existingConversationMessage = convRef.GetPostToUserMessage();
            existingConversationMessage.Text =
                $"Hi, I've completed that long-running job and emailed it to you.";

            conversations.SendToConversation(existingConversationMessage);
        }

        static void StartNewConversation(ConversationReference convRef, IConversations conversations)
        {
            ConversationResourceResponse convResponse =
```

```
            conversations.CreateDirectConversation(convRef.Bot, convRef.User);

            var notificationMessage = convRef.GetPostToUserMessage();
            notificationMessage.Text =
                $"Hi, I haven't heard from you in a while. Want to play?";
            notificationMessage.Conversation = new ConversationAccount(id: convResponse.Id);

            conversations.SendToConversation(notificationMessage);
        }

        static ConversationReference GetConversationReference()
        {
            string convRefJson = File.ReadAllText(path: @"..\..\ConversationReference.json");
            ConversationReference convRef = JsonConvert.DeserializeObject<ConversationReference>(convRefJson);

            return convRef;
        }
    }
}
```

To run, this program requires data for the current conversation, user, and chatbot. You would normally use some type of database to share this information that the chatbot stores and this program reads. However, this program simulates the data store by copying a JSON message from the current conversation. Here are the steps for making this happen:

1. Run the RockPaperScissors4 chatbot.

2. Run the Bot Emulator, select the chatbot URL, and click Connect.

3. Play at least one round of the RockPaperScissors game. e.g. type **scissors**.

4. This created a *ConversationReference.json* file, described later, which you can find in the base folder of the *RockPaperScissors4* project through Windows File Explorer. Open this file and copy its contents.

5. Open the *ConversationReference.json* file in the *RockPaperScissorsNotifier1* project and replace the entire contents of that file with the JSON copied from the *RockPaperScissors4* project. Note: Make sure you have the opening and closing curly braces and that you didn't accidentally copy extra text.

6. Save the *ConversationReference.json* file.

7. Right-click the *RockPaperScissorsNotifier1* project and select Debug | Start new instance.

8. You have a choice to select *1* or *2*. Select *1* and the program will close.

9. Open the Bot Emulator and observe that there is a new message on the screen, as shown in Figure 4-5.

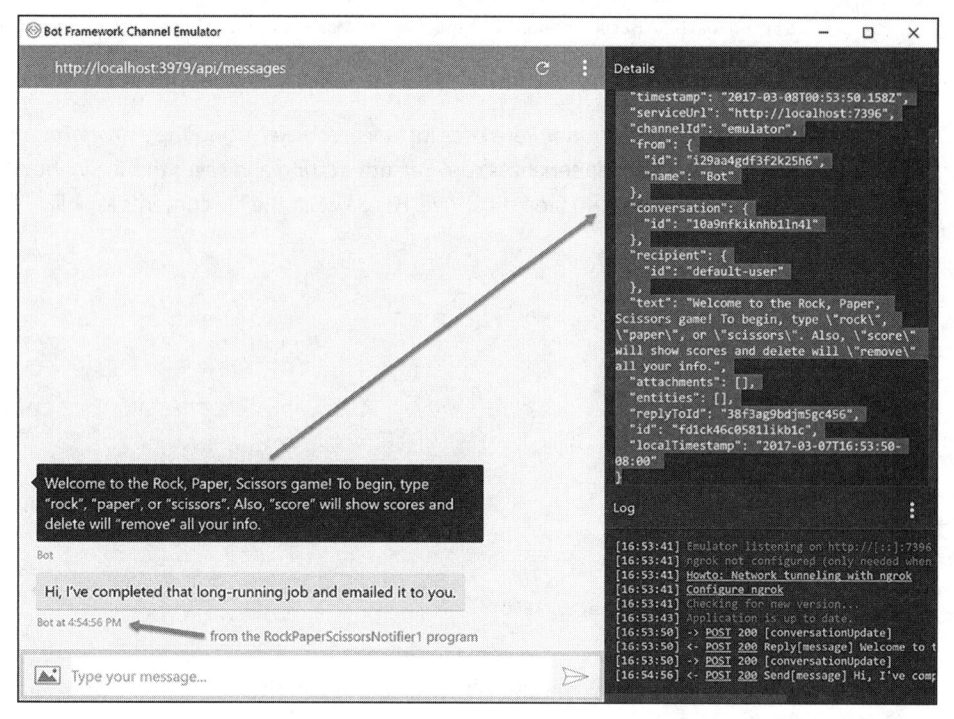

FIGURE 4-5 Sending a Message to a Conversation.

Figure 4-5 shows what the Bot Emulator might look like after the previous steps. Clicking the Hello message shows the JSON in *Details*. When running the project and selecting *1*, a new message appears below the Hello message.

Step #4 discussed a *ConversationReference.json* file created by playing a round of the game. *MessagesController* created this file when handling the user *Message* activity, shown below. It uses a *ConversationReference*, which is a Bot Builder type to help with saving and resuming conversations.

```
if (activity.Type == ActivityTypes.Message)
{
    string message = await GetMessage(connector, activity);
    Activity reply = activity.BuildMessageActivity(message);
    await connector.Conversations.ReplyToActivityAsync(reply);

    SaveActivity(activity);
}
```

The first few lines handle the user message and reply. What's most interesting here is the call to *SaveActivity*, which creates the `ConversationReference.json* file, shown below.

```
void SaveActivity(Activity activity)
{
    ConversationReference convRef = activity.ToConversationReference();
    string convRefJson = JsonConvert.SerializeObject(convRef);
```

CHAPTER 4   Fine-Tuning Your Chatbot   **83**

```
            string path = HttpContext.Current.Server.MapPath(@"..\ConversationReference.json");
            File.WriteAllText(path, convRefJson);
        }
```

SaveActivity creates a *ConversationReference* instance, *convRef*, using the *activity.ToConversationReference* method. Then the code serializes *convRef* into a string and saves the string into the *ConversationReference.json* file, described in Step #4 above. Here's what the file contents look like.

```
{
  "user": {
    "id": "default-user",
    "name": "User"
  },
  "bot": {
    "id": "jn125aajg2ljbg4gc",
    "name": "Bot"
  },
  "conversation": {
    "id": "35hn6jf29di2"
  },
  "channelId": "emulator",
  "serviceUrl": "http://localhost:31750"
}
```

This is a JSON formatted file, with various fields that are identical to what you would find in the JSON representation of an *Activity* class. That was how the file is created.

> **Tip** If you run *RockPaperScissorsNotifier1* and first select option #2 to start a new conversation, that will work. However, that also clears the previous conversation and running the program. Thereafter, selecting option #1 for an existing conversation won't work because the emulator state doesn't reflect the old conversation. To work around this, go back to Step #1, start the conversation, copy the file and try again. While this is an artificiality of the demo environment, it's also an opportunity to understand more about the nature of activities and conversations.

Next, let's look at how the *RockPaperScissorsNotifier1* program uses *ConversationReference*. The following *GetConversationReference*, from Listing 4-5, shows how to do that:

```
        static ConversationReference GetConversationReference()
        {
            string convRefJson = File.ReadAllText(path: @"..\..\ConversationReference.json");
            ConversationReference convRef = JsonConvert.DeserializeObject<ConversationReference>(convRefJson);

            return convRef;
        }
```

The first line of *GetConversationReference* reads all of the text from that file. Then it deserializes the string into a *ConversationReference*, *convRef*. The following *Main* method, from Listing 4-5, shows what the program does with the results of *GetConversationParameters*:

84 PART I Getting Started

```csharp
public static string MicrosoftAppId { get; set; }
    = ConfigurationManager.AppSettings["MicrosoftAppId"];
public static string MicrosoftAppPassword { get; set; }
    = ConfigurationManager.AppSettings["MicrosoftAppPassword"];

static void Main()
{
    ConversationReference convRef = GetConversationReference();

    var serviceUrl = new Uri(convRef.ServiceUrl);

    var connector = new ConnectorClient(serviceUrl, MicrosoftAppId, MicrosoftAppPassword);

    Console.Write(value: "Choose 1 for existing conversation or 2 for new conversation: ");
    ConsoleKeyInfo response = Console.ReadKey();

    if (response.KeyChar == '1')
        SendToExistingConversation(convRef, connector.Conversations);
    else
        StartNewConversation(convRef, connector.Conversations);
}
```

As in previous examples, you need a *ConnectorClient* instance to communicate with the Bot Connector. This time, the *ConnectorClient* constructor overload includes the *MicrosoftAppId* and *MicrosoftAppPassword*, which are read from the *.config* file *appSettings*, just like *Web.config* for the chatbot. The *ConnectorClient* constructor also uses the *ServiceUrl* as its first parameter, which comes from the call to *GetConversationReference*, discussed earlier.

With the *convRef* and *connector*, the program asks the user what they would like to do and passes those as arguments to subsequent methods. The next sections explains how to use those parameters in sending messages to existing and new conversations.

Continuing a Conversation

Normally, there's a continuous back and forth interaction in a conversation where the user sends a message, the chatbot responds, and this repeats until the user stops communicating or you've established a protocol to indicate an end to a given session. There are times though when you might want to continue a conversation later because of a need to do some processing or wait on an event. For these situations, the chatbot sends a message to the user later on, continuing the original conversation. The following *SendToExistingConversation* method, from Listing 4-5, shows how to send a message when you want to participate in an existing conversation.

```csharp
static void SendToExistingConversation(ConversationReference convRef, IConversations conversations)
{
    var existingConversationMessage = convRef.GetPostToUserMessage();
    existingConversationMessage.Text =
        $"Hi, I've completed that long-running job and emailed it to you.";

    conversations.SendToConversation(existingConversationMessage);
}
```

SendToExistingConversation, as its name implies, sends a message to an existing conversation. It uses the *convRef* parameter, calling *GetPostToUserMessage* to convert from a *ConversationReference* to an activity, *existingConversationMessage*. The *Text* indicates a hypothetical scenario where the user might have requested a long running task that has now been completed.

Use the *conversations* parameter, from *connector.Conversations*, to send that *Message Activity* from the chatbot to the user. Figure 4-5 shows how the Bot Emulator might look after calling *SendToConversation*. The next section is very similar, but the semantics are different because instead sending a message to an existing conversation, it explains how the chatbot can initiate a new conversation.

Starting a New Conversation

Another scenario for a conversation is when a chatbot wants to initiate a brand new conversation. This could happen if the purpose of the chatbot is to notify the user when something happens. e.g. What if an appointment chatbot notified a user of a meeting and the user wanted to order a taxi to get to the place of meeting, could ask for an agenda, or might want to reschedule. Another scenario might be if the chatbot notified the user of an event, like breaking news or a daily quote. The following code, from Listing 4-5 shows the *StartNewConversation* and how it can originate a new conversation with the user:

```
static void StartNewConversation(ConversationReference convRef, IConversations conversations)
{
    ConversationResourceResponse convResponse =
        conversations.CreateDirectConversation(convRef.Bot, convRef.User);

    var notificationMessage = convRef.GetPostToUserMessage();
    notificationMessage.Text =
        $"Hi, I haven't heard from you in a while. Want to play?";
    notificationMessage.Conversation = new ConversationAccount(id: convResponse.Id);

    conversations.SendToConversation(notificationMessage);
}
```

At first look, you might notice that the new *Activity*, *notificationMessage*, is built very much the same as the previous section. One difference is the *Text*, that suggests this is a notification that you might send after a period of inactivity from the user. The other difference is the fact that this *Message Activity* is sent to a brand new conversation.

At the top of the method, the call to *CreateDirectConversation* on the *conversations* parameter receives a *ConversationResourceResponse* return value. The *convResponse* supplies the *Id* for a new *ConversationAccount* instance that is passed to the *Conversation* property of *notificationMessage*. Starting a new conversation clears out the existing conversation in the Bot Emulator and only shows this message, but the actual behavior is dependent on the channel the chatbot communicates with.

> **Note** Another way to send a message is to address multiple recipients for group scenarios. As of this writing, the only channels supporting groups is email and Skype. You can check with other channels in the future to see if they eventually support groups.

Summary

This chapter was a conglomeration of subjects on testing, *Activity* management, and advanced conversation techniques. You should become familiar with the Bot Emulator because it offers many features for testing, including message details, logging, and simulating *Activities*.

The section on *Activities* went in-depth on *Activities* other than *Messages*. You learned how to handle when the user adds a chatbot to their friends list, when a user first communicates, and how to detect when the user is typing. Other *Activities* include *Ping* for when a channel wants to know if it can communicate with your chatbot and *Typing* to let a chatbot know that the user is typing.

Finally, you learned about different ways to send messages other than passively waiting and replying to something the user said. This includes the ability to send a *Typing* message to the user, sending a message to an existing conversation, and starting a new notification/alert style conversation with the user.

PART II

Bot Builder

CHAPTER 5 Building Dialogs 91

CHAPTER 6 Using FormFlow. 119

CHAPTER 7 Customizing FormFlow 153

CHAPTER 8 Using Natural Language Processing (NLP) with LUIS 183

CHAPTER 9 Managing Advanced Conversation. 203

As you learned in Part I, the promise of chatbots go beyond a set of platforms for surfacing a new type of app. It has a lot to do with building software that can engage in a conversation with users. This goes beyond a command-type interface to being able to code and manage multi-turn interactions. This is where the Bot Builder excels, giving you the tools to manage different types of conversations and freeing your time to concentrate on the purpose of the chatbot, rather than the mechanics of the conversation itself.

This part of the book contains five chapters, detailing various aspects of the Bot Builder to help you manage conversations. Chapter 5, *Building Dialogs*, explains the fundamental type, IDialog<T>, that shows how to manage a conversation. Chapter 6, *Using FormFlow* introduces a different type of dialog, called FormFlow, allowing you to build quick question and answer chatbots. Because FormFlow is so sophisticated, Chapter 7, *Customizing FormFlow* continues the story by showing you how to customize FormFlow. In Chapter 8, *Using Natural Language*

Processing (NLP) with LUIS , you learn how to make a chatbot understand human sentences. Finally, in Chapter 9, *Managing Advanced Conversation* you'll learn about how the dialog stack works, how to use chaining for even more dynamic conversations, and a few other advanced topics that show the power and flexibility of the Bot Builder library.

After reading this part of the book, you'll understand how dialogs work and what's available. Additionally, you'll have a suite of useful tools for building and managing conversations.

CHAPTER 5

Building Dialogs

Previous chapters built a game chatbot named Rock, Paper, Scissors. It was a simple chatbot in that it mostly responded to single word commands. Interactions were largely limited to a single request and response pair. That worked well for that situation, but there are times when a user needs a more in-depth conversation with a chatbot. For example, filling out forms, getting help, or ordering a product or service can often lead to a series of interactions to share information. A more sophisticated set of tools for conversations is Bot Builder dialogs.

This chapter shows how to use dialogs for conversations. You'll learn the essential parts of a dialog class and the reasons each part exists. You'll learn how to manage a conversation. You'll also learn about convenience methods for prompting the user and receiving their responses.

Introducing WineBot

This chapter shifts away from a game theme to a type of chatbot that supports ordering or storefront interfaces. The particular example is a chatbot named WineBot. The purpose of WineBot is to let the user perform searches for a wine that they're interested in. To perform searches, the chatbot needs to ask the user several questions to learn what they want. WineBot serves as an example of how you can use Bot Builder dialogs to manage a conversation.

Figure 5-1 shows the beginning of a typical session with WineBot. After an initial welcome message, the user kicks off the conversation and answers a series of questions, resulting in a list of wines matching the criteria based on the user's answers to the questions.

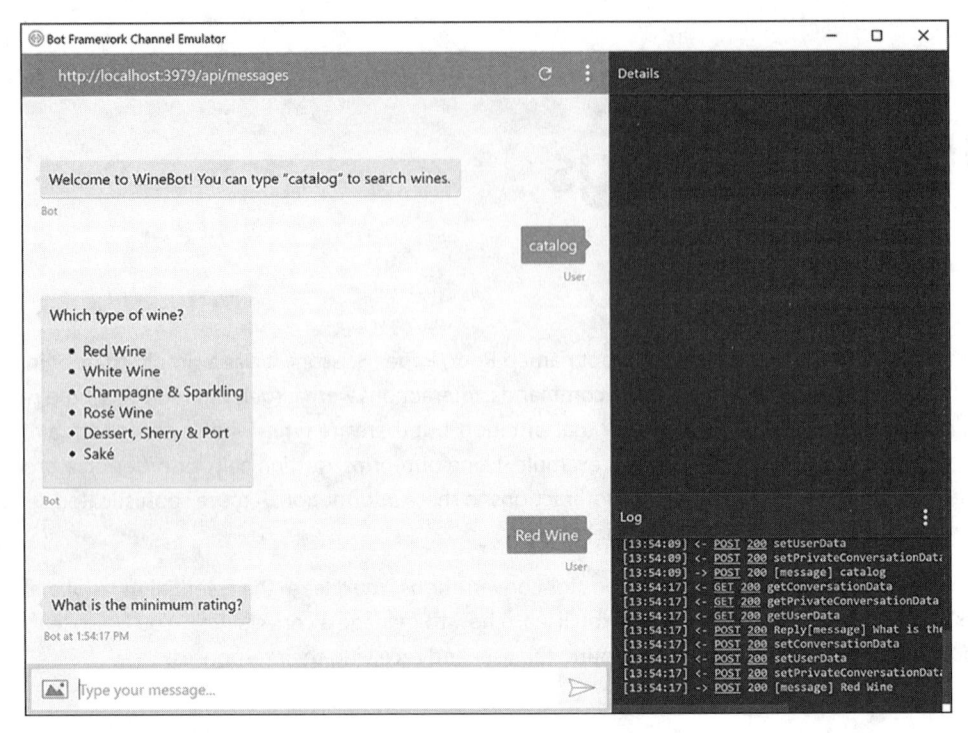

FIGURE 5-1 Chatting with WineBot.

The menu options, shown in Figure 5-1, and search results come from the Wine.com API. This API allows retrieving catalogs, performing searches for wine, and more. WineBot uses a subset of the Wine.com API, which the next section discusses.

Using the Wine.com API

We'll get to how the chatbot works soon, but let's first look at how WineBot gets its data. The Wine.com API uses a REST interface and WineBot communicates with it through HTTP GET requests. Listing 5-1 through Listing 5-4 show the *Status*, *WineCategories*, *WineProducts*, and *WineApi* classes that handle all of this communication.

> **Note** To run this program, you need an API key from Wine.com. You can find the Wine.com API documentation at *https://api.wine.com/*.

LISTING 5-1 WineBot – *Status* Class

```
namespace WineBot
{
    public class Status
    {
```

```
            public object[] Messages { get; set; }
            public int ReturnCode { get; set; }
        }
    }
```

LISTING 5-2 WineBot – *WineCategories* Class

```
using System;

namespace WineBot
{
    public class WineCategories
    {
        public Status Status { get; set; }
        public Category[] Categories { get; set; }
    }

    public class Category
    {
        public string Description { get; set; }
        public int Id { get; set; }
        public string Name { get; set; }
        public Refinement[] Refinements { get; set; }
    }

    [Serializable]
    public class Refinement
    {
        public string Description { get; set; }
        public int Id { get; set; }
        public string Name { get; set; }
        public string Url { get; set; }
    }
}
```

LISTING 5-3 WineBot – *WineProducts* Class

```
namespace WineBot
{
    public class WineProducts
    {
        public Status Status { get; set; }
        public Products Products { get; set; }
    }

    public class Products
    {
        public List[] List { get; set; }
        public int Offset { get; set; }
        public int Total { get; set; }
        public string Url { get; set; }
```

```csharp
}
public class List
{
    public int Id { get; set; }
    public string Name { get; set; }
    public string Url { get; set; }
    public Appellation Appellation { get; set; }
    public Label[] Labels { get; set; }
    public string Type { get; set; }
    public Varietal Varietal { get; set; }
    public Vineyard Vineyard { get; set; }
    public string Vintage { get; set; }
    public Community Community { get; set; }
    public string Description { get; set; }
    public Geolocation1 GeoLocation { get; set; }
    public float PriceMax { get; set; }
    public float PriceMin { get; set; }
    public float PriceRetail { get; set; }
    public Productattribute[] ProductAttributes { get; set; }
    public Ratings Ratings { get; set; }
    public object Retail { get; set; }
    public Vintages Vintages { get; set; }
}

public class Appellation
{
    public int Id { get; set; }
    public string Name { get; set; }
    public string Url { get; set; }
    public Region Region { get; set; }
}

public class Region
{
    public int Id { get; set; }
    public string Name { get; set; }
    public string Url { get; set; }
    public object Area { get; set; }
}

public class Varietal
{
    public int Id { get; set; }
    public string Name { get; set; }
    public string Url { get; set; }
    public Winetype WineType { get; set; }
}

public class Winetype
{
    public int Id { get; set; }
    public string Name { get; set; }
    public string Url { get; set; }
}
```

```csharp
public class Vineyard
{
    public int Id { get; set; }
    public string Name { get; set; }
    public string Url { get; set; }
    public string ImageUrl { get; set; }
    public Geolocation GeoLocation { get; set; }
}

public class Geolocation
{
    public int Latitude { get; set; }
    public int Longitude { get; set; }
    public string Url { get; set; }
}

public class Community
{
    public Reviews Reviews { get; set; }
    public string Url { get; set; }
}

public class Reviews
{
    public int HighestScore { get; set; }
    public object[] List { get; set; }
    public string Url { get; set; }
}

public class Geolocation1
{
    public int Latitude { get; set; }
    public int Longitude { get; set; }
    public string Url { get; set; }
}

public class Ratings
{
    public int HighestScore { get; set; }
    public object[] List { get; set; }
}

public class Vintages
{
    public object[] List { get; set; }
}

public class Label
{
    public string Id { get; set; }
    public string Name { get; set; }
    public string Url { get; set; }
}

public class Productattribute
{
```

CHAPTER 5 Building Dialogs

```csharp
        public int Id { get; set; }
        public string Name { get; set; }
        public string Url { get; set; }
        public string ImageUrl { get; set; }
    }
}
```

LISTING 5-4 WineBot – *WineApi* Class

```csharp
using System;
using System.Configuration;
using System.Linq;
using System.Net.Http;
using System.Threading.Tasks;
using Newtonsoft.Json;

namespace WineBot
{
    public class WineApi
    {
        const string BaseUrl = "http://services.wine.com/api/beta2/service.svc/json/";

        static HttpClient http;

        public WineApi()
        {
            http = new HttpClient();
        }

        string ApiKey => ConfigurationManager.AppSettings["WineApiKey"];

        public async Task<Refinement[]> GetWineCategoriesAsync()
        {
            const int WineTypeID = 4;
            string url = BaseUrl + "categorymap?filter=categories(490+4)&apikey=" + ApiKey;

            string result = await http.GetStringAsync(url);

            var wineCategories = JsonConvert.DeserializeObject<WineCategories>(result);

            var categories =
                (from cat in wineCategories.Categories
                 where cat.Id == WineTypeID
                 from attr in cat.Refinements
                 where attr.Id != WineTypeID
                 select attr)
                .ToArray();

            return categories;
        }

        public async Task<List[]> SearchAsync(int wineCategory, long rating, bool inStock, string searchTerms)
```

```csharp
        {
            string url =
                $"{BaseUrl}catalog" +
                $"?filter=categories({wineCategory})" +
                $"+rating({rating}|100)" +
                $"&inStock={inStock.ToString().ToLower()}" +
                $"&apikey={ApiKey}";

            if (searchTerms != "none")
                url += $"&search={Uri.EscapeUriString(searchTerms)}";

            string result = await http.GetStringAsync(url);

            var wineProducts = JsonConvert.DeserializeObject<WineProducts>(result);
            return wineProducts?.Products?.List ?? new List[0];
        }

        public async Task<byte[]> GetUserImageAsync(string url)
        {
            var responseMessage = await http.GetAsync(url);
            return await responseMessage.Content.ReadAsByteArrayAsync();
        }
    }
}
```

To use the Wine.com API, you need a key, which is what the *ApiKey* reads from the *Web.config* file, shown below. You can obtain a key by visiting *https://api.wine.com*.

```xml
<appSettings>
  <add key="WineApiKey" value="YourWineDotComApiKey"/>

  <!-- update these with your BotId, Microsoft App Id and your Microsoft App Password-->
  <add key="BotId" value="YourBotId" />
  <add key="MicrosoftAppId" value="" />
  <add key="MicrosoftAppPassword" value="" />
</appSettings>
```

> **Note** To make the demo code simple, we added keys to the *Web.config* file. However, this isn't recommended practice and you should visit *https://docs.microsoft.com/en-us/aspnet/* and review more secure techniques for working with code secrets.

The Wine.com API returns JSON object responses and Listing 5-1 through Listing 5-3 are C# class representations of those objects. WineBot uses Newtonsoft Json.NET to deserialize the JSON objects into C# classes. The Microsoft.Bot.Builder NuGet package has a dependency on Json.NET, which is included with the Bot Application template. Anywhere you see *JsonConvert.DeserializeObject*, that's Json.NET transforming JSON into a C# class.

> **Tip** You might notice that the WineApi class in Listing 5-4 contains a static *HttpClient* instance, *http*. At first glance, this might seem odd because *HttpClient* implements *IDisposable* and we've all had the mantra of wrapping *IDisposable* objects with a *using* statement burned into our brains. In the case of *HttpClient*, wrapping in a *using* statement hurts performance and scalability. For more information, visit the Microsoft Patterns and Practices Guidance at *https://github.com/mspnp/performance-optimization/blob/master/ImproperInstantiation/docs/ImproperInstantiation.md* to learn more on why wrapping *HttpClient* in a *using* statement is more of an anti-pattern.

The WineApi class handles all the communication with the Wine.com API. It uses the .NET *HttpClient* library to make HTTP GET requests. *GetWineCategoriesAsync* adds the proper URL segment and parameters to the *BaseUrl*, performs the GET request, and deserializes the results into an instance of *WineCategories*. As Listing 5-2 shows, *WineCategories* contains an array of *Category* and each *Category* contains an array of *Refinement*. WineBot only cares about the Wine category, which is represented by *WineTypeID*. The *Refinements* array inside of the Wine category contains the different types of Wine that WineBot shows to the user. *GetWineCategoriesAsync* reads those *Refinements* via the *SelectMany* LINQ query, and returns them to the caller.

After the user answers all questions, the chatbot calls *SearchAsync* to get a list of wines that match the given criteria, represented by the *wineCategory*, *rating*, *inStock*, and *searchTerms* parameters. Just like *GetWineCategoriesAsync*, *SearchAsync* builds the URL, performs the GET request, and deserializes the results – this time into an instance of *WineProducts* as defined in Listing 5-3. Then *SearchAsync* returns the *List* array, containing the wines found so the chatbot can display those wines to the user.

Though *GetUserImageAsync* doesn't interact with the Wine.com API, it uses *HttpClient* to perform some utility work. When a user uploads an image or document to the chatbot, it contains an *Attachment*, which you'll learn about in the upcoming *Dialog Prompt Options* section of this chapter. The *Attachment* contains a URL for where the file resides. The purpose of *GetUserImageAsync* is to perform an HTTP Get request to obtain an image located at a URL and return a byte array of that image to the caller.

Implementing a Dialog

A *dialog* is a class that contains state and a set of methods that guide a conversation. The chatbot hands over control to a dialog and the dialog manages the entire conversation. In this section, we'll examine *WineSearchDialog* – a dialog class that collects information from a user and then does a search of the wines matching the criteria that the user provided in their answers. Listing 5-5 shows *WineSearchDialog*.

LISTING 5-5 WineBot – *WineSearchDialog* Class

```csharp
using System;
using System.Collections.Generic;
using System.IO;
using System.Linq;
using System.Threading.Tasks;
using Microsoft.Bot.Builder.Dialogs;
using Microsoft.Bot.Connector;
using System.Web.Hosting;

namespace WineBot
{
    [Serializable]
    class WineSearchDialog : IDialog<object>
    {
        public Refinement[] WineCategories { get; set; }
        public string WineType { get; set; }
        public long Rating { get; set; }
        public bool InStock { get; set; }
        public string SearchTerms { get; private set; }

        public async Task StartAsync(IDialogContext context)
        {
            context.Wait(MessageReceivedAsync);
        }

        async Task MessageReceivedAsync(
            IDialogContext context, IAwaitable<IMessageActivity> result)
        {
            var activity = await result;

            if (activity.Text.Contains("catalog"))
            {
                WineCategories = await new WineApi().GetWineCategoriesAsync();
                var categoryNames = WineCategories.Select(c => c.Name).ToList();

                PromptDialog.Choice(
                    context: context,
                    resume: WineTypeReceivedAsync,
                    options: categoryNames,
                    prompt: "Which type of wine?",
                    retry: "Please select a valid wine type: ",
                    attempts: 4,
                    promptStyle: PromptStyle.AutoText);
            }
            else
            {
                await context.PostAsync(
                    "Currently, the only thing I can do is search the catalog. " +
                    "Type \"catalog\" if you would like to do that");
            }
        }
```

```csharp
async Task WineTypeReceivedAsync(
    IDialogContext context, IAwaitable<string> result)
{
    WineType = await result;

    PromptDialog.Number(
        context: context,
        resume: RatingReceivedAsync,
        prompt: "What is the minimum rating?",
        retry: "Please enter a number between 1 and 100.",
        attempts: 4);
}

async Task RatingReceivedAsync(
    IDialogContext context, IAwaitable<long> result)
{
    Rating = await result;

    PromptDialog.Confirm(
        context: context,
        resume: InStockReceivedAsync,
        prompt: "Show only wines in stock?",
        retry: "Please reply with either Yes or No.");
}

async Task InStockReceivedAsync(
    IDialogContext context, IAwaitable<bool> result)
{
    InStock = await result;

    PromptDialog.Text(
        context: context,
        resume: SearchTermsReceivedAsync,
        prompt: "Which search terms (type \"none\" if you don't want to add search terms)?");
}


async Task SearchTermsReceivedAsync(
    IDialogContext context, IAwaitable<string> result)
{
    SearchTerms = (await result)?.Trim().ToLower() ?? "none";

    PromptDialog.Confirm(
        context: context,
        resume: UploadConfirmedReceivedAsync,
        prompt: "Would you like to upload your favorite wine image?",
        retry: "Please reply with either Yes or No.");
}

async Task UploadConfirmedReceivedAsync(
    IDialogContext context, IAwaitable<bool> result)
{
    bool shouldUpload = await result;
```

```csharp
    if (shouldUpload)
        PromptDialog.Attachment(
            context: context,
            resume: AttachmentReceivedAsync,
            prompt: "Please upload your image.");
    else
        await DoSearchAsync(context);
}

async Task AttachmentReceivedAsync(
    IDialogContext context, IAwaitable<IEnumerable<Attachment>> result)
{
    Attachment attachment = (await result).First();

    byte[] imageBytes =
        await new WineApi().GetUserImageAsync(attachment.ContentUrl);

    string hostPath = HostingEnvironment.MapPath(@"~/");
    string imagePath = Path.Combine(hostPath, "images");
    if (!Directory.Exists(imagePath))
        Directory.CreateDirectory(imagePath);

    string fileName = context.Activity.From.Name;
    string extension = Path.GetExtension(attachment.Name);
    string filePath = Path.Combine(imagePath, $"{fileName}{extension}");

    File.WriteAllBytes(filePath, imageBytes);

    await DoSearchAsync(context);
}

async Task DoSearchAsync(IDialogContext context)
{
    await context.PostAsync(
        $"You selected Wine Type: {WineType}, " +
        $"Rating: {Rating}, " +
        $"In Stock: {InStock}, and " +
        $"Search Terms: {SearchTerms}");

    int wineTypeID =
        (from cat in WineCategories
         where cat.Name == WineType
         select cat.Id)
        .FirstOrDefault();

    List[] wines =
        await new WineApi().SearchAsync(
            wineTypeID, Rating, InStock, SearchTerms);

    string message;

    if (wines.Any())
        message = "Here are the top matching wines: " +
                  string.Join(", ", wines.Select(w => w.Name));
    else
        message = "Sorry, No wines found matching your criteria.";
```

```
            await context.PostAsync(message);

            context.Wait(MessageReceivedAsync);
        }
    }
}
```

As Listing 5-5 shows, *WineSearchDialog* has several members and the following sections drill down and explain what each member means.

Creating a Dialog Class

A dialog class is serializable, has state, and implements *IDialog<T>*. We'll refer to this type of dialog throughout the book as *IDialog<T>* to make it clear what type of dialog we're discussing. The following snippet, from Listing 5-5 shows how *WineSearchDialog* implements these things:

```
[Serializable]
class WineSearchDialog : IDialog<object>
{
    public Refinement[] WineCategories { get; set; }
    public string WineType { get; set; }
    public long Rating { get; set; }
    public bool InStock { get; set; }
    public string SearchTerms { get; private set; }
}
```

Dialogs must be serializable, which is why the *Serializable* attribute decorates *WineSearchDialog*. Alternatively, you could implement the *ISerializable* interface and define a serialization constructor, which is a way for .NET developers to perform custom serialization. You can visit the .NET Framework documentation for more information on serialization because it isn't a feature specific to the Bot Framework.

The requirement for a dialog to be serializable is interesting because it highlights the fact that the Bot Framework transfers dialog state across the Internet to the Bot State Service. This allows a dialog to not only keep track of your custom data state, but also keep track of where the user is in the conversation, conversation state, that the dialog is managing.

On data state, notice that *WineSearchDialog* has several public properties. Bot Builder persists these properties in the Bot State Service. Remember, this is a web API, which is inherently stateless, and this is the mechanism dialogs use to re-populate state when receiving the next message from the user.

Later, you'll see how *WineCategories* helps read the category ID for a selected wine type, so the code populates *WineCategories* and then re-uses it again later to find the ID associated with a category. *WineType*, *Rating*, *InStock*, and *SearchTerms* all hold answers from the user and are arguments to guide the search for matching wines.

WineSearchDialog implements *IDialog<object>*, which is a Bot Framework type, shown below:

```
public interface IDialog<out TResult>
{
    Task StartAsync(IDialogContext context);
}
```

IDialog has an *out TResult* type parameter, indicating the object type it will return. The return value is used in some advanced scenarios, that you'll learn more about in Chapter 9, Managing Advanced Conversation, where code can call this dialog and receive a result of its operation, which will be type *TResult*. For *WineSearchDialog*, the implemented *IDialog* type is *object* because WineBot doesn't require a return value.

This section explained the details of serialization, state, and implementing *IDialog*. You'll also notice that *IDialog* has a *StartAsync* member and you'll learn about that in the next section.

Dialog Initialization and Workflow

Implementing *IDialog*, *WineSearchDialog* has a *StartAsync* method. You can think of *StartAsync* as the entry point for a dialog because it's the first method the Bot Builder calls when starting a dialog. Here's the *StartAsync* method:

```
public async Task StartAsync(IDialogContext context)
{
    context.Wait(MessageReceivedAsync);
}
```

StartAsync is async, doesn't return a value and has an *IDialogContext* parameter. The call to *context. Wait* suspends the *StartAsync* method and sets *MessageReceivedAsync* as the next method to call in the dialog.

The call to *context.Wait* is important because it tells the Bot Framwork which method to call next. Every scenario doesn't involve passing an *IMessageActivity* to the dialog and that means *StartAsync* might have more logic. In other cases, like *WineSearchDialog*, the dialog always receives an *IMessageActivity* and it's important to call *context.Wait* on a method, *MessageReceivedAsync*, with the logic to handle that input. In the upcoming *Dialog Conversation Flow* section, you'll see how *MessageReceivedAsync* handles the dialog input. Chapter 9, Managing Advanced Conversation, discusses the dialog stack in more detail, illuminating some of the internal behaviors of how the Bot Framework manages flow of control through dialogs.

The next section drills down into the *StartAsync* parameter type *IDialogContext*.

Examining *IDialogContext*

The *IDialogContext* parameter is important because it implements interfaces with members you need during various steps of a conversation, as shown below:

```
public interface IDialogContext : IDialogStack, IBotContext
{
}
```

IDialogContext doesn't have it's own members, but derives from other interfaces, *IDialogStack* and *IBotContext*, which do. Figure 5-2 shows the relationship between *IDialogContext* and the interfaces it derives from.

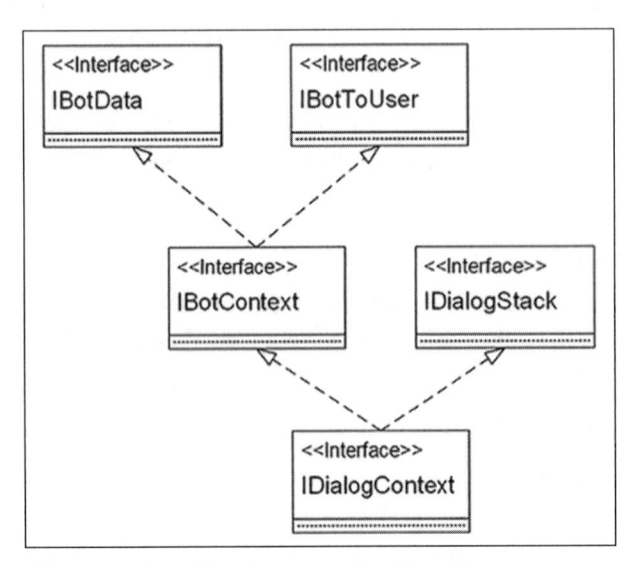

FIGURE 5-2 The *IDialogContext* Interface Hierarchy

Here's *IDialogStack*:

```
public interface IDialogStack
{
    void Wait<R>(ResumeAfter<R> resume);
}
```

IDialogStack has many members, not shown here, to manage a stack of dialogs, which I'll discuss in more detail in Chapter 9, Managing Advanced Conversation. The *Wait* method is relevant to the current discussion because *StartAsync* calls *context.Wait*. This suspends the dialog conversation and sets *MessageReceivedAsync* as the method to resume on.

```
public interface IBotContext : IBotData, IBotToUser
{
    CancellationToken CancellationToken { get; }
    IActivity Activity { get; }
}
```

IBotContext and the interfaces it derives from are part of Bot Builder. It has a *CancellationToken* for async cancellation support and a reference to the current *IActivity*. You've worked with *Activities* in previous chapters, and this is the same *Activity* that is the parameter to *MessagesController Post* method.

The *IBotData* has convenience members for accessing the Bot State Service and we discuss this more in Chapter 9, Managing Advanced Communication. *IBotToUser* has members supporting chatbot communication to the user, as shown below:

```
public interface IBotToUser
{
    Task PostAsync(IMessageActivity message, CancellationToken cancellationToken = default(CancellationToken));
    IMessageActivity MakeMessage();
}
```

IBotToUser lets you post a message to the user with *PostAsync* and you'll see an example of using it in the next section. *MakeMessage* is a convenience method for creating a new *Activity* that implements *IMessageActivity*.

Dialog Conversation Flow

Dialogs move a conversation forward by setting the current state of a conversation to the next method to call when the next *IMessageActivity* arrives from the user. The state machine in Figure 5-3 shows how this works.

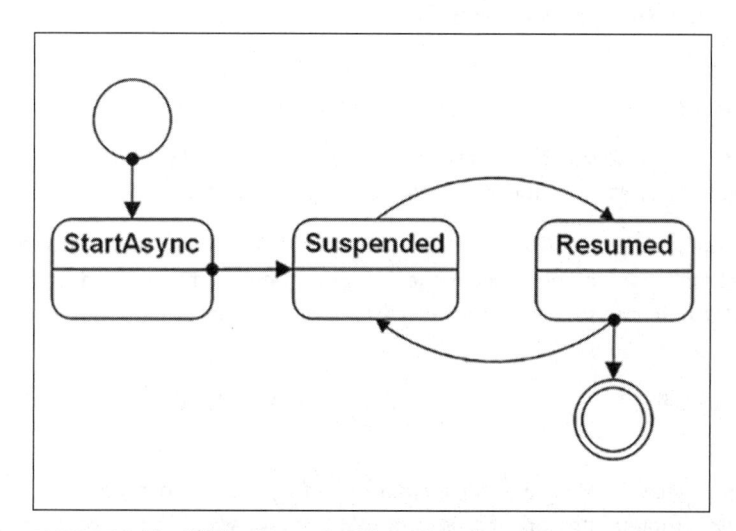

FIGURE 5-3 The *IDialogContext* Interface Hierarchy.

As shown in Figure 5-3, when the dialog starts, Bot Builder calls *StartAsync*, which suspends by calling *Wait* on the *IDialogContext* instance, context. At that point, the dialog is in the *Suspended* state. When a new *IMessageActivity* arrives, Bot Builder moves the dialog state to *Resumed* and executes the next method, which would be *MessageReceivedAsync*. The exception is that if *StartAsync* is called with an *Activity* to process, it immediately resumes on the method it called *Wait* upon – *MessageReceivedAsync* in this example. Each dialog method that is part of a conversation either leaves itself as the next *Resumed* method, or sets another method to be the next *Resumed* method.

The following list shows how conversation state management works with WineBot. The sequence describes when each method, from Listing 5-5, executes and how the conversation flows. We explain *PromptDialog* in the upcoming *Dialog Prompt Options* section of this chapter, but the primary bit of information you need to know is that *PromptDialog* methods have a parameter that sets the next method to resume on.

1. Bot Builder calls *StartAsync*, which calls *context.Wait*, setting the next method to *MessageReceivedAsync*, and suspends.

2. Message arrives to resume on *MessageReceivedAsync*. When message isn't *catalog*, code sends message to user, does not change next method, leaving *MessageReceivedAsync* as the next method to resume on, and suspends. When message is *catalog*, calls *PromptDialog.Choice*, setting *WineTypeReceivedAsync* as the next method to resume on, and suspends. Notice that not setting the next message to resume on, as in the case of user input not being *catalog*, the current method remains as the next method to resume on.

3. Message arrives to resume on *WineTypeReceivedAsync*, which calls *PromptDialog.Number* to set *RatingReceivedAsync* as the next method to resume on, and suspends.

4. Message arrives to resume on *RatingReceivedAsync*, which calls *PromptDialog.Confirm* to set *InStockReceivedAsync* as the next method to resume on, and suspends.

5. Message arrives to resume on *InStockReceivedAsync*, which calls *PromptDialog.Text* to set *SearchTermsReceivedAsync* as the next method to resume on, and suspends.

6. Message arrives to resume on *SearchTermsReceivedAsync*, which calls *PromptDialog.Confirm* to set *UploadConfirmedReceivedAsync* as the next method to resume on, and suspends.

7. Message arrives to resume on *UploadConfirmedReceivedAsync* and goes straight to *DoSearchAsync* if the user doesn't want to upload an image. If the user does want to upload an image, call *PromptDialog.Attachment* to set *AttachmentReceivedAsync* as the next method to resume on, and suspends.

8. If the user wanted to upload an image, message arrives to resume on *AttachmentReceivedAsync* and calls *DoSearchAsync*.

9. Either *UploadConfirmedReceivedAsync* or *AttachmentReceivedAsync* calls *DoSearchAsync*. When *DoSearchAsync* completes its work, it calls *context.Wait* to set *MessageReceivedAsync* as the next method to resume on, and suspends. This is how the conversation re-starts from the beginning.

The point of all of this is that by knowing how a dialog works, via its state machine, you can understand how to design and implement the chatbot conversation flow. In the current discussion, you've seen the *StartAsync* method, it called *Wait* to suspend, and then the dialog resumes with the next message, calling the *MessageReceivedAsync* method, from Listing 5-5, below:

```
async Task MessageReceivedAsync(
    IDialogContext context, IAwaitable<IMessageActivity> result)
{
    var activity = await result;

    if (activity.Text.Contains("catalog"))
    {
        WineCategories = await new WineApi().GetWineCategoriesAsync();
        var categoryNames = WineCategories.Select(c => c.Name).ToList();

        PromptDialog.Choice(
            context: context,
            resume: WineTypeReceivedAsync,
            options: categoryNames,
            prompt: "Which type of wine?",
            retry: "Please select a valid wine type: ",
            attempts: 4,
            promptStyle: PromptStyle.AutoText);
    }
    else
    {
        await context.PostAsync(
            "Currently, the only thing I can do is search the catalog. " +
            "Type \"catalog\" if you would like to do that");
    }
}
```

As discussed in the previous section, all dialog methods participating in the conversation have an *IDialogContext* parameter. Additionally, The *IAwaitable* parameter contains the message that the user sent. The *IAwaitable* is a Bot Builder type that has a type parameter specifying the type of the parameter sent to the method. Notice in the first line of *MessageReceivedAsync* that awaits *result*. The *IAwaitable* type is an async awaitable type and awaiting it makes a call to the Bot State Service to read the value.

When the user sends a message saying *catalog*, *MessageReceivedAsync* starts a conversation to request information from the user that eventually results in a search for wine. It calls *GetWineCategoriesAsync*, as discussed in the previous section and creates a list of wine types for the user to choose from.

When the user types something other than *catalog*, the chatbot responds with a method with more clear instructions on how to get started. As discussed in the previous section, *PostToUser* is a method from the *IBotToUser* interface that sends to the user.

PromptDialog.Choice is one of many prompt methods for interacting with the user and setting the method to resume when the user responds, and we discuss that in detail in the next section.

Dialog Prompt Options

The *PromptDialog* class offers serveral convenience methods, supporting question and answer conversations with the user. Listing 5-5 shows *PromptDialog* in action and we've briefly discussed *PromptDialog.Choice* in the previous section. In this section, we cover all the *PromptDialog* methods. Before doing so, let's review Table 5-1, outlining several common *PromptDialog* method parameters.

TABLE 5-1 Common *PromptDialog* Method Parameters

Parameter Name	Description
context	IDialogContext that was passed into the method. Discussed in a previous section of this chapter.
resume	ResumeAfter<T> delegate specifying the method to resume on when the next IMessageActivity arrives.
prompt	Text (string) to show the user. Typically a question to show the user before displaying options, if any.
retry	Text (string) to display when a user enters an invalid option. Defaults to prompt message.
attempts	Number of times to retry prompt. Bot Builder will show the retry message until the number of attempts occur. Defaults to 3.

When thinking about how a chatbot interacts with a user, the *prompt*, *retry*, and *attempts* parameters come into play. Done right, these can reduce friction for the user. One example might be to define a *retry* message that is worded differently than the *prompt*, assuming the user didn't fully understand what the *prompt* message really wanted. Another option is the *attempts*, which defaults to 3. In some cases, you might want to give the user several more chances to try, but an excessive number of retries might make them walk away. On the other hand, what if you needed a *Confirm* response that needed a *yes* or *no* and any response other than *yes* would mean *no*, meaning that you just want to read the user's response with 0 retries.

> **Tip** When all attempts expire, Bot Builder resumes on the method specified by the *resume* parameter. When awaiting the *IAwaitable* parameter, result in Listing 5-5, Bot Builder throws a *TooManyAttemptsException*. Wrap *await result* in a *try/catch* block to handle the situation. You can learn what the user typed in by examining *context.Activity*, which is the *IMessageActivity* from the user.

A couple of the *PromptDialog* methods have a *promptOptions* parameter with a constructor that takes all of the parameters described in Table 5-1. So, you could instantiate a *PromptOptions* instance, shown below, with those parameters and pass that instance as a parameter to one of the *PromptDialog* methods that accept it instead of coding the parameters to the method itself.

```
public class PromptOptions<T>
{
    public readonly string Prompt;
    public readonly string Retry;
    public readonly IReadOnlyList<T> Options;
    public readonly IReadOnlyList<string> Descriptions;
```

```
    public readonly string TooManyAttempts;
    public readonly PromptStyler PromptStyler;
    public int Attempts { get; set; }
    public string DefaultRetry { get; set; }
    protected string DefaultTooManyAttempts { get; }

    public PromptOptions(
        string prompt, string retry = null, string tooManyAttempts = null,
        IReadOnlyList<T> options = null, int attempts = 3,
        PromptStyler promptStyler = null, IReadOnlyList<string> descriptions = null);
}
```

In the normal case, just using *PromptDialog* parameters works fine, but there is one parameter in *PromptOptions* that you won't find in any of the *PromptDialog* method overloads and that is *tooManyAttempts*, which you can use to specify the message to show the user when they've exceeded the number of specified attempts. *PromptOptions* also has public properties that are synonymous with its constructor parameters.

The following sections follow the conversation flow in WineBot and drill down on *PromptDialog* usage from Listing 5.5.

Choice

When *StartAsync* calls *context.Wait* with its *ResumeAfter<IMessageActivity>* delegate parameter set to *MessageReceivedAsync*; it starts the conversation. *MessageReceivedAsync* handles the condition, where the user sends *catalog*, which a *PromptDialog.Choice* method. The goal is to learn the type of wine the user wants to search for, which is one of several options, and the code below shows how to do this:

```
async Task MessageReceivedAsync(
    IDialogContext context, IAwaitable<IMessageActivity> result)
{
    var activity = await result;

    if (activity.Text.Contains("catalog"))
    {
        WineCategories = await new WineApi().GetWineCategoriesAsync();
        var categoryNames = WineCategories.Select(c => c.Name).ToList();

        PromptDialog.Choice(
            context: context,
            resume: WineTypeReceivedAsync,
            options: categoryNames,
            prompt: "Which type of wine?",
            retry: "Please select a valid wine type: ",
            attempts: 4,
            promptStyle: PromptStyle.AutoText);
    }
    else
    {
        await context.PostAsync(
            "Currently, the only thing I can do is search the catalog. " +
            "Type \"catalog\" if you would like to do that");
    }
}
```

The *PromptDialog.Choice* has all of the common parameters, discussed in the previous section. It also has a *promptStyle* parameter, of type *PromptStyle* enum, specifying how each item in the list of choices appears. Table 5-2 describes what the *PromptStyle* enum members mean.

TABLE 5-2 *PromptStyle* Enum Members

Enum Member	Description
Auto	Select a style based on channel capabilities. Can vary by channel based on the channel's preferences. Bullet list in Bot Emulator.
Keyboard	Displays as keyboard card, mapping to a Hero Card, which you'll learn about in Chapter 10, Attaching Cards.
AutoText	Shows options as text on either separate lines or in the same line, depending on number of choices.
Inline	Show all options as text on a single line. Unlike AutoText, this is regardless of number of choices.
PerLine	Show all options as text with each option on a separate line. Unlike AutoText, this is regardless of number of choices.
None	Don't show any choices. Might be useful if you felt it made more sense to add options to prompt/retry parameter text. E.g. two or three choices.

The method naming convention in Listing 5-5 is *TReceivedAsync*, where *T* is the name of the answer received from the user in the result. After *StartAsync*, the chatbot waits for the next message and that goes to *MessageReceivedAsync*. When *MessageReceivedAsync* prompts for a wine type, it specifies *WineTypeReceivedAsync* for its resume parameter, where it performs a *Number* prompt that we discuss in the next section.

Number

After the user answers with the type of wine they want, the next step is to ask for the minimum rating, which is an integer. *PromptDialog* has a *Number* method that works for this, as shown below:

```
async Task WineTypeReceivedAsync(
    IDialogContext context, IAwaitable<string> result)
{
    WineType = await result;

    PromptDialog.Number(
        context: context,
        resume: RatingReceivedAsync,
        prompt: "What is the minimum rating?",
        retry: "Please enter a number between 1 and 100.",
        attempts: 4);
}
```

Notice how the first statement of *WineTypeReceivedAsync* awaits the *result* parameter, assigning the *string* value to the *WineType* property. The *result* parameter is a Bot Builder *IAwaitable<T>*, allowing you to asynchronously request the result of the previous dialog. Since the *PromptDialog.Choice*, in *MessageReceivedAsync*, resulted in type *string* from the user, the *IAwaitable<T>* type parameter, *T*, must also be type *string*. You could also use an *IAwaitable<object>*, but then the code would need to perform the conversion from *object* to *string* and *IAwaitable<string>* lets Bot Builder do it for us.

WineType is a property of *WineSearchDialog*. If you recall, *WineSearchDialog* is serializable, as all dialogs must be. That means their properties must also be serializable. Further, when Bot Builder serializes *WineSearchDialog*, it also serializes instance state (properties and fields). When a new user message arrives for *WineSearchDialog*, Bot Builder deserializes *WineSearchDialog*, including all of its state. The benefit is that by assigning user input, when received, to properties, as *WineTypeReceivedAsync* does for *WineType*, that state persists for subsequent operations inside the dialog instance. You'll see this later in the *Performing the Search* section.

PromptDialog.Number accepts a *ResumeAfter* delegate with a type parameter of either *double* or *long*. In this example, we need an integer, so the resume delegate, *RatingReceivedAsync* has an *IAwaitable<long>* parameter, as you'll see in the next section.

Confirm

Yes or no answers are another common type of question, which are supported. In this case, you would use a *PromptDialog.Confirm*, as shown below:

```
async Task RatingReceivedAsync(
    IDialogContext context, IAwaitable<long> result)
{
    Rating = await result;

    PromptDialog.Confirm(
        context: context,
        resume: InStockReceivedAsync,
        prompt: "Show only wines in stock?",
        retry: "Please reply with either Yes or No.");
}
```

PromptDialog.Confirm accepts a *bool* value, where a Yes answer is *true* and a No answer is *false*, as described in the next section.

Text

Some answers require text – it's a chatbot after all. In the current situation, WineBot needs to collect some search terms from the user so that the search can filter by those terms. So, if the user types *cabernet*, the search will likely return Cabernet Sauvignon that also matches the other criteria. The following shows how *PromptDialog.Text* works:

```
async Task InStockReceivedAsync(
    IDialogContext context, IAwaitable<bool> result)
{
    InStock = await result;

    PromptDialog.Text(
        context: context,
        resume: SearchTermsReceivedAsync,
        prompt: "Which search terms (type \"none\" if you don't want to add search terms)?");
}
```

PromptDialog.Text expects an answer of type string, which is what it's resume parameter, *Search-TermsReceivedAsync* handles, shown below:

```
async Task SearchTermsReceivedAsync(
    IDialogContext context, IAwaitable<string> result)
{
    SearchTerms = (await result)?.Trim().ToLower() ?? "none";

    PromptDialog.Confirm(
        context: context,
        resume: UploadConfirmedReceivedAsync,
        prompt: "Would you like to upload your favorite wine image?",
        retry: "Please reply with either Yes or No.");
}
```

For the search terms, the *PromptDialog* asked the user to type *none* if they didn't have any terms. You can see how *SearchTermsReceivedAsync* processes the *result*, trimming, lower casing, and setting to *none* if the input doesn't contain a valid term.

> **Warning** Users sometimes say anything and everything you never expected when communicating with a chatbot. Notice how *SearchTermsReceivedAsync* has special handling for the user input. You should code defensively to handle any random input a user provides.

SearchTermsRecievedAsync asks a user if they would like to upload their favorite wine image and sets its resume parameter to *UploadConfirmedReceived*, which is covered next.

Attachment

During the conversation, WineBot asks the user if they would like to upload a picture of their favorite wine image. While this is an image, the user could also upload any type of file. E.g. what if the chatbot took a formatted file like JSON, Excel, or CSV with data to process? Here's how WineBot asks for an attachment:

```
async Task UploadConfirmedReceivedAsync(
    IDialogContext context, IAwaitable<bool> result)
{
    bool shouldUpload = await result;

    if (shouldUpload)
        PromptDialog.Attachment(
            context: context,
            resume: AttachmentReceivedAsync,
            prompt: "Please upload your image.");
    else
        await DoSearchAsync(context);
}
```

When the user confirms that they want to upload, WineBot calls *PromptDialog.Attachment* to let the user know that they should send their attachment. Otherwise, WineBot performs the search.

As explained in Chapter 4, Bot Emulator has an attachment button on the left of the text input box that the user can click to select an image (or any other file) to upload. When the user does this, Bot Builder calls the *AttachmentReceivedAsync* method, shown below, specified as the *PromptDialog.Attachment resume* parameter:

```
async Task AttachmentReceivedAsync(
    IDialogContext context, IAwaitable<IEnumerable<Attachment>> result)
{
    Attachment attachment = (await result).First();

    byte[] imageBytes =
        await new WineApi().GetUserImageAsync(attachment.ContentUrl);

    string hostPath = HostingEnvironment.MapPath(@"~/");
    string imagePath = Path.Combine(hostPath, "images");
    if (!Directory.Exists(imagePath))
        Directory.CreateDirectory(imagePath);

    string fileName = context.Activity.From.Name;
    string extension = Path.GetExtension(attachment.Name);
    string filePath = Path.Combine(imagePath, $"{fileName}{extension}");

    File.WriteAllBytes(filePath, imageBytes);

    await DoSearchAsync(context);
}
```

The *PromptDialog.Attachment resume* parameter takes a *ResumeAfter* delegate with a *IEnumerable<Attachment>* type parameter. Because it's a single attachment, *AttachmentReceivedAsync* takes the first item. *Attachment* has several properties, shown below:

```
public class Attachment
{
    public Attachment();
    public Attachment(
        string contentType = null, string contentUrl = null, object content = null,
        string name = null, string thumbnailUrl = null);
    public string ContentType { get; set; }
    public string ContentUrl { get; set; }
    public object Content { get; set; }
    public string Name { get; set; }
    public string ThumbnailUrl { get; set; }
}
```

An *Attachment* has various properties that are populated according to the context in which the *Attachment* is used. You'll learn more about *Attachment* in Chapter 10, Attaching Cards. In the current scenario, the *Attachment* has a *ContentUrl*, specifying the location of where the file resides.

As discussed when examining Listing 5-4, *WineApi* has a *GetUserImageAsync* method that accepts a URL and returns the *byte[]* with the file data, which *AttachmentReceivedAsync* uses. The code then builds a file path located under the current web API instance and names the file after the user with the file extension. Then it saves the file in that location. This program doesn't do anything significant with that file, but you might imagine saving that file in a folder or database of your choice or processing it in some way that makes sense to your chatbot.

After processing the file, WineBot has all the information it needs to perform the search, which the next section discusses.

Performing the Search

To this point, WineBot hasn't done any extra work to save the answers from the user, other than to populate *WineSearchDialog* properties. It didn't have to because, as described in a previous section, *WineSearchDialog*, like all dialogs must be, is serializable and this allows Bot Builder to save dialog state in the Bot State Service. That means the properties are populated and ready when we need them, like in the *DoSearchAsync* method shown below:

```
async Task DoSearchAsync(IDialogContext context)
{
    await context.PostAsync(
        $"You selected Wine Type: {WineType}, " +
        $"Rating: {Rating}, " +
        $"In Stock: {InStock}, and " +
        $"Search Terms: {SearchTerms}");

    int wineTypeID =
        (from cat in WineCategories
         where cat.Name == WineType
         select cat.Id)
        .FirstOrDefault();

    List[] wines =
        await new WineApi().SearchAsync(
            wineTypeID, Rating, InStock, SearchTerms);

    string message;

    if (wines.Any())
        message = "Here are the top matching wines: " +
                    string.Join(", ", wines.Select(w => w.Name));
    else
        message = "Sorry, No wines found matching your criteria.";

    await context.PostAsync(message);

    context.Wait(MessageReceivedAsync);
}
```

The first thing *DoSearchAsyc* does is send a message to the user with the choices they made. This might be another opportunity for a *PromptDialog.Confirm* to make sure this is what they wanted and this implementation skips that because you've already seen all of the *PromptDialog* options in previous sections.

If you recall *MessageReceivedAsync* stored the wine choices in *WineCategories*. This is useful because now we can take the text *WineType* description and query for it's matching ID. The code calls the *SearchAsync* method from the *WineApi* class in Listing 5-4 and displays the wine list to the user.

Finally, calling *context.Wait* sets the dialog state to *MessageReceivedAsync* as the next method to resume upon. This starts the conversation over again, at the beginning, but the user is still in the same conversation. An important point here is that the code doesn't call *context.Wait* on *StartAsync*. That's because *StartAsync* is only called when a brand new conversation starts. After *StartAsync* calls *context. Wait*, the Bot Builder serializes dialog state, including where the dialog is to resume upon the next message arriving from the user. Bot Builder sends the serialized state to the Bot State Service and the Bot State Service uses private state for the dialog. As you recall, from Chapter 3, private state is for a user in a conversation. So, when the same user continues a conversation, Bot Builder can deserialize and set the next method to resume on from wherever the dialog left off. *StartAsync* doesn't work as a method to resume on because, unlike *MessageReceivedAsync*, it doesn't have an *IAwaitable<T>* to handle the message result – it's only the entry point to a new conversation. That's why *MessageReceivedAsync* is the first method to resume on.

This completes the description of how a dialog works and the next section shows you how to use this dialog in a chatbot.

Calling a Dialog

The last step in using a dialog is telling a chatbot which dialog to use. WineBot does this in the *MessagesController*, shown in Listing 5-6.

LISTING 5-6 WineBot – *MessagesController* Class

```
using System;
using System.Linq;
using System.Net;
using System.Net.Http;
using System.Threading.Tasks;
using System.Web.Http;
using Microsoft.Bot.Builder.Dialogs;
using Microsoft.Bot.Connector;

namespace WineBot
{
    [BotAuthentication]
    public class MessagesController : ApiController
    {
        public async Task<HttpResponseMessage> Post([FromBody]Activity activity)
        {
            if (activity.Type == ActivityTypes.Message)
                await Conversation.SendAsync(activity, () => new WineSearchDialog());
            else
                await HandleSystemMessageAsync(activity);
```

```
            var response = Request.CreateResponse(HttpStatusCode.OK);
            return response;
        }

        async Task HandleSystemMessageAsync(Activity message)
        {
            if (message.Type == ActivityTypes.ConversationUpdate)
            {
                const string WelcomeMessage =
                    "Welcome to WineBot! You can type \"catalog\" to search wines.";

                Func<ChannelAccount, bool> isChatbot =
                    channelAcct => channelAcct.Id == message.Recipient.Id;

                if (message.MembersAdded?.Any(isChatbot) ?? false)
                {
                    Activity reply = message.CreateReply(WelcomeMessage);

                    var connector = new ConnectorClient(new Uri(message.ServiceUrl));
                    await connector.Conversations.ReplyToActivityAsync(reply);
                }
            }
        }
    }
}
```

If the incoming activity is *ActivityType.Message*, WineBot calls *Conversation.SendAsync*. The second parameter to *SendAsync* is a lambda with a new instance of *WineSearchDialog*. This is essentially handing off all processing of that message to *WineSearchDialog*. As you've seen throughout this chapter, *WineSearchDialog* manages all data state and conversation state to keep track of what data it has and where it's at in the conversation.

Of additional note is the *HandleSystemMessageAsync* method that handles *ActivityType.ConversationUpdate*, which we discussed in Chapter 4. When the user first connects with their chatbot, it's polite to send them a Hello or Welcome message, letting them know how to interact with the chatbot. If you wanted to be clever, you could save the user's ID to compare against subsequent connections and vary the message, e.g. welcome back User1. This is a best practice and you'll see this regularly throughout the book.

Summary

This chapter was all about building a dialog, henceforth referred to as *IDialog<T>*. The example was WineBot, a chatbot that lets users answer questions that result in a search for wines matching the given answers. We used the Wine.com API as a data source.

You learned how to specify an *IDialog<T>* and how the *IDialog<T>* manages data state via serialization and interaction with the Bot State Service. This chapter covered the *IDialog<T>* entry point and how a dialog manages conversation state.

The *Implementing a Dialog* section explained parameters to *IDialog<T>* methods, including the *IDialogContext* hierarchy and capabilities. You can use several *PromptDialog* methods to ask the user questions resulting in answers from a list of options, yes or no, text, and numeric. You can even accept file attachments from the user.

Finally you saw how easy it is to use an *IDialog<T>*, instructing *MessagesController* to hand control to the *IDialog<T>* anytime it received a message from the user.

CHAPTER 6

Using FormFlow

In Chapter 5, you learned how to manage conversations with basic dialogs, *IDialog<T>*. This chapter continues the dialog discussion by introducing a different type of dialog, FormFlow. With an *IDialog<T>* dialog, you specify the flow of a conversation through a chain of methods where each method accepts an answer from its antecedent, and then asks another question until you're finally at the end of the conversation. A FormFlow dialog is different, because rather than procedural logic, it's more declarative. Essentially, you specify what you want the conversation to be, rather than how to perform the conversation.

To support the declarative nature of a FormFlow dialog, this chapter starts with the basics of how to create a FormFlow class. Next, you'll learn how to apply attributes to control the chatbot-to-user interface. Some of these attributes affect how the chatbot asks a question, whereas others define the layout of the options, and there are other attributes to help in easy customization. FormFlow also has a pattern language that you'll learn about to help guide the output of some attributes.

Out-of-the-Box Features

This chapter continues with the WineBot chatbot concept introduced in Chapter 5, Building Dialogs. FormFlow has several out-of-the-box features that streamline the appearance and user interface of a chatbot, which is different than the default text of *IDialog<T>*. This section walks you through the user experience for a FormFlow chatbot. One of the first things you'll notice is that menus appear as cards with buttons, as shown in Figure 6-1.

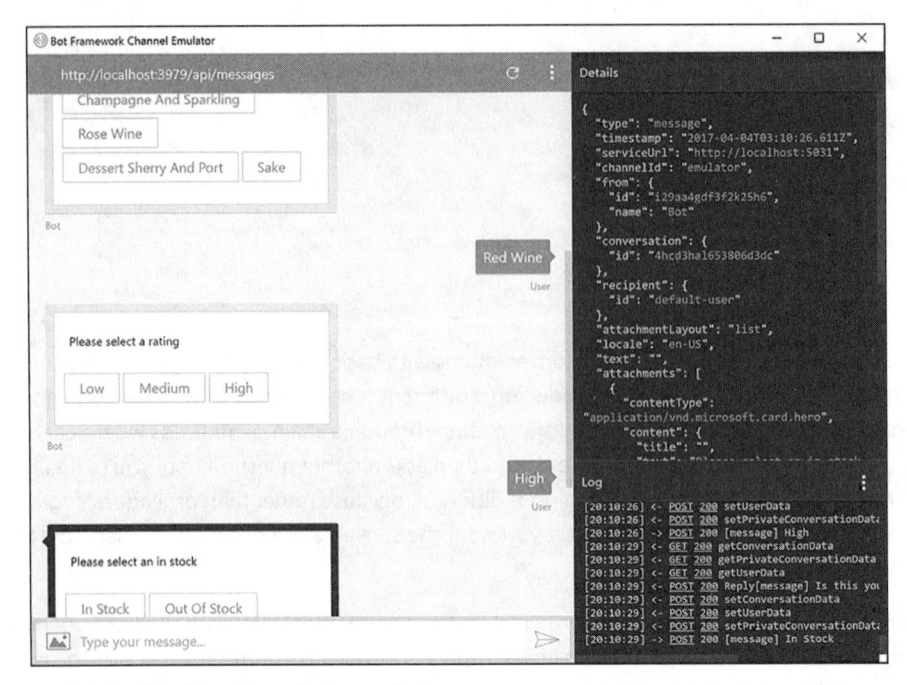

FIGURE 6-1 FormFlow Menu Options as Buttons.

Figure 6-1 shows a subset of the menu options. In this case they appear as button lists because that's the default behavior of FormFlow in the Bot Emulator. Clicking a button results in the text associated with that button being sent to the chatbot – notice the blue responses.

> **Tip** Buttons are a nice user interface feature, because they let users click through options quickly instead of typing. The choice between buttons and text is yours and what you think works best for your chatbot. You'll learn more about making this choice later in the chapter.

Instead of clicking buttons, the user is welcome to type, as shown in Figure 6-2.

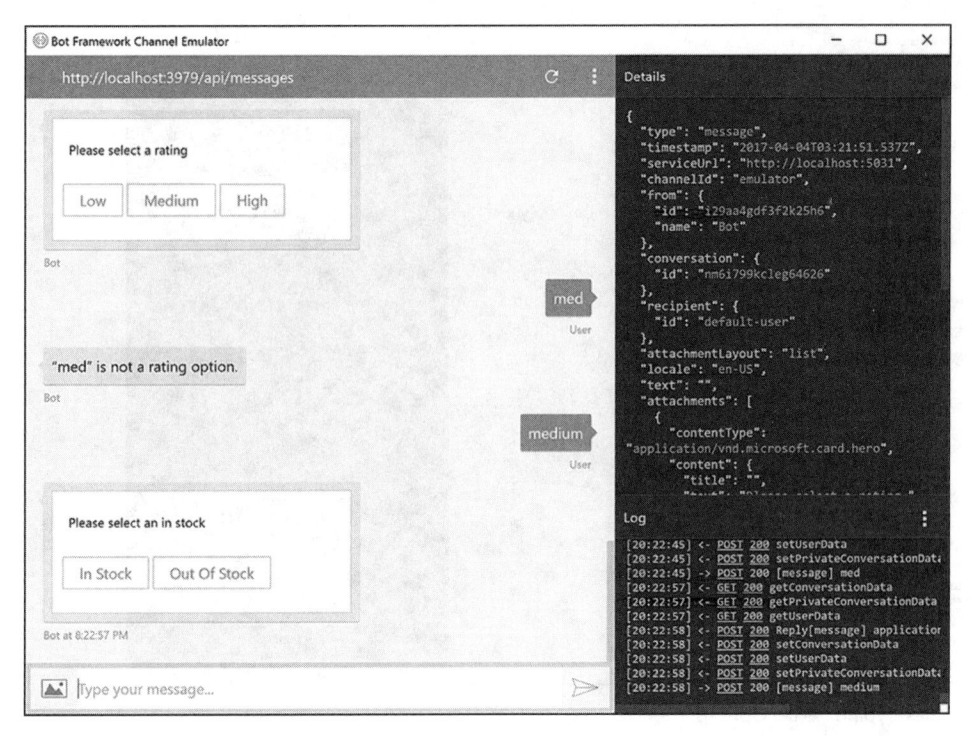

FIGURE 6-2 FormFlow Input Error Correction.

Notice how in Figure 6-2 the user input, **med**, resulted in an error message, explaining "med is not a rating option." Later in this chapter, you'll learn how to customize messages. FormFlow also lets users type in any case and it will still work. For example, notice how FormFlow recognized lower case **medium** as the *Medium* menu option.

> **Note** As the FormFlow developer, you don't have to write code to respond to the user when they enter incorrect input. Much of the FormFlow interaction is out-of-the-box and automatic.

Another subtlety is when users type a single word matching one or two options, as shown in Figure 6-3.

CHAPTER 6 Using FormFlow   **121**

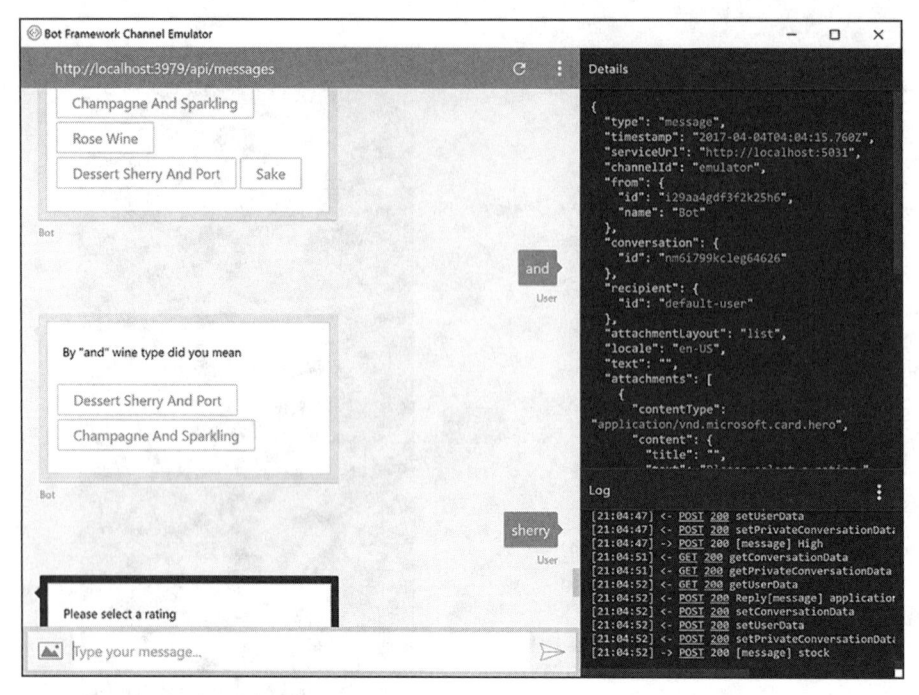

FIGURE 6-3 Typing Partial Options.

In Figure 6-3, both *Champaign And Sparkling* and *Desert Sherry And Port* have the word *And* in their titles. Typing **and** makes FormFlow respond with a choice among the two matching menu items. FormFlow has no idea what the word *and* means, except that it matches two menu options.

Also, notice that typing a single unique word, matching an entire word in a menu item, results in a match. That's what happened when typing **sherry**, matching the *Dessert Sherry And Port* option. The reason typing a partial menu name didn't work in Figure 6-3 is because **med** was not an entire word, but **sherry** is. Later in this chapter, you'll learn how to specify terms, which can be parts of an answer, to match a menu item.

In addition to menu options, users can ask for help anywhere during the operation of a FormFlow, as shown in Figure 6-4.

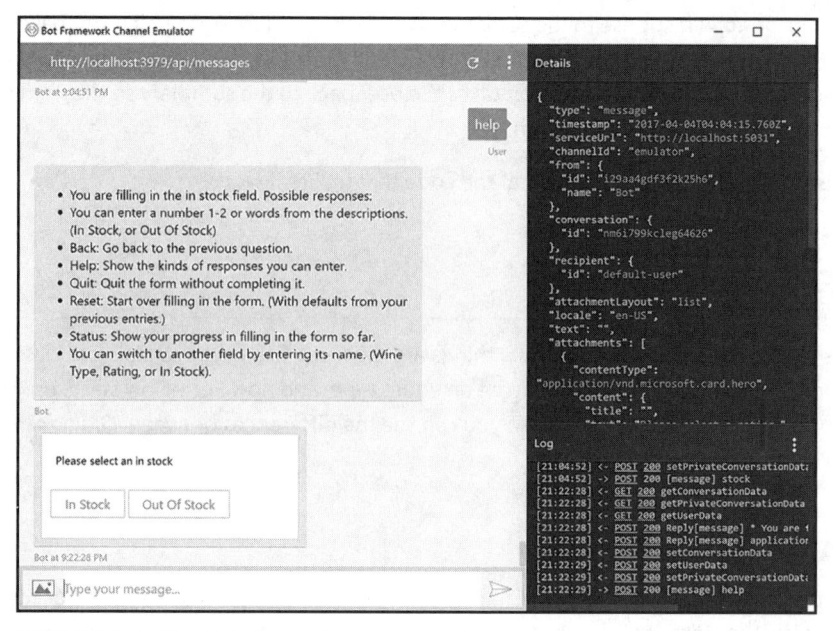

FIGURE 6-4 Getting Help From a FormFlow Chatbot.

Figure 6-4 shows that typing **help** gives a menu of options. Many of the options, like *Back*, *Reset*, or typing an option name support navigation. Users can even type **go to wine** or another field name to navigate. When the user is done filling out the form, FormFlow prompts with a summary of choices, as shown in Figure 6-5.

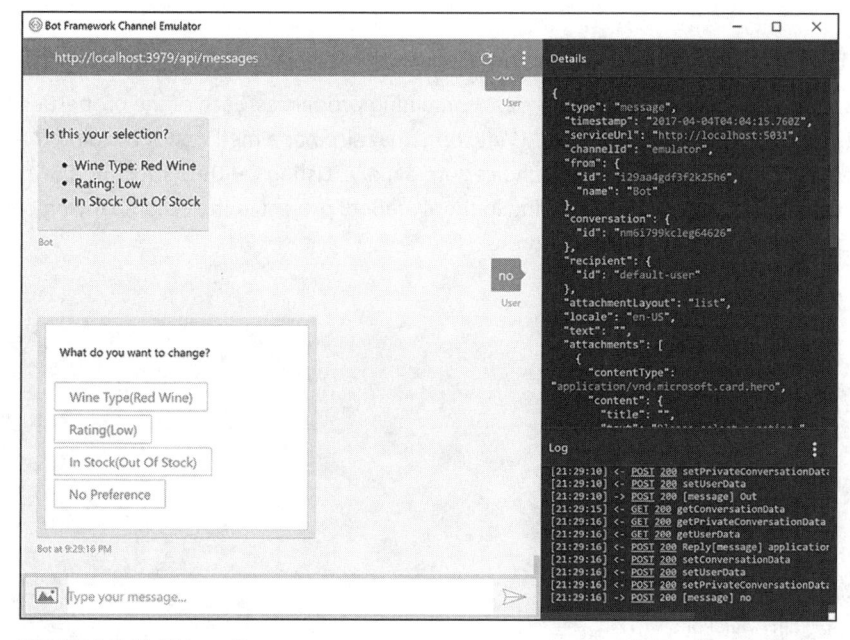

FIGURE 6-5 Finishing a Form.

CHAPTER 6   Using FormFlow   123

Users can type **yes** or **no** in response to the summary. As shown in Figure 6-5, Typing **no** makes FormFlow prompt for which item to change and takes the user there to make a new choice for that item. Typing or clicking the *No Preference* button takes the user back to the summary so they can eventually say **yes** and WineBot does the wine search.

That was the user experience. Now let's look at the code that makes it work.

A Basic FormFlow Chatbot

Since FormFlow is a different type of dialog, a few things are different, like having a plain old class that doesn't derive from an interface, properties that are an enum type, and how FormFlow starts up. This section walks through the basic WineBot code so you can see the differences between *IDialog<T>* and FormFlow code.

The Wine API Interface

This chapter uses the same WineApi class from Chapter 5. Because of that, we won't need to describe it again. You can review the *WineApi* listing in Chapter 5 to get up to speed on what it does. Essentially, it's an abstraction that makes a REST call to the Wine.com Wine API to perform serarches.

> **Note** You need your own Wine.com API key to run the code. You can obtain an API key from *https://api.wine.com/* for free.

WineForm: A FormFlow Form

At its most basic level, a FormFlow form is a C# class containing properties. Each of the properties are enum types. This chapter creates a form, called *WineForm*, that also has a method for building the form. The properties hold the state, which are choices users make. Listing 6-1 through Listing 6-4 show the enums and the *WineForm*, which you can find in the *WineBot2* project in the accompanying code. You'll see how to use *WineForm* in a later section.

LISTING 6-1 WineBot – *WineType* Enum

```
namespace WineBot2
{
    public enum WineType
    {
        None = 0,
        RedWine = 124,
        WhiteWine = 125,
        ChampagneAndSparkling = 123,
        RoseWine = 126,
        DessertSherryAndPort = 128,
        Sake = 134
```

```
        }
    }
```

Listing 6-1 contains the *WineType* enum, where each member has an explicit value. These values correspond to category numbers for the Wine API. While class state holds the enum type, the search logic can cast the enum to *int* to get the category.

LISTING 6-2 WineBot – *RatingType* Enum

```
namespace WineBot2
{
    public enum RatingType
    {
        None = 0,
        Low = 25,
        Medium = 50,
        High = 75
    }
}
```

Listing 6-2 breaks *RatingType* into 3 values: *Low*, *Medium*, and *High*. The Wine API takes a rating between *1* and *100*, but I designed WineBot to work with fewer values. Like *WineType*, the code can cast the value to an *int* for search input.

LISTING 6-3 WineBot – *StockingType* Enum

```
namespace WineBot2
{
    public enum StockingType
    {
        None = 0,
        InStock,
        OutOfStock
    }
}
```

StockingType, in Listing 6-3 handles a bool condition where a wine is either in stock or not. The enum values are there to give the user a choice.

None is a common member of all the enums and explicitly set to *0* for documentation. FormFlow expects the *None*, with value *0*, to be present. The *None* is optional, but if you choose to omit it, or are using an enum from a library that you can't change, FormFlow requires that you make the property type nullable in the form. The *DoSearch* method, from Listing 6-4, uses this value to filter the search based on whether a matching wine is in stock.

LISTING 6-4 WineBot – *WineForm* Class

```csharp
using System;
using System.Linq;
using System.Threading.Tasks;
using Microsoft.Bot.Builder.Dialogs;
using Microsoft.Bot.Builder.FormFlow;

namespace WineBot2
{
    [Serializable]
    public class WineForm
    {
        public WineType WineType { get; set; }
        public RatingType Rating { get; set; }
        public StockingType InStock { get; set; }

        public IForm<WineForm> BuildForm()
        {
            return new FormBuilder<WineForm>()
                .Message(
                    "I have a few questions on your wine search. " +
                    "You can type \"help\" at any time for more info.")
                .OnCompletion(DoSearch)
                .Build();
        }

        async Task DoSearch(IDialogContext context, WineForm wineInfo)
        {
            List[] wines =
                await new WineApi().SearchAsync(
                    (int)WineType,
                    (int)Rating,
                    InStock == StockingType.InStock,
                    "");

            string message;

            if (wines.Any())
                message = "Here are the top matching wines: " +
                            string.Join(", ", wines.Select(w => w.Name));
            else
                message = "Sorry, No wines found matching your criteria.";

            await context.PostAsync(message);
        }
    }
}
```

FormFlow reads enum values to build menus. It adds spaces between words of individual enum members automatically. For example, *RedWine* becomes "Red Wine" in two words. FormFlow breaks words by capitalization. FormFlow also breaks words on underscores, so *Red_Wine* becomes "Red Wine" too. However, standard C# naming conventions prefer PascalCase, as you see in this book.

> **Note** C# has a couple common naming conventions where identifiers use either PascalCase or camelCase. PascalCase is used for identifiers like classes, properties, methods, and other class members. Just as in the name, PascalCase, the each word of the identifier is upper case, including the first. Camel case is for private fields, parameters, and variables. The camelCase format lower-cases the first word and capitalizes subsequent workds in the identifier. These are the only two types of identifier conventions you'll see throughout the book.

WineForm, in Listing 6-4, is *Serializable*. Just like *IDialog<T>*, FormFlow types must be serializable so the Bot Framework can persist state on the Bot State Service between user activities. The properties; *WineType*, *Rating*, and *InStock*; hold the user's choices. Notice that their types are the enum types from Listing 6-1 through Listing 6-3.

> **Note** Form state can also be fields, instead of properties. It's your choice.

Essentially, FormFlow reads these properties, via reflection, and dynamically generates a question and menu based on the property name and type. The order of properties in the class determines the order FormFlow uses to present a question to the user. The logic to make this happen occurs in the *BuildForm* method, repeated below for your convenience:

```
public IForm<WineForm> BuildForm()
{
    return new FormBuilder<WineForm>()
        .Message(
            "I have a few questions on your wine search. " +
            "You can type \"help\" at any time for more info.")
        .OnCompletion(DoSearch)
        .Build();
}
```

BuildForm uses the *FormBuilder* class, which offers a fluent interface for building a form. The only requirement here is to instantiate *FormBuilder* and call *Build*, which performs all of the internal work to prepare the form as a dialog and returns an *IForm<WineForm>*. Later, you'll see how WineBot wraps this *IForm<WineForm>* in a dialog to handle user input.

FormBuilder has several methods to customize how the form asks questions and you'll see how to use each of these methods during a deep dive in Chapter 7, Customizing FormFlow. WineBot uses the *Message* and *OnCompletion* methods to produce a minimal, but working chatbot. *Message* defines the text to show the user at the beginning of interaction with *WineForm*, before asking the first question. This text displays in addition to the *Welcome* message WineBot displays on the *ConversationUpdate Activity*.

After the text, which is specified in *Message* displays, WineBot asks all of the questions corresponding to *WineForm* properties. After WineBot asks all questions and the user confirms that their choices are correct, *FormBuilder* executes the method referenced by *OnCompletion*, whose type is the *OnCompletionAsyncDelegate* delegate below:

```
public delegate Task OnCompletionAsyncDelegate<T>(IDialogContext context, T state);
```

According to *OnCompletionAsyncDelegate*, the referenced method receives an *IDialogContext* reference. As discussed in Chapter 5, Building Dialogs, *IDialogContext* gives access to several convenience methods that chatbots commonly need when responding to user input. In the current implementation the *T* in state will be *WineForm*, because that's the instantiated type of *FormBuilder*. Here's the *DoSearch* method, specified as the *OnCompletionAsyncDelegate* type argument that *OnCompletion* refers to:

```
async Task DoSearch(IDialogContext context, WineForm wineInfo)
{
    List[] wines =
        await new WineApi().SearchAsync(
            (int)WineType,
            (int)Rating,
            InStock == StockingType.InStock,
            "");

    string message;

    if (wines.Any())
        message = "Here are the top matching wines: " +
                  string.Join(", ", wines.Select(w => w.Name));
    else
        message = "Sorry, No wines found matching your criteria.";

    await context.PostAsync(message);
}
```

WineBot uses the same *WineApi* class as the one used in Chapter 5, but look at the *SearchAsync* arguments. Both *WineType* and *Rating* use a cast operator to convert their enum type values to an *int*, passing values that the Wine API recognizes. The *inStock* argument is *bool*, so a comparison operator works there. To keep this version of WineBot simple, the *searchTerms* is empty, but you'll see an example of how to obtain text input later in this chapter.

DoSearch reads parameters directly from it's containing type, because they're available in the same instance. However, if *DoSearch* were *static* or resided in another class, the *wineInfo* parameter would have supplied access to properties. This might be a good argument for properties over fields if you prefer the encapsulation benefits.

> **Tip** You can also allow a user to select multiple items from a single option by specifying the type as a List. For example, *List<WineType> WineTypes { get; set; }* would allow a user to choose multiple wine types. In this case, you could read each value of *WineTypes* in the *DoSearch* method, Listing 6-4, to read each value and create a *WineApi SearchAsync* overload that accepts multiple values for *wineType*.

Finally, *IDialogContext* comes in handly by letting *DoSearch* communicate results with the user via *PostAsync*.

After *DoSearch* returns, *WineForm* is ready to go again from the beginning. Until then, FormFlow maintains the state of the conversation, as long as that state persists in the Bot State Service, as discussed in Chapter 4, Fine-Tuning Your Chatbot. This means that if the user stops communicating and then returns after a while, FormFlow will resume where it left off. Then the user has the option to request status, resume by answering remaining questions, or simply resetting.

That's the basics of how a FormFlow form works. Next, you'll learn how to use *WineForm* as the main dialog to WineBot.

Using *WineForm* as a Dialog

If you recall, from Listing 6-4, *BuildForm* returns an *IForm<WineForm>*. The next task for *WineForm* is to wrap that *IForm<WineForm>* in a dialog to handle user input activities. This section introduces the *MessagesController* for WineBot and shows how to use *WineForm* as a dialog, which you can see in Listing 6-5.

LISTING 6-5 WineBot – *MessagesController* Class

```
using System;
using System.Linq;
using System.Net;
using System.Net.Http;
using System.Threading.Tasks;
using System.Web.Http;
using Microsoft.Bot.Builder.Dialogs;
using Microsoft.Bot.Builder.FormFlow;
using Microsoft.Bot.Connector;

namespace WineBot2
{
    [BotAuthentication]
    public class MessagesController : ApiController
    {
        public async Task<HttpResponseMessage> Post([FromBody]Activity activity)
        {
            if (activity.Type == ActivityTypes.Message)
            {
                try
                {
                    await Conversation.SendAsync(activity, BuildWineDialog);
                }
                catch (FormCanceledException ex)
                {
                    HandleCanceledForm(activity, ex);
                }
            }
            else
            {
                await HandleSystemMessageAsync(activity);
            }

            return Request.CreateResponse(HttpStatusCode.OK);
        }
```

```csharp
IDialog<WineForm> BuildWineDialog()
{
    return FormDialog.FromForm(
        new WineForm().BuildForm());
}

async Task HandleSystemMessageAsync(Activity message)
{
    if (message.Type == ActivityTypes.ConversationUpdate)
    {
        const string WelcomeMessage =
            "Welcome to WineBot! " +
            "Through a series of questions, WineBot can do a " +
            "search and return wines that match your answers. " +
            "You can type \"start\" to get started.";

        Func<ChannelAccount, bool> isChatbot =
            channelAcct => channelAcct.Id == message.Recipient.Id;

        if (message.MembersAdded?.Any(isChatbot) ?? false)
        {
            Activity reply = message.CreateReply(WelcomeMessage);

            var connector = new ConnectorClient(new Uri(message.ServiceUrl));
            await connector.Conversations.ReplyToActivityAsync(reply);
        }
    }
}

void HandleCanceledForm(Activity activity, FormCanceledException ex)
{
    string responseMessage =
        $"Your conversation ended on {ex.Last}. " +
        "The following properties have values: " +
        string.Join(", ", ex.Completed);

    var connector = new ConnectorClient(new Uri(activity.ServiceUrl));
    var response = activity.CreateReply(responseMessage);
    connector.Conversations.ReplyToActivity(response);
}
}
```

The *MessagesController.Post* method in Listing 6-5 calls *SendAsync* if the activity type is *ActivityTypes.Message*. As in the examples of previous chapters, the first parameter of *SendAsync* is the activity passed to the *Post* method and the second is an *IDialog<T>*. The difference this time is that the previous examples created *IDialog<object>*, but now we're working with an *IDialog<WineForm>*—the same *WineForm* shown in Listing 6-4.

> **Note** Whether derived directly from *IDialog<T>* or wrapped in an *IDialog<T>* derived type, like FormFlow, all dialogs are ultimately *IDialog<T>*. Because the term "dialog" is so generic, we still use the *IDialog<T>* in this book to differentiate the discussion of the difference of FormFlow and a method-based dialog of Chapter 6, Building Dialogs.

The *BuildWineDialog* method returns an *IDialog<WineForm>* that *SendAsync* uses. Inside of *BuildWineDialog* are a couple nested method calls that result in the *IDialog<WineForm>*, repeated here:

```
return FormDialog.FromForm(
    new WineForm().BuildForm));
```

FormDialog implements *IFormDialog<T>*, which derives from *IDialog<T>*. As you learned in Chapter 5, *IDialog<T>* has a *StartAsync* member, which is the dialog entry point. So, *FormDialog* is managing all of the logic to manage dialog conversation. *FromForm* takes a *BuildFormDelegate* argument type, shown bhere:

```
public delegate IForm<T> BuildFormDelegate<T>() where T : class;
```

This is why *BuildForm* returns an *IForm<T>*, just like the *BuildFormDelegate*–so you can pass a *BuildForm* reference to *FromForm*. Internally, *FromForm* instantiates a new *FormDialog*, passing the *BuildForm* reference and invokes *BuildForm* along with other tasks to configure and manage conversation state. Coming back full-circle, this is how *BuildWineDialog* returns the *IDialog<WineForm>* that *SendAsync* needs.

> **Note** Listing 6-5 handles the *ActivityTypes.ConversationUpdate* activity, which was described in detail in Chapter 4, Fine-Tuning Your Chatbot. In fact, you'll see this as a common practice throughout this book. The message recommends that you type **start** to begin the conversation. In reality, the user can type anything and this particular code will launch the FormFlow form as a dialog. In previous chapters, you saw how to intercept and handle commands, which is another way to handle this situation. In this case, WineBot gives the user specific guidance in order to be helpful.

Finally, one of the commands that a user can give to FormFlow is *Quit*, for when they want to exit before completing the form. When the user types **quit**, FormFlow throws *FormCanceledException*, and Table 6-1 shows available *FormCanceledException* members.

TABLE 6-1 FormCanceledException Members

Member	Description
Completed	Names of steps that the user has already provided answers for
Last	Step that the user was on
LastForm	Reference to the FormFlow form that the user was using

Table 6-1 uses the term *Steps* to refer to the state of the form when the user has sen the *Quit* command. In *WineForm*, those steps correspond to the properties that hold the user's answers. The following *HandleCanceledForm* method, from Listing 6-5, shows one way to handle the *FormCanceledException*:

```
void HandleCanceledForm(Activity activity, FormCanceledException ex)
{
    string responseMessage =
        $"Your conversation ended on {ex.Last}. " +
        "The following properties have values: " +
        string.Join(", ", ex.Completed);

    var connector = new ConnectorClient(new Uri(activity.ServiceUrl));
    var response = activity.CreateReply(responseMessage);
    connector.Conversations.ReplyToActivity(response);
}
```

When building the *responseMessage* string, *HandleCanceledForm* accesses *ex.Last* and *ex.Completed* properties. The *ex.Last* property tells where the user was at in the form. The *ex.Completed* contains a list of property names that the user already provided answers for, at the point in time that the user *Quit*. When handling *FormCanceledException*, you can also reference the form, like *WineForm*, instance to access the property values via the *ex.LastForm* property.

Now that the basics of FormFlow have been covered, let's move on to more about attributes and other features that you can use with forms to enhance the user experience.

Enhancing FormFlow Conversations

In FormFlow, you can decorate the form class and its members with attributes. These attributes, outlined in Table 6-2, manage the appearance and layout of text for improving the chatbot user experience. The examples use *WebBot3* in the accompanying source code, which is largely the same code as *WebBot2*, except for enhancements discussed in this section.

TABLE 6-2 FormFlow Attributes

Attribute	Description	Applies To
Describe	Overrides FormFlow's default description	Enum, field, or property
Numeric	Specifies a min and max number	Field or property
Optional	Indicates that a value is not required	Field or property
Pattern	Uses a regular expression to validate value	Field or property
Prompt	Overrides default question	Field or property
Template	Defines text and patterns for how to prompt the user	Class, struct, field, or property
Terms	Allows alternate terms that can match choices	Field or property

The following sections explain each of the attributes from Table 6-2.

> **Note** The code in this section uses *WineBot3* in the accompanying source code. It uses port *3978* to avoid confusion with *WineBot2* that might have cached on port *3979*.

The *Describe* Attribute

Each field, property, or enum value has a default description. FormFlow creates default descriptions by using capitalization, or underscores, to separate words of an identifier to use in help text, menu items, and questions. An example of a property is *WineType*, which becomes "Wine Type" in the Bot Emulator question. An example of an enum member is *WineTypes.DessertSherryAndPort*, which becomes "Dessert Sherry And Port" as a menu item.

Often, the default description is inadequate or you might want to show the user the description in a different way. This is where the *Description* attribute helps. The following example changes the default *WineType* property in *WineForm*:

```
[Describe(Description="Type of Wine")]
public WineType WineType { get; set; }
```

Now, the user sees "Type of Wine" instead of "Wine Type." You can also change enum members, as shown in the following *WineType* enum example:

```
public enum WineType
{
    None = 0,
    RedWine = 124,
    WhiteWine = 125,
    ChampagneAndSparkling = 123,
    RoseWine = 126,
    [Describe("Dessert, Sherry, and Port")]
    DessertSherryAndPort = 128,
    Sake = 134
}
```

The *Describe* attribute decorates the *DessertSherryAndPort* member, adding punctuation and lower casing *and*. When the user communicates with the chatbot, they'll see the text from *Describe* attributes, as shown in Figure 6-6.

CHAPTER 6   Using FormFlow   **133**

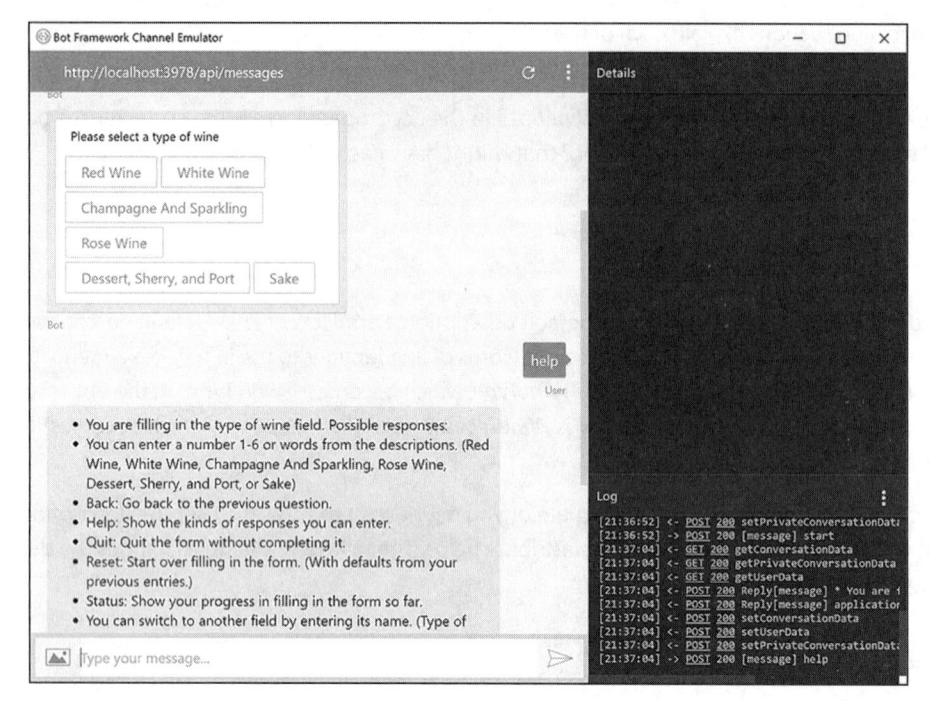

![Figure 6-6 screenshot of Bot Framework Channel Emulator]

FIGURE 6-6 Viewing Results of the *Describe* Attribute.

In Figure 6-6, I typed **help** at the menu for the *WineForm.WineType* property. You can see "type of wine" in both the help text and the menu title. You can also see the differences in the *WineType* members, decorated with the *Describe* attribute.

> **Note** *Describe* has other properties for *Title*, *Subtitle*, *Image*, and *Message* that are used for a more graphical presentation of the menu, which are covered in Chapter 10, Attaching Cards.

The *Numeric Attribute*

WineBot2 defined *Rating* as an enum with three choices, making it easier to integrate with default FormFlow behavior. This, however, doesn't give the user sufficient flexibility because the Wine API takes a value between *1* and *100*. This is where the *Numeric* attribute helps, as shown in the refactored *Rating* property, shown here:

```
[Numeric(1, 100)]
public int Rating { get; set; }
```

The *Numeric* attribute has two parameters: *Min* and *Max*. Here, you can see that *Min* is *1* and *Max* is *100*. What this does is change from a button menu in the emulator to a text string, shown below, that let's the user enter a number:

Please enter a number between 1 and 100 for rating (current choice: 0).

Now the user can enter a real number and FormFlow validates that the number is greater than or equal to *1* or less than or equal to *100*. Also, notice how the current choice, above, is *0*. That's because the *Rating* property hasn't been set and *0* is the default value for an *int*. You can initialize the property to a default value to avoid any problems with this.

The *Optional* Attribute

FormFlow normally performs validation, making all fields/properties required. There might be times, however, when you won't require the user to enter a value. That's the case with *InStock*, shown here:

```
[Optional]
public StockingType InStock { get; set; }
```

Here the *Optional* attribute allows the user to bypass entering a value. When you do this, FormFlow shows an additional *No Preference* option. Selecting *No Preference* returns the enum option with a value of *0* (*None* in WineBot) or *null* if the property type is nullable.

The *Pattern* Attribute

You can allow the user to enter text by using the *Pattern* attribute. The *Pattern* attribute accepts a single required parameter, which is a regular expression string. The following property demonstrates how to use the *Pattern* attribute:

```
[Pattern(@".+@.+\..+")]
public string EmailAddress { get; set; }
```

The *EmailAddress* property should receive a properly formatted email address. While the regular expression used in this example isn't very sophisticated, it demonstrates how you add a string parameter to the *Pattern* attribute. Sometimes, you might want a purely free-form string field, which you can do with the following example:

```
[Pattern(".*")]
public string SearchTerms { get; set; }
```

In the *SearchTerms* example, the input must not have any constraints, and the regular expression reflects that a user can provide any input they desire.

The *Prompt* Attribute

Sometimes the default question doesn't necessarily make sense for the current entry, so you want to add additional info, or use something that reflects the personality of the chatbot. The *Prompt* attribute lets you do this by changing the question text, as the following example demonstrates:

```
[Numeric(1, 100)]
[Prompt(
    "Please enter a minimum rating (from 1 to 100).",
```

```
        "What rating, 1 to 100, would you like to search for?")]
public int Rating { get; set; }
```

The Prompt attribute for *Rating* has two arguments, representing different ways to ask the same question. This lets a chatbot vary the conversation, which might be more interesting for the user. You have a choice to add only one string or multiple strings, beyond the two in the *Rating* example.

The *Terms Attribute*

FormFlow out-of-the-box is very flexible and has a lot of options. When the user types, however, they must spell everything correctly or receive an error message. Sometimes, you might want to select an item if the user types part of the answer, an abbreviation, or a common misspelling. The *Terms* attribute demonstrated here, shows how to give the user flexibility in the answers they provide:

```
public enum WineType
{
    None = 0,
    RedWine = 124,
    WhiteWine = 125,

    [Terms("[S|s|C|c]hamp.*", MaxPhrase=2)]
    ChampagneAndSparkling = 123,

    RoseWine = 126,

    [Terms(
        "desert", "shery", "prt",
        "dessert", "sherry", "port",
        "Dessert, Sherry, and Port")]
    [Describe("Dessert, Sherry, and Port")]
    DessertSherryAndPort = 128,

    Sake = 134
}
```

In the example, there are two *WineType* enum members decorated with the *Terms* attribute: *ChampagneAndSparkling* and *DessertSherryAndPort*. The *Terms* attribute on *DessertSherryAndPort* has a list of seven different terms, and the user's input will pass if it matches any of those terms. As you can see, there are a few misspellings along with the full spelling that matches the *Describe* attribute, just in case the user has trouble and wants to make sure they type the exact words.

The *Terms* attribute for *ChampagneAndSparkling* has two parameters. The first is a regular expression that assumes it might be easy for the user to misspell champagne. The second parameter, *MaxPhrase*, supports auto-generation capability where FormFlow creates multiple phrases that would match the value. In this example, *MaxPhrase* is 2, indicating that the user has two additional variations that could potentially match their input.

Both regular expressions and MaxPhrase increase the chances that a given option will be selected. However, you might want to adjust this based on what other options are available. When a user's input matches multiple items, FormFlow shows potential matches and asks the user which item they really wanted. e.g. notice that I didn't have *Terms* attributes, with *MaxPhrase* on *RedWine*, *WhiteWine*, and *RoseWine* properties because typing **wine** would result in FormFlow querying the user for which wine, which is half the list. Imagine if there were even more matches in a long list. You could probably see where too many matches of this type might introduce friction for the user. So, testing and limiting options appropriately might improve the user experience.

Combined, alternate terms, regular expressions, and auto-generation give a chatbot a lot of flexibility in being able to recognize what the user wants.

> **Tip** This might seem like a lot of fuss about allowing the user to type in text when buttons are more efficient. However, remember that one of the great features of the Bot Framework is in giving you the opportunity to write a chatbot one time and deploy to multiple platforms. There are some channels, like SMS and email that don't offer buttons, which means that text is the only option. Another consideration is that an entire generation of users has grown up communicating with messaging applications and text is second nature to them, as well as other power users who have embraced the conversational nature of messaging applications. While buttons might be the default interface when it's available, it's best practice to also offer the most pleasant user experience via text. In the longer term, the advent of voice will make this obvious.

The remaining attribute that I have yet to cover is *Template*. Both *Prompt* and *Template* have sophisticated options, including the ability to use a pattern language, which you'll learn about next.

Advanced Templates and Patterns

FormFlow has an advanced templating system that supports overriding its default messaging by targeting where a message can be used, and employing a pattern language to customize a message. In this section, you'll learn about the FormFlow pattern language and how both *Prompt* and *Template* attributes use it. You'll also learn about the advanced features of *Template* attributes and various options that enhance the chatbot user experience.

> **Note** When discussing patterns, I'll use the term *field* to refer to either C# fields or properties, both of which work the same for FormFlow attributes and associated patterns.

Pattern Language

The FormFlow pattern language is a set of formatting items for a string that specifies items like substitution options or question layout. The patterns fall into a couple different categories: field question display and template usage placeholders. This section describes the field question display patterns, leaving template usage placeholders to a later section.

The following *Prompt* attribute for the *WineType* property shows how pattern language can be used, and starts off with the description and value of the current field for *WineType*:

```
[Describe(Description="Type of Wine")]
[Prompt("Which {&} would you like? (current value: {})")]
public WineType WineType { get; set; }
```

In the pattern string for the *Prompt* attribute above, the *{&}* is a placeholder for the property description, which is "type of wine" as specified in the *Describe* attribute. Even though the *Description* property for the *Describe* attribute is upper case, FormFlow lower cases the description when it's in the middle of a sentence. Without the *Describe* attribute, FormFlow would have substituted the default "wine type" into the placeholder. The *{}* is a placeholder for the value of the current field. Now, the question "Which type of wine would you like? (current value: Unspecified)" displays instead the default "Please select a type of wine.". The first time through, the current value is unspecified, but if the user picks *Red Wine* and then re-displays (by typing **WineType**), the new message is "Which type of wine would you like? (current value: Red Wine)." Here again, the *Describe* attribute of an enum member will override the default FormFlow generated description.

In addition to the current field, there are patterns to display the description and value of a separate field. The patterns for the *Rating* field below combine all of the description/value patterns:

```
[Numeric(1, 100)]
[Prompt(
    "Please enter a minimum rating (from 1 to 100).",
    "What rating, 1 to 100, would you like to search for?",
    "Minimum {&} (selected {&WineType}: {WineType}, current rating: {:000})")]
public int Rating { get; set; }
```

In this example, the third prompt string is "Minimum {&} (selected {&WineType}: {WineType}, current rating: {:000})". As mentioned earlier, *{&}* is the description of the current field, but *{&WineType}* is the description of the *WineType* field, demonstrating how to display the description of a field other than the current field. For displaying values, *{WineType}* is the value of the *WineType* field and *{:000}* is the value of the current, *Rating*, field. In the previous example, for *WineType*, the current field value is empty braces, *{}*. However, contents of the value pattern for the rating field has a format specifier *:000*, indicating that the format is 3 digits, left-padded with *0*. Also, notice the colon prefix to the pattern, which is required. To summarize, the ampersand is a placeholder for field description and no ampersand is a placeholder for field value.

In the previous description, field description/value patterns replaced the default FormFlow questions. However, they also replaced the menus with nothing. Essentially, FormFlow asked the question without showing any options. The user could type **help** and see the options, but that might not be the best user experience. The next example shows how to add field members back to the question and how to change the appearance of those options:

```
[Describe(Description="Type of Wine")]
[Prompt("Which {&} would you like? (current value: {}) {||}")]
public WineType WineType { get; set; }
```

This example is nearly identical to the *WineType* Prompt above, except for the *{||}* appended to the end of the string, indicating that the UI should display menu options.

> **Note** The menu pattern *{||}* shows buttons in the Bot Emulator and most messaging channels, but that isn't guaranteed. Remember that the Bot Framework is multi-platform and it's up to the channel platform on whether buttons appear. Additionally, some channels, like SMS, can't display buttons and the user sees a text menu instead.

The remaining patterns are designed to work with Templates, which you'll learn about next.

Basic Templates

FormFlow has an extensive set of standard messaging that covers questions, help, status, and more. As you learned in earlier sectons, the *Prompt* attribute is designed to help customize messaging, but this was focused on fields. There's another attribute, *Template*, that allows even further customization of FormFlow messaging. *Template* attributes, like *Prompt*, can decorate fields, but also classes and structs. Here's an example of how a *Template* attribute can decorate a class:

```
[Serializable]
[Template(TemplateUsage.String, "What {&} would you like to enter?")]
public class WineForm
{
}
```

In this example, the *Template* attribute has two parameters, the *TemplateUsage* enum and pattern string. The *TemplateUsage.String* means that this attribute applies to all fields in this class whose type is *string*. The pattern string replaces the default FormFlow question for *string* fields. Because both *SearchTerms* and *EmailAddress* are type *string*, this customizes their question text. To be complete, any other fields of different type are not affected by a *Template* attribute with *TemplateUsage.String*. The next sections explain more of the template usage and other available options.

Template Usage

The *TemplateUsage* enum, Listing 6-6, shows that the Bot Framework has an extensive set of messages available for customization. At first glance, this list might appear quite daunting. However, there are some common patterns that can help break down and categorize members to help digest what is available.

First, notice that there are several members associated with .NET types, such as *Bool*, *DateTime*, *Double*, and more. These members help customize questions associated with a type – just like *TemplateUsage.String* customized all fields of type *string* in the previous example. Additionally, each of the members for a type has a corresponding member with a *Help* suffix, *<type>Help*, such as in *BoolHelp*, *DateTimeHelp*, *DoubleHelp*, and more. Each of the *<type>Help* members help customize the FormFlow *Help* message (when the user types **help**) for questions whose type is *<type>*.

Another set of members customize FormFlow navigation and commands, such as *Help* and *Status-Format*.

LISTING 6-6 The Bot Framework *TemplateUsage* Enum

```
/// <summary>
/// All of the built-in templates.
/// </summary>
/// <remarks>
/// A good way to understand these is to look at the default templates defined in <see cref="FormConfiguration.Templates"/>
/// </remarks>
public enum TemplateUsage
{
    /// <summary>   An enum constant representing the none option. </summary>
    None,

    /// <summary>
    /// How to ask for a boolean.
    /// </summary>
    Bool,

    /// <summary>
    /// What you can enter when entering a bool.
    /// </summary>
    /// <remarks>
    /// Within this template {0} is the current choice if any and {1} is no preference if optional.
    /// </remarks>
    BoolHelp,

    /// <summary>
    /// Clarify an ambiguous choice.
    /// </summary>
    /// <remarks>This template can use {0} to capture the term that was ambiguous.</remarks>
    Clarify,

    /// <summary>
    /// Default confirmation.
    /// </summary>
    Confirmation,

    /// <summary>
    /// Show the current choice.
    /// </summary>
    /// <remarks>
    /// This is how the current choice is represented as an option.
    /// If you change this, you should also change <see cref="FormConfiguration.CurrentChoice"/>
    /// so that what people can type matches what you show.
    /// </remarks>
    CurrentChoice,

    /// <summary>
```

```csharp
/// How to ask for a <see cref="DateTime"/>.
/// </summary>
DateTime,

/// <summary>
/// What you can enter when entering a <see cref="DateTime"/>.
/// </summary>
/// <remarks>
/// Within this template {0} is the current choice if any and {1} is no preference if optional.
/// </remarks>
/// <remarks>
/// This template can use {0} to get the current choice or {1} for no preference if field is optional.
/// </remarks>
DateTimeHelp,

/// <summary>
/// How to ask for a double.
/// </summary>
/// <remarks>
/// Within this template if numerical limits are specified using <see cref="NumericAttribute"/>,
/// {0} is the minimum possible value and {1} is the maximum possible value.
/// </remarks>
Double,

/// <summary>
/// What you can enter when entering a double.
/// </summary>
/// <remarks>
/// Within this template {0} is the current choice if any and {1} is no preference if optional.
/// If limits are specified through <see cref="NumericAttribute"/>, then {2} will be the minimum possible value
/// and {3} the maximum possible value.
/// </remarks>
/// <remarks>
/// Within this template, {0} is current choice if any, {1} is no preference for optional
/// and {1} and {2} are min/max if specified.
/// </remarks>
DoubleHelp,

/// <summary>
/// What you can enter when selecting a single value from a numbered enumeration.
/// </summary>
/// <remarks>
/// Within this template, {0} is the minimum choice. {1} is the maximum choice
/// and {2} is a description of all the possible words.
/// </remarks>
EnumOneNumberHelp,

/// <summary>
///  What you can enter when selecting multiple values from a numbered enumeration.
```

```csharp
        /// </summary>
        /// <remarks>
        /// Within this template, {0} is the minimum choice. {1} is the maximum choice
        /// and {2} is a description of all the possible words.
        /// </remarks>
        EnumManyNumberHelp,

        /// <summary>
        /// What you can enter when selecting one value from an enumeration.
        /// </summary>
        /// <remarks>
        /// Within this template, {2} is a list of the possible values.
        /// </remarks>
        EnumOneWordHelp,

        /// <summary>
        /// What you can enter when selecting mutiple values from an enumeration.
        /// </summary>
        /// <remarks>
        /// Within this template, {2} is a list of the possible values.
        /// </remarks>
        EnumManyWordHelp,

        /// <summary>
        /// How to ask for one value from an enumeration.
        /// </summary>
        EnumSelectOne,

        /// <summary>
        /// How to ask for multiple values from an enumeration.
        /// </summary>
        EnumSelectMany,

        /// <summary>
        /// How to show feedback after user input.
        /// </summary>
        /// <remarks>
        /// Within this template, unmatched input is available through {0}, but it should be wrapped in
        /// an optional {?} in \ref patterns in case everything was matched.
        /// </remarks>
        Feedback,

        /// <summary>
        /// What to display when asked for help.
        /// </summary>
        /// <remarks>
        /// This template controls the overall help experience.  {0} will be recognizer specific help and {1} will be command help.
        /// </remarks>
        Help,

        /// <summary>
        /// What to display when asked for help while clarifying.
        /// </summary>
        /// <remarks>
```

```
        /// This template controls the overall help experience.  {0} will be recognizer
specific help and {1} will be command help.
        /// </remarks>
        HelpClarify,

        /// <summary>
        /// What to display when asked for help while in a confirmation.
        /// </summary>
        /// <remarks>
        /// This template controls the overall help experience.  {0} will be recognizer
specific help and {1} will be command help.
        /// </remarks>
        HelpConfirm,

        /// <summary>
        /// What to display when asked for help while navigating.
        /// </summary>
        /// <remarks>
        /// This template controls the overall help experience.  {0} will be recognizer
specific help and {1} will be command help.
        /// </remarks>
        HelpNavigation,

        /// <summary>
        /// How to ask for an integer.
        /// </summary>
        /// <remarks>
        /// Within this template if numerical limits are specified using <see cref="NumericAttribute"/>,
        /// {0} is the minimum possible value and {1} is the maximum possible value.
        /// </remarks>
        Integer,

        /// <summary>
        /// What you can enter while entering an integer.
        /// </summary>
        /// <remarks>
        /// Within this template, {0} is current choice if any, {1} is no preference for optional
        /// and {1} and {2} are min/max if specified.
        /// </remarks>
        IntegerHelp,

        /// <summary>
        /// How to ask for a navigation.
        /// </summary>
        Navigation,

        /// <summary>
        /// Help pattern for navigation commands.
        /// </summary>
        /// <remarks>
        /// Within this template, {0} has the list of possible field names.
        /// </remarks>
        NavigationCommandHelp,
```

```csharp
        /// <summary>
        /// Navigation format for one line in navigation choices.
        /// </summary>
        NavigationFormat,

        /// <summary>
        /// What you can enter when navigating.
        /// </summary>
        /// <remarks>
        /// Within this template, if numeric choies are allowed {0} is the minimum possible choice
        /// and {1} the maximum possible choice.
        /// </remarks>
        NavigationHelp,

        /// <summary>
        /// How to show no preference in an optional field.
        /// </summary>
        NoPreference,

        /// <summary>
        /// Response when an input is not understood.
        /// </summary>
        /// <remarks>
        /// When no input is matched this template is used and gets {0} for what the user entered.
        /// </remarks>
        NotUnderstood,

        /// <summary>
        /// Format for one entry in status.
        /// </summary>
        StatusFormat,

        /// <summary>
        /// How to ask for a string.
        /// </summary>
        String,

        /// <summary>
        /// What to display when asked for help when entering a string.
        /// </summary>
        /// <remarks>
        /// Within this template {0} is the current choice if any and {1} is no preference if optional.
        /// </remarks>
        StringHelp,

        /// <summary>
        /// How to represent a value that has not yet been specified.
        /// </summary>
        Unspecified
    };
```

Another example of *TemplateUsage* is in handling situations where the chatbot doesn't understand what the user wants. The following template shows how to use *TemplateUsage.NotUnderstood*:

```
[Template(TemplateUsage.NotUnderstood,
    "Sorry, I didn't get that.",
    "Please try again.",
    "My apologies, I didn't understand '{0}'.",
    "Excuse me, I didn't quite get that.",
    "Sorry, but I'm a chatbot and don't know what '{0}' means.")]
```

Notice that this *Template* attribute has several pattern strings. Imagine a user interacting with a chatbot and continuously receiving the same message. After a few times, the user might become frustrated and stop using the chatbot, but by varying the message, they might not feel like they're stuck and pay attention to the message.

Notice the *{0}* specifier in a couple of the strings. This holds the value that the user typed in. For example, if the user was entering a response for the *WineType* field and typed **rse**, they might receive the *My apologies, "I didn't understand 'rse'."* message, highlighting that they misspelled *rose*. To know what values are available, examine the *TemplateUsage* enum in Listing 6-6 for the *NotUnderstood* member:

```
/// <summary>
/// Response when an input is not understood.
/// </summary>
/// <remarks>
/// When no input is matched this template is used and gets {0} for what the user entered.
/// </remarks>
NotUnderstood,
```

The comments explain that *{0}* is the placeholder for what the user typed in. Additionally, this message will appear in the Visual Studio IDE Intellisense when typing the dot after *TemplateUsage*.

> **Tip** To learn what default message matches a *TemplateUsage* member, visit the open source *BotBuilder* project code at *https://github.com/Microsoft/BotBuilder/*. Then find *FormConfiguration.Templates* and look for the mapping of the *TemplateUsage* to resource name. Then map the resource name to the *.resx* file in the *BotBuilder* source code *Resource* folder.

Now that you know about *TemplateUsage* and placeholders, here's a useful pattern, demonstrating how to customize a help message:

```
[Pattern(".*")]
[Template(
    TemplateUsage.StringHelp,
    "Additional words to filter search {?{0}}")]
public string SearchTerms { get; set; }
```

Notice the *{?{0}}* appended to the *Template* attribute string. This is a conditional placeholder, symbolized by the leading question mark. The *{0}* placeholder in *TemplateUsage* for *StringHelp* is for the current choice. If this is the first time the user encounters this question, there is not a current choice

CHAPTER 6 Using FormFlow **145**

and the conditional placeholder hides the current choice. However, if the user has specified a value for *SearchTerms*, they will see the current choice. Since this is a *<type>Help TemplateUsage* member, the user sees this message when they are on the *SearchTerms* question and they type **help**.

Also, the previous example demonstrated how to apply a *Template* attribute to a single field, rather than a whole class.

Template Options

In addition to patterns and *TemplateUsage*, FormFlow offers a set of options to customize messages even further. Listing 6-7 shows available options via the *TemplateBaseAttribute* class. Both *Prompt* and *Template* attributes derive from *TemplateBaseAttribute* and inherit these options.

LISTING 6-7 The Bot Framework *TemplateBaseAttribute* Class

```csharp
    /// <summary>
    /// Abstract base class used by all attributes that use \ref patterns.
    /// </summary>
    public abstract class TemplateBaseAttribute : FormFlowAttribute
    {
        /// <summary>
        /// When processing choices {||} in a \ref patterns string, provide a choice for the default value if present.
        /// </summary>
        public BoolDefault AllowDefault { get; set; }

        /// <summary>
        /// Control case when showing choices in {||} references in a \ref patterns string.
        /// </summary>
        public CaseNormalization ChoiceCase { get; set; }

        /// <summary>
        /// Format string used for presenting each choice when showing {||} choices in a \ref patterns string.
        /// </summary>
        /// <remarks>The choice format is passed two arguments, {0} is the number of the choice
        /// and {1} is the field name.</remarks>
        public string ChoiceFormat { get; set; }

        /// <summary>
        /// When constructing inline lists of choices using {||} in a \ref patterns string, the string used before the last choice.
        /// </summary>
        public string ChoiceLastSeparator { get; set; }

        /// <summary>
        /// When constructing inline choice lists for {||} in a \ref patterns string
        /// controls whether to include parentheses around choices.
        /// </summary>
        public BoolDefault ChoiceParens { get; set; }
```

```csharp
        /// <summary>
        /// When constructing inline lists using {||} in a \ref patterns string,
        /// the string used between all choices except the last.
        /// </summary>
        public string ChoiceSeparator { get; set; }

        /// <summary>
        /// How to display choices {||} when processed in a \ref patterns string.
        /// </summary>
        public ChoiceStyleOptions ChoiceStyle { get; set; }

        /// <summary>
        /// Control what kind of feedback the user gets after each input.
        /// </summary>
        public FeedbackOptions Feedback { get; set; }

        /// <summary>
        /// Control case when showing {&amp;} field name references in a \ref patterns string.
        /// </summary>
        public CaseNormalization FieldCase { get; set; }

        /// <summary>
        /// When constructing lists using {[]} in a \ref patterns string, the string used before the last value in the list.
        /// </summary>
        public string LastSeparator { get; set; }

        /// <summary>
        /// When constructing lists using {[]} in a \ref patterns string, the string used between all values except the last.
        /// </summary>
        public string Separator { get; set; }

        /// <summary>
        /// Control case when showing {} value references in a \ref patterns string.
        /// </summary>
        public CaseNormalization ValueCase { get; set; }
    }
```

Each of the options are named parameters for the *Prompt* or *Template* attributes. Here's an example that customizes the *WineType* options format:

```csharp
[Describe(Description="Type of Wine")]
[Prompt(
    "Which {&} would you like? (current value: {}) {||}",
    ChoiceStyle = ChoiceStyleOptions.PerLine)]
public WineType WineType { get; set; }
```

The *ChoiceStyleOptions* enum has several members for formatting the menu options. In this example, *ChoiceStyleOptions.PerLine* causes the output to be text, where each entry is numbered and is on a new line, as shown in Figure 6-7.

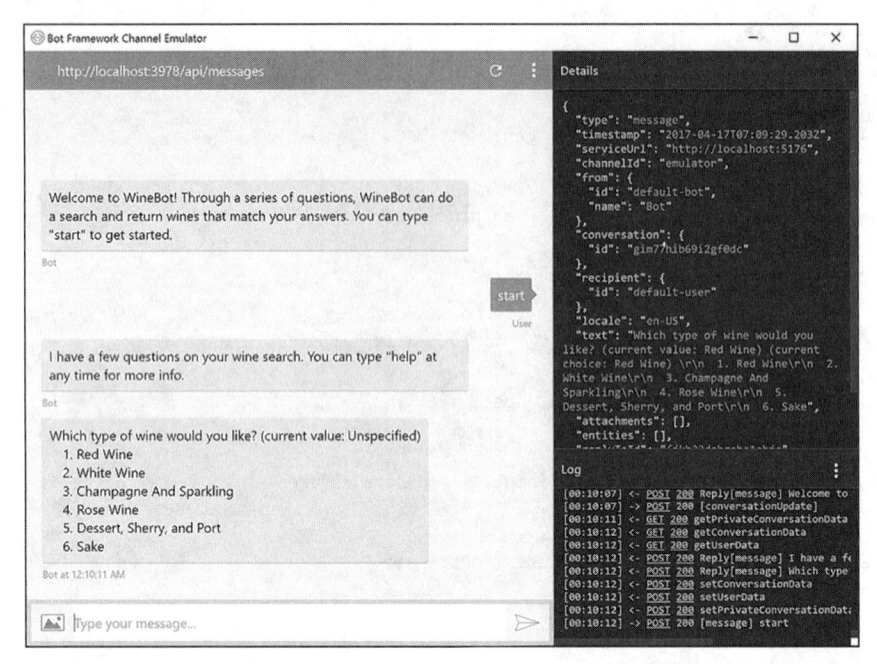

FIGURE 6-7 *ChoiceStyleOptions.PerLine Prompt* Attribute Option Showing Text Options on Separate Lines.

Listing 6-8 has the entire update of WineForm that includes all of the attributes and patterns demonstrated in this section.

LISTING 6-8 *The WineForm* Class with Attributes and Patterns

```
using Microsoft.Bot.Builder.Dialogs;
using Microsoft.Bot.Builder.FormFlow;
using System;
using System.Linq;
using System.Threading.Tasks;

namespace WineBot3
{
    [Serializable]
    [Template(TemplateUsage.String, "What {&} would you like to enter?")]
    [Template(TemplateUsage.NotUnderstood,
        "Sorry, I didn't get that.",
        "Please try again.",
        "My apologies, I didn't understand '{0}'.",
        "Excuse me, I didn't quite get that.",
        "Sorry, but I'm a chatbot and don't know what '{0}' means.")]
    public class WineForm
    {
        [Describe(Description="Type of Wine")]
        [Prompt(
            "Which {&} would you like? (current value: {}) {||}",
            ChoiceStyle = ChoiceStyleOptions.PerLine)]
        public WineType WineType { get; set; }
```

```
    [Numeric(1, 100)]
    [Prompt(
        "Please enter a minimum rating (from 1 to 100).",
        "What rating, 1 to 100, would you like to search for?",
        "Minimum {&} (selected {&WineType}: {WineType}, current rating: {:000})")]
    public int Rating { get; set; }

    [Optional]
    public StockingType InStock { get; set; }

    [Pattern(".*")]
    [Template(
        TemplateUsage.StringHelp,
        "Additional words to filter search {?{0}}")]
    public string SearchTerms { get; set; }

    [Pattern(@".+@.+\..+")]
    public string EmailAddress { get; set; }

    public IForm<WineForm> BuildForm()
    {
        return new FormBuilder<WineForm>()
            .Message(
                "I have a few questions on your wine search. " +
                "You can type \"help\" at any time for more info.")
            .OnCompletion(DoSearch)
            .Build();
    }

    async Task DoSearch(IDialogContext context, WineForm wineInfo)
    {
        List[] wines =
            await new WineApi().SearchAsync(
                (int)wineInfo.WineType,
                wineInfo.Rating,
                wineInfo.InStock == StockingType.InStock,
                wineInfo.SearchTerms);

        string message;

        if (wines.Any())
            message = "Here are the top matching wines: " +
                        string.Join(", ", wines.Select(w => w.Name));
        else
            message = "Sorry, No wines found matching your criteria.";

        await context.PostAsync(message);
    }
  }
}
```

While this is a lot of information on FormFlow customization, Listing 6-8 shows how straight forward it is to customize the user experience for a chatbot with attributes and patterns.

Choosing Between FormFlow and *IDialog<T>*

At this point in time, you might be wondering how to choose between FormFlow and *IDialog<T>*. This section examines the characteristic of each to help you decide which is the best tool for the job.

As you've just seen, FormFlow is very easy for building Q&A chatbots. It's very simple to get started and relatively easy to customize with attributes. This is a very powerfull capability compared to the small amount of associated coding. FormFlow excels in situations where the chatbot needs to collect information from a user. This chapter showed how to collect filters for a wine search. Other examples are when you need a user to fill out a form to order food or another service, take a survey, or complete an application for a job.

An *IDialog<T>* is good for more free-form interaction. There isn't a specified sequence for how the user interacts with the chatbot. One minute they could be asking about one thing and the next minute move to another subject. You can put logic inside of callbacks to take the logical path through a conversation that the user wants to go. Examples might be browsing a store catalog, playing a game, or asking a vendor what type of services they provide and their hours of operation.

There isn't always a definitive rule on when to use one over the other and you'll no doubt find gray areas in between and wonder which way to go. e.g. In the case of WineBot, you could go either way without too much trouble. However, lets also examine some potentially extreme cases of when not to use a specific dialog type. One example of when a *IDialog<T>* might not work is for a survey. Imagine writing a dialog where there's a separate method for each question. The chatbot asks a question, the user responds, the chatbot obtains the result and saves in a property, and then moves on to ask the next question. This would result in a lot of repetitive code doing the same thing, taking more time than required, and potentially introducing bugs. In this example, FormFlow would have been the better choice.

In Chapter 7, Customizing FormFlow, you'll learn how to specify options where you don't want to ask a user a specific question, how to validate input, and how to dynamically present input. While those will be useful in scenarios where FormFlow does well, imagine writing a toy store chatbot. The dialog would ask what type of toy the user is interested in, show a list, put the selected toy in a cart, and take the user to checkout. If the user wants to modify their cart or select another toy, they could use the built-in FormFlow navigation. You might be able to write some adaptive logic for all of these conversation paths, but it would soon become unwieldy. Unless the process of buying something is so simple, such as when there's a small number of options and the user is expected to only buy one and move forward in sequence, the amount of work to get FormFlow to do this would not be worth it. In this example, an *IDialog<T>* would have been a better choice.

You'll want to look at the task your chatbot is designed to accomplish and use the right tool for the right job.

Summary

FormFlow offers an easy way to quickly build a chatbot for question and answer scenarios. This chapter started by showing the FormFlow user experience and how it provides a lot of default functionality out-of-the-box.

A FormFlow form is relatively simple to construct as a class with fields and properties. You need to add *FormBuilder* code, that the *Post* method configures to handle user activities. That's what tells the chatbot to hand over the conversation to your FormFlow class.

You'll probably need to customize the user experience, and FormFlow offers several attributes to do so in a declarative manner. You learned how the *Describe* attribute changes a field name, the *Numeric* attribute specifies bounds for integers and floating point numbers, the *Optional* attribute allows users to skip a field, and the *Pattern* attribute constrains a field to match a regular expression. There's even more customization available through the *Prompt* and *Template* attributes. In this chapter you also learned how to apply patterns and additional options to enhance the user experience.

This chapter provided the essentials of how to create a powerful user experience with FormFlow, and the next chapter builds on this with more advanced FormFlow features.

CHAPTER 7

Customizing FormFlow

FormFlow offers a quick way to build conversations and Chapter 6, Using FormFlow, explained the essentials for how FormFlow works. For all the great standard features that FormFlow offers, in practical use, you'll want additional customizations. This chapter moves beyond the basics and explains a few customization options.

One of the customizations you'll need is to build dynamic menus, rather than relying on a static enum. In this chapter you'll learn how to modify the form build process to dynamically build a menu for a field. You'll also learn how to add custom validation to that field. Another customization is a reuse strategy and you'll learn how to build a custom field type. Finally, a useful technique to make the user's experience better is to pre-populate fields.

This chapter uses the WineForm concept from Chapter 6, with customizations. We'll focus on the customizations in this chapter (but see Chapter 6 for more information on how WineBot works if needed). The first customization in this chapter uses the FormFlow fluent interface.

Understanding the FormFlow Fluent Interface

A fluent interface is one where the return value of one class member is the instance used to call another method. Visually, this starts by instantiating a class, following the class instantiation with a dot operator, calling another method, and continuing with the dot method pattern, eventually ending in a method that returns a type you want to operate on. The Chapter 6 listings for *WineForm* used the FormFlow fluent interface like the code shown here:

```
public IForm<WineForm> BuildForm()
{
    return new FormBuilder<WineForm>()
        .Message(
            "I have a few questions on your wine search. " +
            "You can type \"help\" at any time for more info.")
        .OnCompletion(DoSearch)
        .Build();
}
```

This is the *BuildForm* method from Chapter 6, Listing 6-4, but you don't need to go back and look at that to understand what is covered here. To summarize, the previous code example shows that the

FormBuilder<T> type has *Message, OnCompletion,* and *Build* methods. *FormBuilder<T>* implements *IFormBuilder<T>* and *Message, OnCompletion,* and *Build* return *IFormBuilder<T>* instances, supporting the ability to build a FormFlow form with a fluent interface. Here's part of the *IFormBuilder<T>* definition, showing these and more members:

```
public interface IFormBuilder<T>
    where T : class
{
    IForm<T> Build(
        Assembly resourceAssembly = null,
        string resourceName = null);

    FormConfiguration Configuration { get; }

    IFormBuilder<T> Message(
        string message,
        ActiveDelegate<T> condition = null,
        IEnumerable<string> dependencies = null);

    IFormBuilder<T> Field(IField<T> field);

    IFormBuilder<T> AddRemainingFields(
        IEnumerable<string> exclude = null);

    IFormBuilder<T> Confirm(
        string prompt = null,
        ActiveDelegate<T> condition = null,
        IEnumerable<string> dependencies = null);

    IFormBuilder<T> OnCompletion(
        OnCompletionAsyncDelegate<T> callback);

    bool HasField(string name);
}
```

All of the methods in *IFormBuilder<T>* return *IFormBuilder<T>*, except for *Build* and *HasField*, which return *IForm<T>* and *bool*, respectively. Some of the methods have multiple overloads, not shown, but the following sections discuss overloads and explain how each *IFormBuilder<T>* member works.

The *Configuration* Property

As explained in Chapter 6, FormFlow has a set of defaults it uses in three areas: commands, responses, and templates. If those defaults don't meet your needs, you can change them through the *Configuration* property as shown in Listing 7-1.

LISTING 7-1 Customizing FormFlow *Configuration*

```csharp
using Microsoft.Bot.Builder.Dialogs;
using Microsoft.Bot.Builder.FormFlow;
using System;
using System.Linq;
using System.Threading.Tasks;
using Microsoft.Bot.Builder.Resource;
using WineBotLib;

namespace WineBotConfiguration
{
    [Serializable]
    public class WineForm
    {
        public WineType WineType { get; set; }
        public RatingType Rating { get; set; }
        public StockingType InStock { get; set; }

        public IForm<WineForm> BuildForm()
        {
            var builder = new FormBuilder<WineForm>();
            ConfigureFormBuilder(builder);

            return builder
                .Message(
                    "I have a few questions on your wine search. " +
                    "You can type \"help\" at any time for more info.")
                .OnCompletion(DoSearch)
                .Build();
        }

        void ConfigureFormBuilder(FormBuilder<WineForm> builder)
        {
            FormConfiguration buildConfig = builder.Configuration;

            buildConfig.Yes = "Yes;y;sure;ok;yep;1;good".SplitList();

            TemplateAttribute tmplAttr = buildConfig.Template(TemplateUsage.EnumSelectOne);
            tmplAttr.Patterns = new[] {"What {&} would you like? {||}"};

            buildConfig.Commands[FormCommand.Quit].Help =
                "Quit: Quit the form without completing it. " +
                "Warning - this will clear your previous choices!";
        }

        async Task DoSearch(IDialogContext context, WineForm wineInfo)
        {
            List[] wines =
                await new WineApi().SearchAsync(
                    (int)WineType,
                    (int)Rating,
                    InStock == StockingType.InStock,
                    "");
```

```
            string message;

        if (wines.Any())
            message = "Here are the top matching wines: " +
                        string.Join(", ", wines.Select(w => w.Name));
        else
            message = "Sorry, No wines found matching your criteria.";

        await context.PostAsync(message);
        }
    }
}
```

The *BuildForm* method in Listing 7-1 is a little different than previous examples because it doesn't continue calling methods off the new *FormBuilder<WineForm>()* instance. It assigns the new instance to *builder* and then passes *builder* as an argument to the *ConfigureFormBuilder* method.

> **Tip** *IFormBuilder<T>* methods rely on default configuration to operate and then consider attributes to override the defaults. Because of this, remember to perform configuration to change the defaults before calling any methods.

Inside *ConfigureFormBuilder*, the code assigns a reference to the *FormConfiguration* instance of the *Configuration* property. The following sections show the three types of configuration actions you can take to customize FormFlow *Configuration*.

Configuring Responses

Having accepted an *IFormBuilder<WineForm>* instance, builder, the *ConfigureFormBuilder* method in Listing 7-1 shows how to customize the three areas of *Configuration*. The first example, repeated next, shows how to modify *FormConfiguration* responses:

```
buildConfig.Yes = "Yes;y;sure;ok;yep;1;good".SplitList();
```

This particular response configuration is for the *Yes* property, which is a *string[]* of potential responses to any situation where FormFlow expects an answer and one of the answers could be *Yes*. The *SplitList* method is a Bot Builder extension method that creates an array, using semi-colons as the default separator, but with overloads allowing you to change to your preferred separator. The following summarization of *FormConfiguration* shows the response properties you can customize:

```
    public class FormConfiguration
    {
        /// <summary>
        /// Enumeration of strings for interpreting a user response as setting an optional field
to be unspecified.
        /// </summary>
```

```
        /// <remarks>
        /// The first string is also used to describe not having a preference for an optional
field.
        /// </remarks>
        public string[] NoPreference = Resources.MatchNoPreference.SplitList();

        /// <summary>
        /// Enumeration of strings for interpreting a user response as asking for the current
value.
        /// </summary>
        /// <remarks>
        /// The first value is also used to describe the option of keeping the current value.
        /// </remarks>
        public string[] CurrentChoice = Resources.MatchCurrentChoice.SplitList();

        /// <summary>
        /// Enumeration of values for a "yes" response for boolean fields or confirmations.
        /// </summary>
        public string[] Yes = Resources.MatchYes.SplitList();

        /// <summary>
        /// Enumeration of values for a "no" response for boolean fields or confirmations.
        /// </summary>
        public string[] No = Resources.MatchNo.SplitList();

        /// <summary>
        /// String for naming the "navigation" field.
        /// </summary>
        public string Navigation = Resources.Navigation;

        /// <summary>
        /// String for naming "Confirmation" fields.
        /// </summary>
        public string Confirmation = Resources.Confirmation;
    };
```

In each property, you can see that FormFlow uses a Bot Builder resource file to set strings. Similarly, you can define your own .NET resource file to localize possible responses to common questions.

Configuring Templates

The next item you can customize is templates. If you recall, from Chapter 6, the *Template* attribute allows you to override the default prompt for anything that FormFlow prints out. Through *Configuration*, you can change these defaults, as shown by the excerpt from Listing 7-1, repeated below:

```
TemplateAttribute tmplAttr = buildConfig.Template(TemplateUsage.EnumSelectOne);
tmplAttr.Patterns = new[] {"What {&} would you like? {||}"};
```

FormConfiguration has a method named *Template*, accepting a *TemplateUsage* argument, which is the same *TemplateUsage* discussed in Chapter 6 for the *Template* attribute. This returns a *TemplateAttribute* instance that you can use to customize the defaults for that particular template. This example replaces the default *string[]* for the *Patterns* used to select a prompt to display to the user. The following *Templates* field shows the available *FormConfiguration* defaults:

```
public class FormConfiguration
{
    public List<TemplateAttribute> Templates = new List<TemplateAttribute>
    {
        new TemplateAttribute(TemplateUsage.Bool, Resources.TemplateBool),
        // {0} is current choice, {1} is no preference
        new TemplateAttribute(TemplateUsage.BoolHelp, Resources.TemplateBoolHelp),

        // {0} is term being clarified
        new TemplateAttribute(TemplateUsage.Clarify, Resources.TemplateClarify),

        new TemplateAttribute(TemplateUsage.Confirmation, Resources.TemplateConfirmation),

        new TemplateAttribute(TemplateUsage.CurrentChoice, Resources.TemplateCurrentChoice),

        new TemplateAttribute(TemplateUsage.DateTime, Resources.TemplateDateTime),
        // {0} is current choice, {1} is no preference
        // new TemplateAttribute(TemplateUsage.DateTimeHelp,
        //          "Please enter a date or time expression like 'Monday' or 'July 3rd'{?, {0}}{?, {1}}."),
        new TemplateAttribute(TemplateUsage.DateTimeHelp, Resources.TemplateDateTimeHelp),

        // {0} is min and {1} is max.
        new TemplateAttribute(TemplateUsage.Double, Resources.TemplateDouble)
                { ChoiceFormat = Resources.TemplateDoubleChoiceFormat },
        // {0} is current choice, {1} is no preference
        // {2} is min and {3} is max
        new TemplateAttribute(TemplateUsage.DoubleHelp, Resources.TemplateDoubleHelp),

        // {0} is min, {1} is max and {2} are enumerated descriptions
        new TemplateAttribute(TemplateUsage.EnumManyNumberHelp, Resources.TemplateEnumManyNumberHelp),
        new TemplateAttribute(TemplateUsage.EnumOneNumberHelp, Resources.TemplateEnumOneNumberHelp),

        // {2} are the words people can type
        new TemplateAttribute(TemplateUsage.EnumManyWordHelp, Resources.TemplateEnumManyWordHelp),
        new TemplateAttribute(TemplateUsage.EnumOneWordHelp, Resources.TemplateEnumOneWordHelp),

        new TemplateAttribute(TemplateUsage.EnumSelectOne, Resources.TemplateEnumSelectOne),
        new TemplateAttribute(TemplateUsage.EnumSelectMany, Resources.TemplateEnumSelectMany),

        // {0} is the not understood term
        new TemplateAttribute(TemplateUsage.Feedback, Resources.TemplateFeedback),

        // For {0} is recognizer help and {1} is command help.
        new TemplateAttribute(TemplateUsage.Help, Resources.TemplateHelp),
        new TemplateAttribute(TemplateUsage.HelpClarify, Resources.TemplateHelpClarify),
```

```
            new TemplateAttribute(TemplateUsage.HelpConfirm, Resources.TemplateHelpConfirm),
            new TemplateAttribute(TemplateUsage.HelpNavigation, Resources.
TemplateHelpNavigation),

            // {0} is min and {1} is max if present
            new TemplateAttribute(TemplateUsage.Integer, Resources.TemplateInteger)
                    { ChoiceFormat = Resources.TemplateIntegerChoiceFormat },
            // {0} is current choice, {1} is no preference
            // {2} is min and {3} is max
            new TemplateAttribute(TemplateUsage.IntegerHelp, Resources.TemplateIntegerHelp),

            new TemplateAttribute(TemplateUsage.Navigation, Resources.TemplateNavigation)
                    { FieldCase = CaseNormalization.None },
            // {0} is list of field names.
            new TemplateAttribute(TemplateUsage.NavigationCommandHelp, Resources.
TemplateNavigationCommandHelp),
            new TemplateAttribute(TemplateUsage.NavigationFormat, Resources.
TemplateNavigationFormat)
                    {FieldCase = CaseNormalization.None },
            // {0} is min, {1} is max
            new TemplateAttribute(TemplateUsage.NavigationHelp, Resources.
TemplateNavigationHelp),

            new TemplateAttribute(TemplateUsage.NoPreference, Resources.TemplateNoPreference),

            // {0} is the term that is not understood
            new TemplateAttribute(TemplateUsage.NotUnderstood, Resources.TemplateNotUnderstood),

            new TemplateAttribute(TemplateUsage.StatusFormat, Resources.TemplateStatusFormat)
                    {FieldCase = CaseNormalization.None },

            new TemplateAttribute(TemplateUsage.String, Resources.TemplateString)
                    { ChoiceFormat = Resources.TemplateStringChoiceFormat },
            // {0} is current choice, {1} is no preference
            new TemplateAttribute(TemplateUsage.StringHelp, Resources.TemplateStringHelp),

            new TemplateAttribute(TemplateUsage.Unspecified, Resources.TemplateUnspecified)
        };
    }
```

Each *TemplateAttribute* constructor accepts arguments of type *TemplateUsage* and *string[]*. As with the response properties, you can assign localizable resources to the patterns argument.

Similar to *TemplateAttribute*, *FormConfiguration* has a *PromptAttribute* that you can replace to set default prompt settings, shown here:

```
    public class FormConfiguration
    {
        public PromptAttribute DefaultPrompt = new PromptAttribute("")
        {
            AllowDefault = BoolDefault.True,
            ChoiceCase = CaseNormalization.None,
            ChoiceFormat = Resources.DefaultChoiceFormat,
            ChoiceLastSeparator = Resources.DefaultChoiceLastSeparator,
```

```
            ChoiceParens = BoolDefault.True,
            ChoiceSeparator = Resources.DefaultChoiceSeparator,
            ChoiceStyle = ChoiceStyleOptions.Auto,
            FieldCase = CaseNormalization.Lower,
            Feedback = FeedbackOptions.Auto,
            LastSeparator = Resources.DefaultLastSeparator,
            Separator = Resources.DefaultSeparator,
            ValueCase = CaseNormalization.InitialUpper
        };
    }
```

PromptAttribute inherits all of those properties from its abstract base class, *TemplateBaseAttribute*. You can also set those properties through a *TemplateAttribute* instance, which also derives from *TemplateBaseAttribute*.

Configuring Commands

The last *Configuration* category is commands. A command is a way to communicate with FormFlow itself, rather than the form class, like WineForm, that the developer creates. As Chapter 6 discusses, these commands include *Back*, *Status*, *Help*, and more. The following code shows how to customize the *Help* command:

```
buildConfig.Commands[FormCommand.Quit].Help =
    "Quit: Quit the form without completing it. " +
    "Warning - this will clear your previous choices!";
```

This sets the *Help* message for the *Quit* command, which as you saw in Chapter 6, will appear where the user types **Help**. In the help message is a **Quit** option and this sets the message that appears.

The *Commands* property is a *Dictionary<FormCommand, CommandDescription>* where *FormCommand* is an enum with a specific set of commands that FormFlow supports and *CommandDescription*, shown here:

```
public class CommandDescription
{
    /// <summary>
    /// Description of the command.
    /// </summary>
    public string Description;

    /// <summary>
    /// Regexs for matching the command.
    /// </summary>
    public string[] Terms;

    /// <summary>
    /// Help string for the command.
    /// </summary>
    public string Help;

    /// <summary>
    /// Construct the description of a built-in command.
    /// </summary>
```

```
/// <param name="description">Description of the command.</param>
/// <param name="terms">Terms that match the command.</param>
/// <param name="help">Help on what the command does.</param>
public CommandDescription(string description, string[] terms, string help)
{
    Description = description;
    Terms = terms;
    Help = help;
}
}
```

As you can see, *CommandDescription* has *Description*, *Terms*, and *Help* fields that can be set directly, or you can replace the entire dictionary value for that command with a new *CommandDescription* instance. Notice that the only fields available customize the appearance of an existing command. You can't add arbitrary commands this way, but can customize any that are members of the *FormCommand* enum, shown below:

```
public enum FormCommand
{
    /// <summary>
    /// Move back to the previous step.
    /// </summary>
    Backup,

    /// <summary>
    /// Ask for help on responding to the current field.
    /// </summary>
    Help,

    /// <summary>
    /// Quit filling in the current form and return failure to parent dialog.
    /// </summary>
    Quit,

    /// <summary>
    /// Reset the status of the form dialog.
    /// </summary>
    Reset,

    /// <summary>
    /// Provide feedback to the user on the current form state.
    /// </summary>
    Status
};
```

> **Tip** Besides just organizing code to separate and clarify custom configuration logic, the *ConfigureFormBuilder* method implies that you might want to move this logic into another method and/or class to reuse and set defaults with common code for multiple forms in a chatbot.

After optionally configuring the form, you can call methods and the next section eases into that by discussing common parameters for *Message* and other methods.

The *Message* Method and Common Parameters

The *Message* method supports sending any type of non-question related information to the user. It has overloads that accept a varying list of parameters that either permit building the message text a specific way or controlling the conditions of whether or not the *Message* method will display text. Listing 7-2, from the *WineFormParams* project in the accompanying source code, shows the different *Message* overloads.

LISTING 7-2 Using the *Message* method

```
using Microsoft.Bot.Builder.Dialogs;
using Microsoft.Bot.Builder.FormFlow;
using System;
using System.Linq;
using System.Threading.Tasks;
using WineBotLib;

namespace WineBotParams
{
    [Serializable]
    public class WineForm
    {
        public WineType WineType { get; set; }
        [Optional]
        public RatingType Rating { get; set; }
        public StockingType InStock { get; set; }

        public IForm<WineForm> BuildForm()
        {
            ActiveDelegate<WineForm> shouldShowSpecial =
                wineForm => DateTime.Now.DayOfWeek == DayOfWeek.Friday;

            var prompt = new PromptAttribute
            {
                Patterns =
                    new[]
                    {
                        "Hi, I have a few questions to ask.",
                        "How are you today? I just have a few questions.",
                        "Thanks for visiting - please answer a few questions."
                    }
            };

            int numberOfBackOrderDays = 15;

            MessageDelegate<WineForm> generateMessage =
                async wineForm =>
                    await Task.FromResult(
```

```csharp
                    new PromptAttribute(
                        $"Note: Delivery back order is {numberOfBackOrderDays} days."));

            return new FormBuilder<WineForm>()
                .Message(prompt)
                .Message(
                    "You can type \"help\" at any time for more info.")
                .Message(
                    "It's your lucky day - 10% discounts on Friday!",
                    shouldShowSpecial)
                .Message(
                    $"Today you get an additional %5 off.",
                    wineForm => wineForm.Rating == RatingType.Low,
                    new[] { nameof(Rating) })
                .Message(
                    generateMessage,
                    wineForm => wineForm.InStock == StockingType.OutOfStock)
                .OnCompletion(DoSearch)
                .Build();
        }

        async Task DoSearch(IDialogContext context, WineForm wineInfo)
        {
            List[] wines =
                await new WineApi().SearchAsync(
                    (int)WineType,
                    (int)Rating,
                    InStock == StockingType.InStock,
                    "");

            string message;

            if (wines.Any())
                message = "Here are the top matching wines: " +
                    string.Join(", ", wines.Select(w => w.Name));
            else
                message = "Sorry, No wines found matching your criteria.";

            await context.PostAsync(message);
        }
    }
}
```

As Listing 7-2 shows, the *BuildForm* method instantiates a new *FormBuilder<WineForm>* and calls several *Message* overloads. The second *Message* overload displays a string, which appears as the first user message when FormFlow starts, before any questions. The following sections discuss other overloads and their associated parameters.

The *condition* Parameter

The condition parameter determines whether the *Message* will display. Its parameter conforms to the *ActiveDelegate* signature, shown here:

```
public delegate bool ActiveDelegate<T>(T state);
```

With *ActiveDelegate* the state type is the FormFlow form, which is *WineForm* in these examples. The code the developer writes for this delegate instance is logic that returns a bool result. When the return value is *true*, the message displays, and won't display if *false*. The following excerpt from Listing 7-2 shows how this works:

```
ActiveDelegate<WineForm> shouldShowSpecial =
    wineForm => DateTime.Now.DayOfWeek == DayOfWeek.Friday;

return new FormBuilder<WineForm>()
    .Message(
        "It's your lucky day - 10% discounts on Friday!",
        shouldShowSpecial)
    .Build();
```

Here, you can see the code assigns a lambda to an *ActiveDelegate<WineForm>* instance, *shouldShowSpecial*. The concept here is that the chatbot offers a 10% discount on Fridays. On any other day, the message won't display.

The *dependencies* Parameter

Some questions are optional and you need a way to display a message only if the user decides to provide a value. That's the case with the *Rating* property in *WineForm*, as shown here:

```
[Optional]
public RatingType Rating { get; set; }

public IForm<WineForm> BuildForm()
{
    return new FormBuilder<WineForm>()
        .Message(
            $"Today you get an additional %5 off.",
            wineForm => wineForm.Rating == RatingType.Low,
            new[] { nameof(Rating) })
        .OnCompletion(DoSearch)
        .Build();
}
```

This excerpt, from Listing 7-2, decorates *Rating* with an *Optional* attribute. The *Message* example in *BuildForm* accepts three parameters: the message *text*, the *condition* (as described in the previous section), and the *dependency* fields. In particular, if the user doesn't choose to complete the *Rating* property, this message won't dispay. If the user does provide a value, and meets the condition that the selection is a *Low* rating, then the message displays. This example defines condition as a lambda, instead of an *ActivieDelegate<WineForm>* delegate reference. The concept here is a potential situation where the chatbot wants to clear inventory that's moving slow because of low ratings and wants to provide an incentive for the user to buy.

> **Note** *WineForm* implements properties for its state, but let's refer to them as fields, which doesn't conform to C# language syntax guidelines. However, here we refer to both properties and fields as FormFlow fields. Since FormFlow is a multi-language library in Bot Builder, it has its own naming idioms, rather than conforming to what a specific language prescribes. We could use C# fields, which might reduce confusion, but let's use properties. However, from a terminology standpoint, we'll cover *Field* methods later in the chapter and discuss the FormFlow perspective, using the term *field*, which has merit.

Notice that dependencies is a *string[]*, indicating that there can be dependencies on multiple fields. Dependencies on required fields, without the *Optional* attribute, prevents the message from appearing.

The point in time of the conversation when a message appears, assuming its matched the dependencies and condition critertia, is after acknowledging the final confirmation of the form. This makes sense because there's no way to know for sure that the value is set until after the user confirms so. Until final confirmation, the user can navigate at any time and change the field to *No Preference*.

The *prompt* Parameter

The *Message* method has an overload accepting the *prompt* parameter, which is type *PromptAttribute*. This is the same *Prompt* attribute that's explained in Chapter 6. Here's an excerpt from Listing 7-2, showing how to define a *prompt* parameter:

```
public IForm<WineForm> BuildForm()
{
    var prompt = new PromptAttribute
    {
        Patterns =
            new[]
            {
                "Hi, I have a few questions to ask.",
                "How are you today? I just have a few questions.",
                "Thanks for visiting - please answer a few questions."
            }
    };

    return new FormBuilder<WineForm>()
        .Message(prompt)
        .OnCompletion(DoSearch)
        .Build();
}
```

Notice that the *PromptAttribute Patterns* property has an array of multiple messages. This is a handy way to vary a message so the user has some variation across multiple interactions with a chatbot.

CHAPTER 7 Customizing FormFlow **165**

The *generateMessage* Parameter

Another *Message* overload accepts a *generateMessage* argument. This is a good idea if you need dynamic logic to build the message. Here's an excerpt from Listing 7-2, showing how to use *generateMessage*:

```
public IForm<WineForm> BuildForm()
{
    int numberOfBackOrderDays = 15;

    MessageDelegate<WineForm> generateMessage =
        async wineForm =>
            await Task.FromResult(
                new PromptAttribute(
                    $"Note: Delivery back order is {numberOfBackOrderDays} days."));

    return new FormBuilder<WineForm>()
        .Message(
            generateMessage,
            wineForm => wineForm.InStock == StockingType.OutOfStock)
        .OnCompletion(DoSearch)
        .Build();
}
```

The *generateMessage* parameter type is *MessageDelegate<WineForm>*, which returns a *Task* type, allowing an async operation. This is useful because, although this example hard codes it, you might want to make an async database call to get the current number of back order days. Combined with the condition for *OutOfStock*, this message attempts to be helpful and let's the user know how long they'll wait for products that aren't in stock.

All of the parameters for *Message* are available in *Confirm* method overloads, which is discussed next.

The *Confirm* Method

The *Confirm* method calls for the user to acknowledge a statement. Positive responses, such as saying **yes**, allow the user to progress with the form and negative responses, such as **no**, will prevent the user from proceeding in the form. Listing 7-3, from the *WineFormConfirm* project in the accompanying source code, shows how to use the *Confirm* method.

LISTING 7-3 Using the *Confirm* Method

```
using Microsoft.Bot.Builder.Dialogs;
using Microsoft.Bot.Builder.FormFlow;
using System;
using System.Linq;
using System.Threading.Tasks;
using WineBotLib;

namespace WineBotConfirm
{
```

```csharp
[Serializable]
public class WineForm
{
    public WineType WineType { get; set; }
    [Optional]
    public RatingType Rating { get; set; }
    public StockingType InStock { get; set; }

    public IForm<WineForm> BuildForm()
    {
        ActiveDelegate<WineForm> shouldShowContest =
            wineForm => DateTime.Now.DayOfWeek == DayOfWeek.Friday;

        var prompt = new PromptAttribute
        {
            Patterns =
                new[]
                {
                    "Hi, May I ask a few questions?",
                    "How are you today? Can I ask a few questions?",
                    "Thanks for visiting - would you answer a few questions?"
                }
        };

        int numberOfBackOrderDays = 15;

        MessageDelegate<WineForm> generateMessage =
            async wineForm =>
                await Task.FromResult(
                    new PromptAttribute(
                        $"Delivery back order is {numberOfBackOrderDays} days. Are you sure?"));

        return new FormBuilder<WineForm>()
            .Confirm(prompt)
            .Confirm(
                "You can type \"help\" at any time for more info. Would you like to proceed?")
            .Confirm(
                "Would you like to enter a contest for free bottle of Wine?",
                shouldShowContest)
            .Confirm(
                $"Low rated wines are limited in stock - are you sure?",
                wineForm => wineForm.Rating == RatingType.Low,
                new[] { nameof(Rating) })
            .Confirm(
                generateMessage,
                wineForm => wineForm.InStock == StockingType.OutOfStock)
            .OnCompletion(DoSearch)
            .Build();
    }

    async Task DoSearch(IDialogContext context, WineForm wineInfo)
    {
        List[] wines =
            await new WineApi().SearchAsync(
```

```
                (int)WineType,
                (int)Rating,
                InStock == StockingType.InStock,
                "");

        string message;

        if (wines.Any())
            message = "Here are the top matching wines: " +
                        string.Join(", ", wines.Select(w => w.Name));
        else
            message = "Sorry, No wines found matching your criteria.";

        await context.PostAsync(message);
    }
  }
}
```

The *BuildForm* method in Listing 7-3 is similar to the *BuildForm* method in Listing 7-2, except it uses the *Confirm* method and different text. The parameters in each of the *Confirm* method overloads are the same parameters used in the previous section on *Message*, where you'll find a more detailed description.

> **Tip** The difference between *Confirm* and *Message* methods is that the *Message* doesn't require user acknowledgement and always moves forward, but *Confirm* prevents a user from moving forward until they positively acknowledge the question.

Working with Fields

Until now, you've seen examples that rely on public properties as FormFlow fields. By default, FormFlow uses reflection to read these fields in the order they appear in the class. Chapter 6 showed how to decorate these fields with attributes for customization, but that's limited. In this section, you'll learn how to take full control of fields with dynamic definition and validation. You'll also learn how to control which fields appear and in what order. The following sections take a deep dive into Listing 7-4, from the *WineBotFields* project in the accompanying source code, which shows different ways to work with fields.

LISTING 7-4 *WineForm* for Working with Fields

```
using Microsoft.Bot.Builder.Dialogs;
using Microsoft.Bot.Builder.FormFlow;
using Microsoft.Bot.Builder.FormFlow.Advanced;
using System;
using System.Linq;
using System.Threading.Tasks;
```

```csharp
using WineBotLib;

namespace WineBotFields
{
    [Serializable]
    public class WineForm
    {
        public string WineType { get; set; }
        public int Rating { get; set; }
        public StockingType InStock { get; set; }
        public int Vintage { get; set; }
        public string SearchTerms { get; set; }

        public Refinement[] WineCategories { get; set; }

        public IForm<WineForm> BuildForm()
        {
            var form = new FormBuilder<WineForm>()
                .Message("Welcome to WineBot!")
                .Field(nameof(InStock), wineForm => DateTime.Now.DayOfWeek == DayOfWeek.Wednesday)
                .Field(new FieldReflector<WineForm>(nameof(WineType))
                    .SetType(null)
                    .SetFieldDescription("Type of Wine")
                    .SetDefine(async (wineForm, field) =>
                    {
                        foreach (var category in WineCategories)
                            field
                                .AddDescription(category.Name, category.Name)
                                .AddTerms(category.Name, Language.GenerateTerms(category.Name, 6));

                        return await Task.FromResult(true);
                    }))
                .Field(
                    name: nameof(Rating),
                    prompt: new PromptAttribute("What is your preferred {&} (1 to 100)?"),
                    active: wineForm => true,
                    validate: async (wineForm, response) =>
                    {
                        var result = new ValidateResult { IsValid = true, Value = response };

                        result.IsValid =
                            int.TryParse(response.ToString(), out int rating) &&
                            rating > 0 && rating <= 100;

                        if (!result.IsValid)
                        {
                            result.Feedback = $"'{response}' isn't a valid option!";
                            result.Choices =
                                new List<Choice>
                                {
                                    new Choice
                                    {
```

```csharp
                                Description = new DescribeAttribute("25"),
                                Value = 25,
                                Terms = new TermsAttribute("25")
                            },
                            new Choice
                            {
                                Description = new DescribeAttribute("50"),
                                Value = 50,
                                Terms = new TermsAttribute("50")
                            },
                            new Choice
                            {
                                Description = new DescribeAttribute("75"),
                                Value = 75,
                                Terms = new TermsAttribute("75")
                            }
                        };

                        return await Task.FromResult(result);
                    })
            .AddRemainingFields(new[] { nameof(Vintage) });

        if (!form.HasField(nameof(Vintage)))
            form.Field(nameof(Vintage));

        form.OnCompletion(DoSearch);

        return form.Build();
    }

    async Task DoSearch(IDialogContext context, WineForm wineInfo)
    {
        int wineType =
            (from refinement in WineCategories
             where refinement.Name == wineInfo.WineType
             select refinement.Id)
            .SingleOrDefault();

        List[] wines =
            await new WineApi().SearchAsync(
                wineType,
                wineInfo.Rating,
                wineInfo.InStock == StockingType.InStock,
                wineInfo.SearchTerms);

        string message;

        if (wines.Any())
            message = "Here are the top matching wines: " +
                      string.Join(", ", wines.Select(w => w.Name));
        else
            message = "Sorry, No wines found matching your criteria.";

        await context.PostAsync(message);
```

```
                context.EndConversation(EndOfConversationCodes.CompletedSuccessfully);
            }
        }
    }
```

In its simplest form, the *Field* method has an overload that only requires the name of a field, as in the following excerpt from Listing 7-4:

```
var form = new FormBuilder<WineForm>()
    .Message("Welcome to WineBot!")
    .Field(nameof(InStock), wineForm => DateTime.Now.DayOfWeek == DayOfWeek.Wednesday);
```

This *Field* defines the *InStock* field. Additionally, it uses an optional *active* parameter, which is an *ActiveDelegate* instance, as described in the previous *Message* method section. Since the default FormFlow behavior is to order properties based on their declared order in the class, this is one way to change the order of fields from the default. In this example, the default ordering is to ask the *WineType* field question first, because it appears first in the class, but using the *Field* method makes the *InStock* field question show first instead.

The first time you use the *Field* method, it overrides the default FormFlow behavior, requiring explicitly using a *Field* method for each field. In the following sections, you'll learn how to dyamically add, exclude, and fill in remaining fields.

Dynamic Field Definition

The *Field* method has several overloads, one of which let's you dynamically define the field. This allows you to specify all aspects of a field to include its type, members, underlying values, and any other customizations to define the field. The following excerpt, from Listing 7-4, shows how to dynamically define a field:

```
var form = new FormBuilder<WineForm>()
    .Message("Welcome to WineBot!")
    .Field(new FieldReflector<WineForm>(nameof(WineType))
        .SetType(null)
        .SetFieldDescription("Type of Wine")
        .SetDefine(async (wineForm, field) =>
        {
            foreach (var category in WineCategories)
                field
                    .AddDescription(category.Name, category.Name)
                    .AddTerms(category.Name, Language.GenerateTerms(category.Name, 6));

            return await Task.FromResult(true);
        }));
```

The *Field* method overload accepts a parameter of type *Field<T>*. The *FieldReflector<WineForm>* instance helps build a new *Field<WineForm>* instance. Like *FormBuilder<T>*, *FieldReflector<T>* is the entry point of a fluent interface where each method returns *Field<T>*. Further, *Field<T>* has methods, such as the *SetType*, *SetFieldDescription*, and *SetDefine* in the example above, in addition to many more.

To understand *Field<T>* members, consider all that you've learned about fields and their attributes from Chapter 6 and this chapter. Each of the *Field<T>* members support everything you've learned so far.

SetType indicates the field type, except when the value should be treated like an enum and is set to *null*. By passing *null* to *SetType*, the previous example tells FormFlow to treat the *WineType* field as an enum. In *WineForm*, the *WineType* property type is *string*, which is used to hold the value that the user responds with, and while the type of the property doesn't need to be an enum, FormFlow gives the user an enum-like experience by showing buttons with the question.

SetFieldDescription is analogous to using a *Description* attribute on a field. Instead of the constructing the default name *Wine Type*, FormFlow will use *Type of Wine*, as specified by the parameter.

SetDefine defines the contents of the question, adding allowable values for the user to choose from. Remember, the whole purpose of using this particular overload is to dynamically define available answers, rather than relying on hard-coded enum values. *SetDefine* takes an async *DefineAsyncDelegate<T>*, shown here:

```
public delegate Task<bool> DefineAsyncDelegate<T>(T state, Field<T> field)
    where T : class;
```

For this example, the type *T* is *WineForm* and the two parameters, *wineForm* and *field*, are type *WineForm* and *Field<WineForm>*, respectively.

WineCategories is a *Refinement[]*–a serializable type returned via repository calls to *WineApi* and you'll see how *WineCategories* gets populated later in this chapter. The call to *AddDescription* adds possible answers, where the parameters are *value* and *description*. *AddTerms*, both analogous to using the *Terms* attribute, where the first parameter is the *value* and the second is *string[]*.

> **Note** We've seen a couple methods whose identifiers contain *Description*. One is *SetFieldDescription*, which operates the same as a *Description* attribute decorating a FormFlow property or field. The *AddDescription* is different because it displays a name for a potential value of a field, instead of the field itself.

Rather than manually specify terms, the example uses the Bot Builder *Language.GenerateTerms* method to automatically create up to six regular expressions that can possibly match this value. Table 7-1 shows what Language.GenerateTerms returns for each category.

TABLE 7-1 *Language.GenerateTerms* Examples

Phrases	Generated Terms
Red Wine	reds?, wines?, reds wines?
White Wine	whites?, wines?, whites? wines?
Champagne & Sparkling	champagnes?, sparklings?, champagnes? & sparklings?
Rosé Wine	rosés?, wines?, rosés? wines?
Dessert, Sherry & Port	dessert,s?, sherrys?, ports?, dessert,s? sherrys?, sherrys? & ports?, dessert,s? sherrys? & ports?
Saké	sakés?

The second parameter to Language.GenerateTerms is set to 6, meaning that it can generate 6 or fewer terms. The generated terms are regular expressions that FormFlow recognizes when validating that user input matches. Table 7-1 shows that longer phrases, like Dessert, Sherry & Port generate up to 6 regular expressions. Shorter phrases don't have enough words to generate 6 regular expressions. The *DefineAsyncDelegate<T>* instance returns *true* if it defined the field.

Field Validation

Field<T> has a *SetValidation* method you can use to perform validation on the user's input. You don't need all of the ceremony of a *FieldReflector<T>* with multiple method calls, however, because the *Field* method has an overload taking a *validation* parameter, as shown in the following excerpt from Listing 7-4:

```
var form = new FormBuilder<WineForm>()
    .Message("Welcome to WineBot!")
    .Field(
        name: nameof(Rating),
        prompt: new PromptAttribute("What is your preferred {&} (1 to 100)?"),
        active: wineForm => true,
        validate: async (wineForm, response) =>
        {
            var result = new ValidateResult { IsValid = true, Value = response };

            result.IsValid =
                int.TryParse(response.ToString(), out int rating) &&
                rating > 0 && rating <= 100;

            result.IsValid =
                int.TryParse(response.ToString(), out int rating) &&
                rating > 0 && rating <= 100;

            if (!result.IsValid)
            {
                result.Feedback = $"'{response}' isn't a valid option!";
                result.Choices =
                    new List<Choice>
                    {
                        new Choice
                        {
```

```
                        Description = new DescribeAttribute("25"),
                        Value = 25,
                        Terms = new TermsAttribute("25")
                    },
                    new Choice
                    {
                        Description = new DescribeAttribute("50"),
                        Value = 50,
                        Terms = new TermsAttribute("50")
                    },
                    new Choice
                    {
                        Description = new DescribeAttribute("75"),
                        Value = 75,
                        Terms = new TermsAttribute("75")
                    }
                };
        }

        return await Task.FromResult(result);
    });
```

This *Field* overload has *name*, *prompt*, *active*, and *validate* parameters. The *name* is the name of the field to store results in, *prompt* is analogous to the *Prompt* attribute on a field, and *active* is the *ActiveDelegate* used in previous examples to determine whether FormFlow should ask a question for that field. The *validate* parameter is a *ValidateAsyncDelegate*, shown here:

```
public delegate Task<ValidateResult> ValidateAsyncDelegate<T>(T state, object value);
```

The first parameter of *ValidateAsyncDelegate*, in the *Field* example, is a reference to the current *WineForm* instance and the second is the user's *response* to the question. The validation logic instantiates a *ValidateResult*, initializing *IsValid* to *true* and *Value* to the user's *response*. The validation you use depends on what makes sense for the field. In this case, the validation is to ensure the user's response is an integer between 1 and 100. The code returns the *ValidateResult* instance, *result*.

When the user's input is not valid, the code populates the *result.Feedback* property to give the user more information on what went wrong. The *result.Choices* property displays a list of buttons the user can click. The *Choice* class has properties for *Description*, *Value*, and *Terms*. The user can read *Description*, *Value* is the input sent back to the chatbot when the user clicks the button, and *Terms* is a list of items that could match that choice. (See the previous discussion on *Language.GenerateTerms* for ideas on how to populate *Terms*.) When you populate *result.Choices*, only the presented choices are valid and the user can't type any other text. Another way to communicate error details to the user is with the *FeedbackCard* property, shown here:

```
            result.FeedbackCard =
                new FormPrompt
                {
                    Prompt = $"'{response}' isn't a valid option!",
                    Buttons =
                        new List<DescribeAttribute>
                        {
```

```
                    new DescribeAttribute("25"),
                    new DescribeAttribute("50"),
                    new DescribeAttribute("75")
                }
        };
```

This is nearly the same as the *Feedback* and *Choice* properties, except we use a single *FormPrompt* instance and the result comes out as a card. The *FormPrompt* has a *Prompt* and *Buttons* properties where the *Prompt* is the text the user explaining what went wrong. The *Buttons* property shows a button for each option, using a *DescribeAttribute* for each option. Unlike the *Choice* property, *FeedbackCard* lets the user type any value, regardless of whether it's specified by a button, making this option more desirable when the buttons aren't the only values a user could type.

The *AddRemainingFields* Method

As mentioned previously, calling a single *Field* method on *FormBuilder<T>* overrides the default behavior of FormFlow to automatically ask questions for each field. When that happens, FormFlow will only ask questions for fields that are explicitly defined. That poses a problem in that it can be cumbersome to explicitly define every field, so *FormBuilder<T>* has a method, *AddRemainingFields*, that does exactly what it sounds like—adds all remaining fields that haven't been explicitly specified. Here's an excerpt from Listing 7-4, showing how to use *AddRemainingFields*:

```
            var form = new FormBuilder<WineForm>()
                .Message("Welcome to WineBot!")
                .Field(nameof(InStock), wineForm => DateTime.Now.DayOfWeek == DayOfWeek.
Wednesday)
                .AddRemainingFields(new[] { nameof(Vintage) });
```

In the previous example, if *InStock* were the only field specified, it would be the only question that FormFlow asks the user. The *AddRemainingFields* method includes all of the other fields in the form, except for a *string[]* with field names to exclude. This example adds all of the fields, except for *Vintage*. *AddRemainingFields* has a parameterless overload in case you don't want to exclude any fields.

The *HasField* Method

FormBuilder<T> offers a *HasField* method, allowing queries to determine if a given field is specified in the *FormBuilder<T>*. Here's an excerpt, from Listing 7-4, showing one way to use *HasField*:

```
            if (!form.HasField(nameof(Vintage)))
                form.Field(nameof(Vintage));
```

This code checks to see if the *FormBuider<T>* defines a *Vintage* field. If not, the code adds the *Vintage* field.

> **Tip** The *HasField* example shows the beauty of using a fluent interface. You can use your own logic to dynamically determine whether to add or remove elements from a *FormBuilder<T>*. The same applies for *Field<T>*.

The *OnCompletion* Method

When FormFlow completes asking questions, you'll want to do something with the results, such as saving in a database, providing an answer, or performing a search. The following excerpt, from Listing 7-4, shows how to use *OnCompletion* to process those results:

```
var form = new FormBuilder<WineForm>()
    .Message("Welcome to WineBot!")
    .Field(nameof(InStock), wineForm => DateTime.Now.DayOfWeek == DayOfWeek.Wednesday)
    .AddRemainingFields(new[] { nameof(Vintage) })
    .OnCompletion(DoSearch);
```

OnCompletion specifies the *DoSearch* method for processing form results. The *DoSearch* method implements the *OnCompletionAsyncDelegate* signature, shown here:

```
public delegate Task OnCompletionAsyncDelegate<T>(IDialogContext context, T state);
```

Type *T* in the *OnCompletionAsyncDelegate* that *OnCompletion* refers to is *WineForm* and the parameter types are *IDialogContext* and *WineForm*. This *IDialogContext* is the same type described in detail in Chapter 5, Building Dialogs, for *IDialog<T>* implementation method parameters. Here's the *DoSearch* method, showing one way to use these parameters to process results:

```
async Task DoSearch(IDialogContext context, WineForm wineInfo)
{
    int wineType =
        (from refinement in WineCategories
         where refinement.Name == wineInfo.WineType
         select refinement.Id)
        .SingleOrDefault();

    List[] wines =
        await new WineApi().SearchAsync(
            wineType,
            wineInfo.Rating,
            wineInfo.InStock == StockingType.InStock,
            wineInfo.SearchTerms);

    string message;

    if (wines.Any())
        message = "Here are the top matching wines: " +
            string.Join(", ", wines.Select(w => w.Name));
    else
        message = "Sorry, No wines found matching your criteria.";

    await context.PostAsync(message);

    context.EndConversation(EndOfConversationCodes.CompletedSuccessfully);
}
```

The *wineInfo* parameter is the *WineForm* instance, containing the user's responses to FormFlow questions. The first thing *DoSearch* does is get the category number via the LINQ query on the *WineCategories* collection. *WineCategories* is a *Refinement[]* from the *WineApi*, containing both the category name and id and *WineApi* needs a category id for its search.

The call to *WineApi.SearchAsync* passes the category id, *wineType*, and the rest of the values it needs from *wineInfo*. The return value is a *List*, which is a type from *WineApi* containing details of wine results. The method creates a message for the user, based on whether the search returned wine.

The *PostAsync* method, from the *IDialogContext* instance, *context*, is the same used many times previously. It sends a message to the user with the search results.

Finally, the method tells Bot Connector that it's done with this form by calling *context.EndConversation*. The parameter to *EndConversation* is a string, named *code*, to communicate the reason why the form is ending the conversation. This example uses *EndOfConversationCodes.CompletedSuccessfully* from the following *EndOfConversationCodes* class to indicate successful completion:

```
public class EndOfConversationCodes
{
    /// <summary>
    /// The conversation was ended for unknown reasons
    /// </summary>
    public const string Unknown = "unknown";

    /// <summary>
    /// The conversation completed successfully
    /// </summary>
    public const string CompletedSuccessfully = "completedSuccessfully";

    /// <summary>
    /// The user cancelled the conversation
    /// </summary>
    public const string UserCancelled = "userCancelled";

    /// <summary>
    /// The conversation was ended because requests sent to the bot timed out
    /// </summary>
    public const string BotTimedOut = "botTimedOut";

    /// <summary>
    /// The conversation was ended because the bot sent an invalid message
    /// </summary>
    public const string BotIssuedInvalidMessage = "botIssuedInvalidMessage";

    /// <summary>
    /// The conversation ended because the channel experienced an internal failure
    /// </summary>
    public const string ChannelFailed = "channelFailed";
}
```

The *Build* Method

After specifying each part of the form, call the *Build* method, as copied here from Listing 7-4:

```
return form.Build();
```

In this example, form is the *FormBuilder<WineForm>* instance. The *Build* method performs all of the internal work to prepare the form, based on how you defined it with *FormBuilder<WineForm>* methods and wraps the result in an *IForm<WineForm>* to return from the *BuildForm* method.

That's it for how to build a FormFlow form. Now, let's look at how properties, like *WineCategories*, get populated and a new way to wrap a FormFlow form in an *IDialog<T>* for Bot Builder consumption.

Initializing FormFlow

Previous examples used the static *FormDialog.FromForm* method to wrap a FormFlow form in an *IDialog<T>*. However, that won't work for our latest form in Listing 7-4 because this example requires pre-populated fields. Therefore, this section shows another way to get an *IDialog<T>*, while providing default values for form fields. Listing 7-5 shows the chatbot code that does this.

LISTING 7-5 Initializing a FormFlow Form

```csharp
using System.Net;
using System.Net.Http;
using System.Threading.Tasks;
using System.Web.Http;
using Microsoft.Bot.Builder.Dialogs;
using Microsoft.Bot.Connector;
using Microsoft.Bot.Builder.FormFlow;
using System;
using WineBotLib;

namespace WineBotFields
{
    [BotAuthentication]
    public class MessagesController : ApiController
    {
        public async Task<HttpResponseMessage> Post([FromBody]Activity activity)
        {
            if (activity.Type == ActivityTypes.Message)
                try
                {
                    IDialog<WineForm> wineDialog = await BuildWineDialogAsync();
                    await Conversation.SendAsync(activity, () => wineDialog);
                }
                catch (FormCanceledException<WineForm> fcEx)
                {
                    Activity reply = activity.CreateReply(fcEx.Message);
                    var connector = new ConnectorClient(new Uri(activity.ServiceUrl));
```

```
                    connector.Conversations.ReplyToActivity(reply);
            }

            return Request.CreateResponse(HttpStatusCode.OK);
        }

        Refinement[] wineCategories;

        async Task<IDialog<WineForm>> BuildWineDialogAsync()
        {
            if (wineCategories == null)
                wineCategories = await new WineApi().GetWineCategoriesAsync();

            var wineForm = new WineForm
            {
                WineCategories =
                    wineCategories,
                InStock = StockingType.InStock,
                Rating = 75,
                Vintage = 2010
            };

            return new FormDialog<WineForm>(
                wineForm,
                wineForm.BuildForm,
                FormOptions.PromptFieldsWithValues);
        }
    }
}
```

In Listing 7-5, the *Post* method calls *BuildWineDialogAsync* to get an *IDialog<WineForm>* for Bot Builder's *Conversation.SendAsync*. In previous examples, this was a simple call to *FormDialog. FromFrom*, but the requirements for this example are different. Essentially, we want to send a list of wine categories, from *WineApi*, to *WineForm*.

The code populates *wineCategories*, which is a *Refinement[]*, from the call to *GetWineCategoriesAsync* in *WineApi*. This is the same *WineApi* code that we've been using since Chapter 5, where you can find a detailed description of how it works.

The code instantiates a *WineForm*, but this isn't the *WineForm* instance that FormFlow will use. Instead, its purpose is to initialize fields. Notice that the code also populates *WineCategories*, which isn't included as a *WineForm* field, but is how we pass those categories to the form.

> **Tip** Imagine if you were keeping track of a user's choices between each visit to WineBot, saving the choices and retrieving those choices on subsequent visits. You can populate a *WineForm* instance like this example, with those saved values. This can save the user's time in filling out the form.

Finally, the code instantiates a new *FormDialog<WineForm>* whose parameters are *state*, *buildForm*, and *options*. The *state* parameter is the *WineForm* instance, *wineForm*. *FormDialog* populates the fields in the FormFlow form with matching values from the *wineForm* instance passed as the *state* parameter, setting new defaults for the user. The *buildForm* parameter refers to the *BuildForm* method in *WineForm*. The default behavior when fields already have values is for FormFlow to ignore the fields with values and only display fields that don't have values. Using *FormOptions.PromptFieldsWithValues* forces FormFlow to abandon the default behavior and ask questions for each field, regardless of whether the field has a value. Here's the FormOptions enum:

```
[Flags]
public enum FormOptions
{
    /// <summary>
    /// No options.
    /// </summary>
    None,

    /// <summary>
    /// Prompt when the dialog starts.
    /// </summary>
    PromptInStart,

    /// <summary>
    /// Prompt for fields that already have a value in the initial state when processing form.
    /// </summary>
    PromptFieldsWithValues
};
```

The *FormOptions* parameter, *options*, defaults to *None*. Also, notice the *Flags* attribute, allowing you to combine *PromptInStart* and *PromptFieldsWithValues*.

As you've seen, the process of initializing a FormFlow form wraps a class inside of FormFlow types, resulting in an object implementing *IDialog<T>*. In particular, you saw how to pass arguments, which is useful when you have full or partial information and want to hand off the next part of the conversation to a FormFlow form. You'll learn how perform these hand-offs and navigate between dialogs in Chapter 9, Managing Advanced Conversation.

Summary

This chapter showed several advanced techniques for customizing FormFlow. It started with an overview of the *FormBuilder<T>* fluent interface and continued with available methods.

You learned how to add messages and confirmations to the form, along with common parameters for customizing their behaviors.

This chapter showed several examples of how to define fields, changing order, dynamic definition, and custom validation. There is also a discussion of how to fill in the rest of the form automatically and determine if a field has been defined.

You also learned how to handle the results of a form when it's complete.

Finally, the chapter showed you how to set default values before the user starts the form and how to determine whether the user should fill in forms, regardless of whether a field has a value or not.

You now have several tools for dynamically creating and customizing FormFlow forms. The next chapter continues the story of Bot Framework dialogs with *LuisDialog*, allowing users to interact with a chatbot using natural language.

CHAPTER 8

Using Natural Language Processing (NLP) with LUIS

In the world of chatbots, there are plenty of opportunities for buttons, cards, and quick commands, but the essential aspect of chatbots is that they have conversational interfaces. With the Microsoft Bot Framework, you're able to target multiple channels, where text conversation is the defacto standard. In fact, some channels like email and SMS are primarily text, where any attempts to make them graphical would look and feel like an old comfortable tool with new technology bolted onto its side. What you'll find is that a lot of applications can naturally fit into this conversational world and often yield a superior solution.

To see the benefit of the conversational interface, this chapter discusses a branch of Artificial Intelligence (AI) known as Natural Language Processing (NLP). While just mentioning sophisticated computer science subjects like AI might sound complex, it really isn't. That's because Microsoft has an NLP service that does all the work for us. It's called Language Understanding Intelligence Service (LUIS). The LUIS API is part of Microsoft's Cognitive Services, a suite of AI services you can use, which I'll talk more about in Chapter 14, Integrating Cognitive Services. LUIS is a NLP API that translates plain language text into objects that a chatbot can recognize and act on.

In this chapter, you'll learn important concepts of how LUIS works. Then you'll learn how to visit the LUIS site to use graphical and text tools to train what is called a model that will recognize commands for a chatbot. With a model, you'll learn how to add code to a chatbot that receives commands from the user and act on those commands. You'll see how easy it is to get a chatbot up and running quickly with LUIS, and will learn how to improve the model afterwards.

Learning Essential LUIS Concepts

The goal of LUIS is to translate human text into something a computer program can understand. As demonstrated in Figure 8-1, chatbots send messages from the user to LUIS, LUIS performs the translation into something the chatbot can understand, and the chatbot processes the LUIS results to formulate a response to the user.

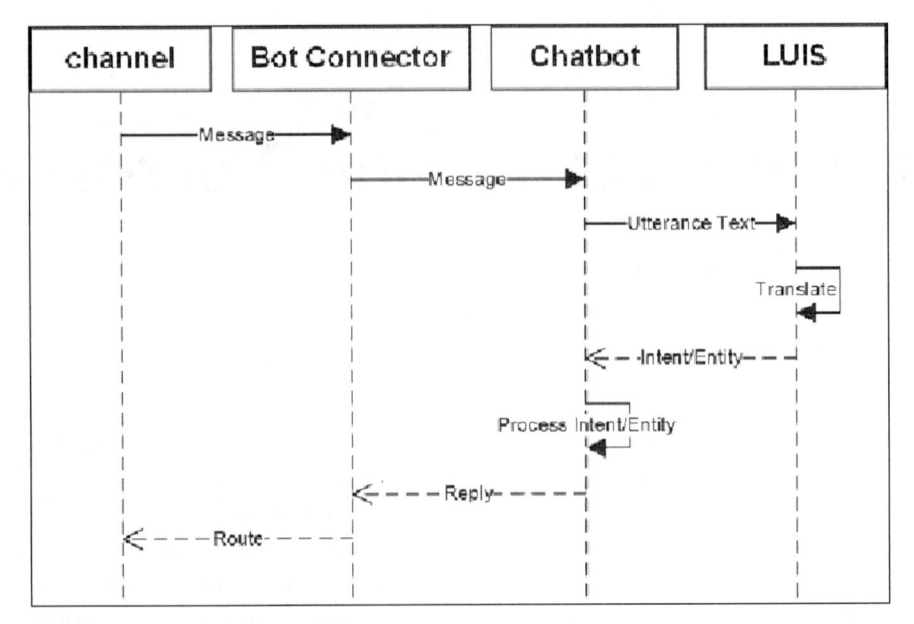

FIGURE 8-1 How Chatbots Use LUIS.

In Figure 8-1, *Channel* represents the client application that the user interacts with to communicate with a chatbot. Since the Bot Framework supports a growing list of channels, this means that it's easier for users to find a preferred client to communicate with a chatbot.

As normal, the *Channel* sends messages, with the user's plain language text, to *Bot Connector*, which then routes that message to a *Chatbot*. At this point in time, the *Chatbot* can handle the user's text as it sees fit. More specifically, in this chapter you'll learn about the last component in Figure 8-1, *LUIS*, which translates utterance text into intents and entities. The *Chatbot* extracts utterance text from the message received from *Bot Connector*.

These intents and entities that *LUIS* returns arrive in the form of JSON text. Bot Framework translates that JSON into objects for it to operate on and provide data in parameters for a chatbot to make decisions upon. The intents map to a goal that the user wants to accomplish. Entities map to facts inside of the utterance text that a chatbot needs as parameters of the intent. Later in this chapter, you're going to see how intents map to methods and entities are extracted from parameters to those methods. Based on intent and entities, the *Chatbot* processes logic to determine what action to take and what response to send to the user.

The next section explains how to train a LUIS model that translates utterances into intents and entities.

Setting up LUIS and Training Models

The existence of NLP doesn't mean that a chatbot will be able to understand anything a user says. Rather, the developer needs to train an NLP model to understand very specific tasks belonging to a chatbot's domain. In the case of this chapter's example, the domain is *searching for wine*. It isn't about farming, shopping, health or any other type of general conversation, just *searching for wine*. Not only is this good because chatbots targeting a specific domain make early success easier, but because it limits the amount of work by the developer to train a model. For this reason, WineBot will limit conversation to its domain, making the process of training an NLP model with LUIS easier.

This section explains how to build, train, and deploy a LUIS model. Let's start by visiting the LUIS Web site and creating a new model.

Creating Models

Working with LUIS, you need a model, which is a set of intents, entities, and utterances that train the model. To get started, visit *https://www.luis.ai*, sign in, and go to the *My apps* tab. Click the button to create a new app and you'll see a screen similar to Figure 8-2.

> **Note** LUIS uses the term *app* and you see me using the term *model*. For the purposes of this text, they're synonymous. I prefer the term *model* because it's based on the underlying implementation, which is a machine learning model.

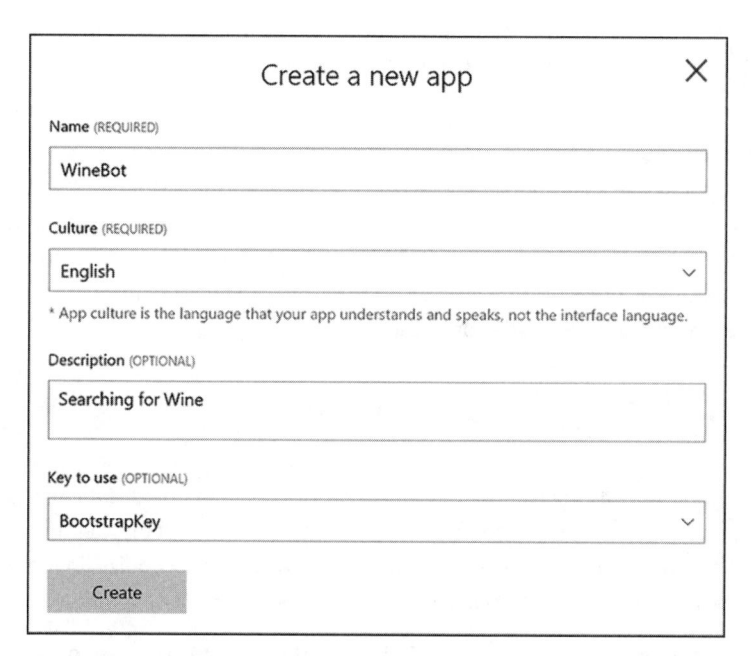

FIGURE 8-2 Creating a New LUIS Model.

In Figure 8-2, add a meaningful name for the chatbot, which will appear in a list of LUIS models for your login. The *Culture* is *English*, but you can change it to one of a growing list of cultures. The optional *Description* lets you explain what the model is for.

The *Key to use* can be any pre-configured key or, since it's optional, no key at all. Click *Create* and observe the page for your new model, shown in Figure 8-3.

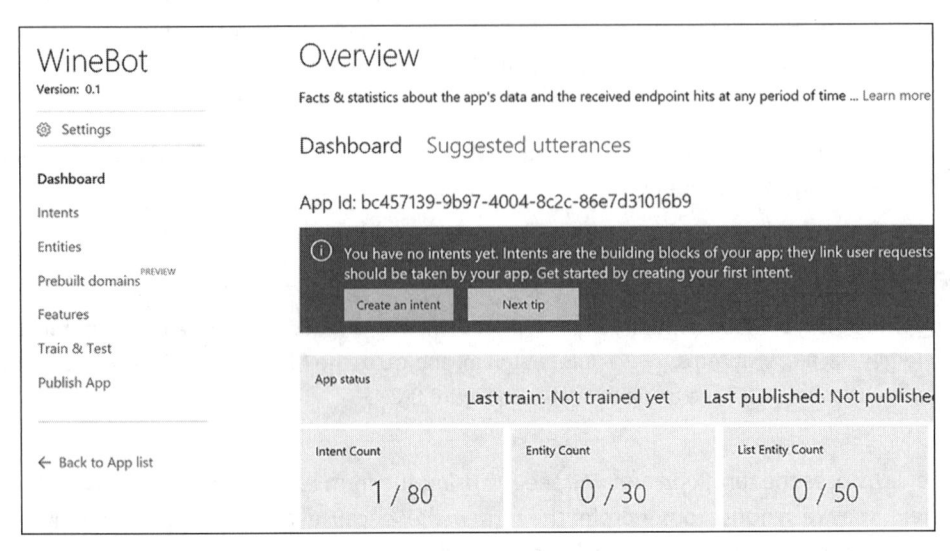

FIGURE 8-3 A New LUIS Model Page.

Figure 8-3 shows a *Dashboard* with a menu on the left side of the page. In that menu are *Intents* and *Entities*. The next couple of sections show how to create intents and entities.

> **Note** As you might already know, Web pages change over time and the LUIS site is no exception. Rather than reviewing the exact makeup of a given page for a snapshot in time, it's more important to have a sense of what you're looking for to accomplish a task. LUIS continues to evolve and add new and interesting features, which you'll want to explore over time. However, for the purposes of this book, intents and entities are the essential features of a LUIS model that you'll learn how to write code for in this chapter.

Building Intents

As its name suggests, an *intent* represents the user's intention. In other words, an intent is the user's goal or a task they wish to accomplish. An intent consists of an intent name and a set of utterances that a user would typically ask or say to accomplish the goal the intent is designed to recognize. In the current example, we want an intent to represent the goal of searching for wine.

To add an intent, click on the *Intents* menu item. Figure 8-4 shows the next steps in how to create an intent.

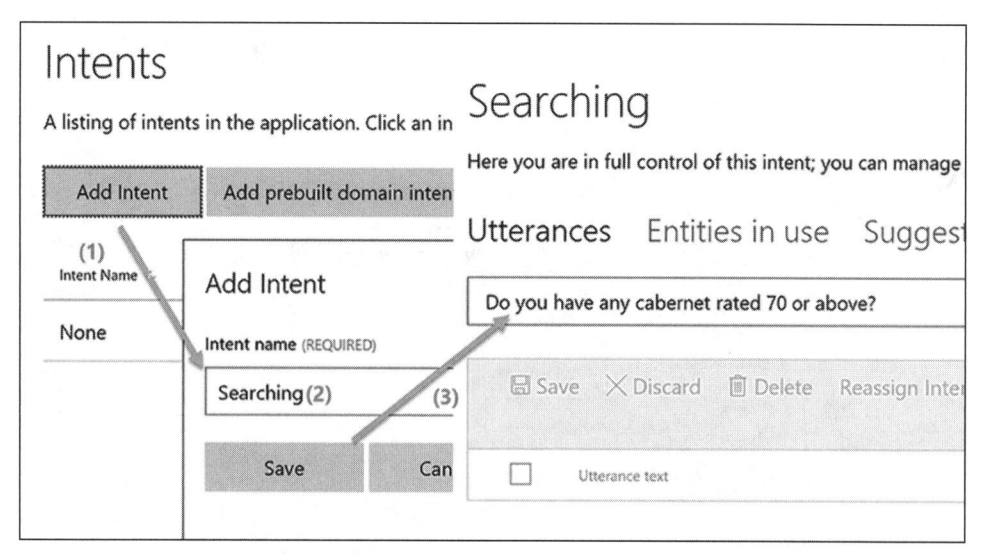

FIGURE 8-4 Creating a New Intent.

The *Intents* page, in Figure 8-4, has a list of intents and a button for creating a new intent. Notice that there is also a *None* intent, which is a default intent that LUIS assigns utterances to whenever it can't figure out which intent a user's utterance maps to.

Clicking *Add Intent* opens a pop-up, *Add Intent*, to type the name of the event. Step 2 of Figure 8-4, shows the *Searching* intent name, because its purpose will be to search for wines.

After adding an *Intent Name*, click *Save* and you'll see a page that allows you to add utterances for the intent. The *Searching* page, Step 3 of Figure 8-4, shows a possible utterance: "Do you have any cabernet rated 70 or above?". On this page, type several possible things that the user can say that can match this intent. For example, "Please search for champaigne rated fifty," or "What white wines do you have rated 80 or above?".

At this point, we're not ready to start adding utterances yet. Besides knowing that the user wants to do a search, we need to know the facts (aka entities) associated with a search request, which is discussed next.

Specifying Entities

Just having the user's intent is not necessarily sufficient to know what the user is asking. The utterance in Figure 8-4, "Do you have any cabernet rated 70 or above?" let's us know that the user wants to do a search, but we really need to know that the type of wine is "cabernet" and the rating is "70." We also don't want to manually write code to parse these values ourselves. This is the role of entities.

Simple Entities

To add a new entity, click on the *Entities* menu item. Figure 8-5 shows how to create an entity.

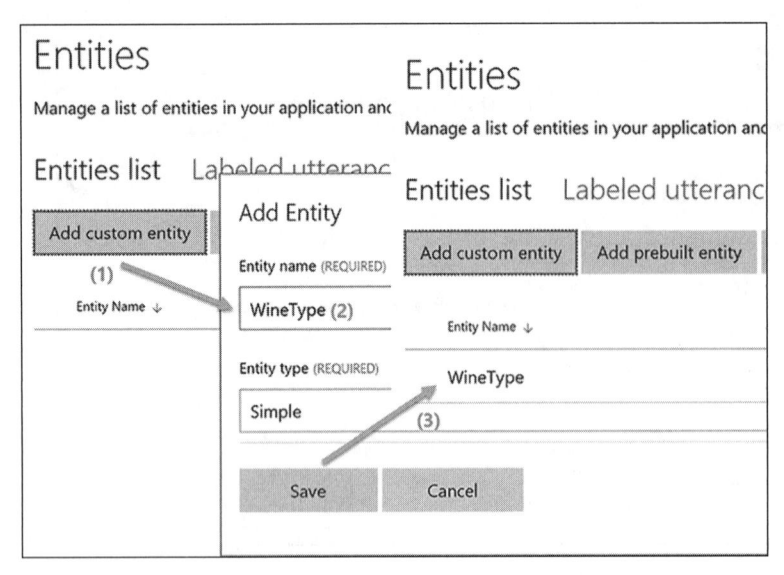

FIGURE 8-5 Creating a New Entity.

As shown in Figure 8-5, you can create an entity by clicking *Add custom entity*, opening the *Add Entity* pop-up.

Type the *Entity name*, which is *WineType* in Figure 8-5. LUIS has different types of entities, but taking the default of *Simple* is sufficient enough to extract a value that a chatbot can act on.

> **Note** LUIS Entity type is an evolving feature with some items in beta. These things can change between the time of this writing and when you read this. Fortunately, once you understand how the *Simple* entity type works and can extract values in code, it will be easier to review and experiment with other entity types as they come online.

Clicking *Save* closes the *Add Entity* pop-up and shows the new entity in the list.

Prebuilt Entites

In addition to custom entities, LUIS has *prebuilt entities*, which are more sophisticated in that they can recognize different ways to type an entity. The prebuilt entity this example uses is *number*, which can not only recognize numbers like "70," but also variations spelled out like "seventy." This saves a lot of time in writing custom code to try to recognize what an entity means. To get started, click the *Add prebuilt entity* button on the *Entities* page, to reveal the *Add prebuilt entities* pop-up, shown in Figure 8-6.

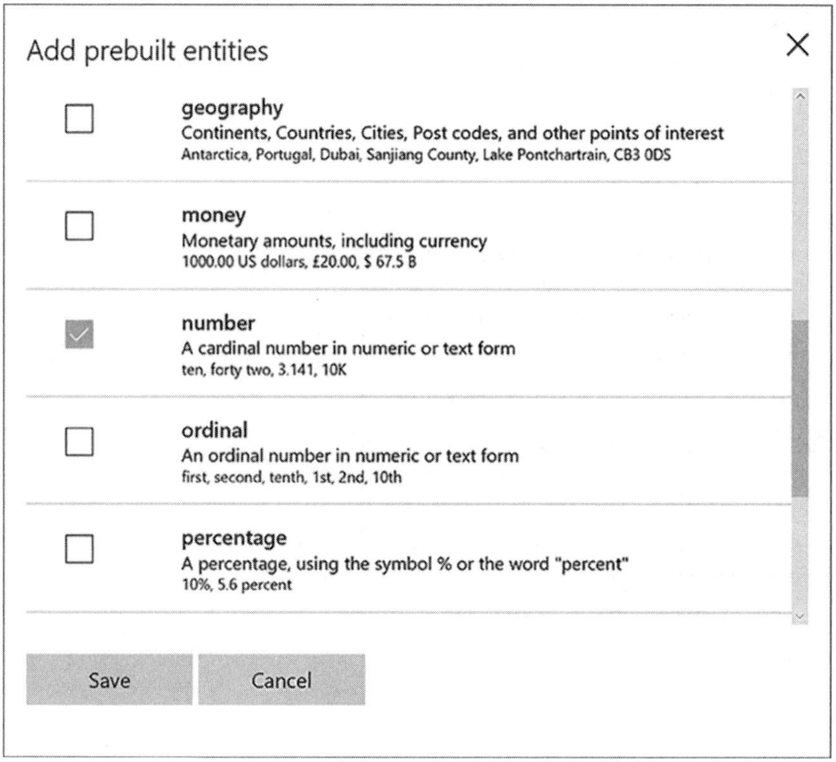

FIGURE 8-6 Creating a New Prebuilt Entity.

Scanning the different options in Figure 8-6: *Geography, Money, Number, Ordinal, Percentage*, and more. You can get a feel for the benefits of using a prebuilt entity—they recognize text in each of their domains in several different forms. Check your choice, *number* in Figure 8-6, and click the *Save* button, closing the pop-up and showing the new number entity in the list.

Now that you can create intents and entities, it's time to pull them together and train the model, which is discussed next.

Training and Deploying

LUIS models are based on machine learning, which you must train by providing data. In the case of LUIS, the data is a list of utterances that a user can possibly ask or say. You classify utterances into intents, representing what the user desires to accomplish. Each of these utterances has zero or more entities, which are placeholders for the facts your code needs to fulfill the user's request. This section explains how to train the model by adding utterances to intents and specifying entities.

Adding Utterances

To train a model, click on the *Intents* menu, click on the intent to train (*Searching* in this example), and ensure you're on the *Utterances* tab. As shown in Figure 8-7, type an utterance in the text box and press **Enter**.

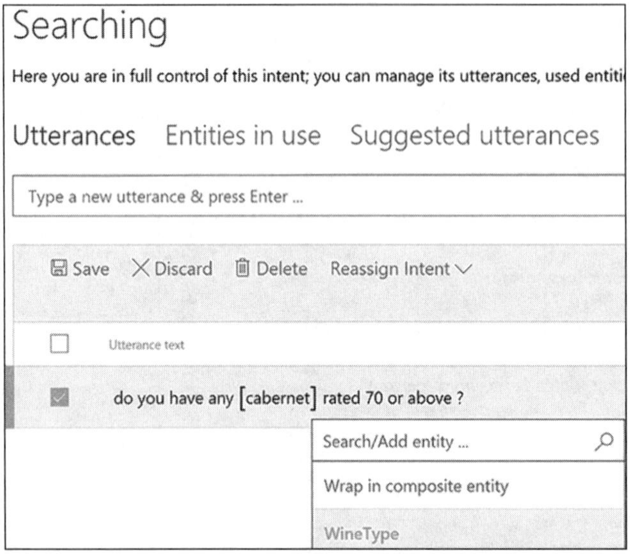

FIGURE 8-7 Adding Utterances.

For the example in Figure 8-7, you must type **Do you have any cabernet rated 70 or above?** LUIS normalized the sentence by lower casing and spacing the ending punctuation. While this isn't proper grammar, it doesn't matter for training. In fact, for better training you'll want to provide different examples that aren't necessarily grammatically correct because users won't use perfect grammar. More examples make a better model.

Sometimes, just an intent is sufficient and you won't need entitites. For example, if you have an *OpeningHours* intent, people might ask "What are your hours of operation?" and the only answer is, "We're open from 8 to 5." That doesn't require entities because the intent is already clear.

For the *Searching* intent, we need to know *WineType* and *Rating*, which is why earlier examples created entities for these values. Figure 8-7 shows what happens when clicking a word, *cabernet*, showing the associated context menu. Clicking *WineType*, designates that word as an entity.

A single utterance is insufficient because people will say things to a chatbot that you never planned for. To increase the chances of recognition, add several more utterance variations, as shown in Figure 8-8.

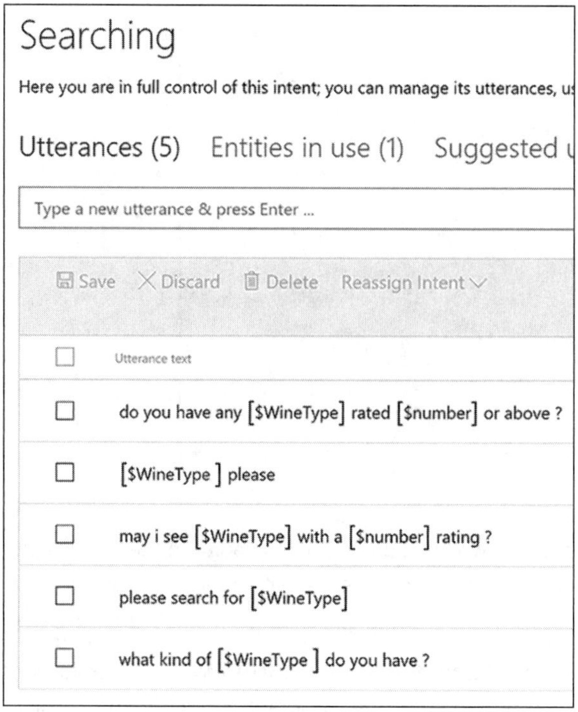

Searching

Here you are in full control of this intent; you can manage its utterances, u[...]

Utterances (5)   Entities in use (1)   Suggested [...]

	Type a new utterance & press Enter ...

🖫 Save   ✕ Discard   🗑 Delete   Reassign Intent ⌄

	Utterance text
☐	do you have any [$WineType] rated [$number] or above ?
☐	[$WineType] please
☐	may i see [$WineType] with a [$number] rating ?
☐	please search for [$WineType]
☐	what kind of [$WineType] do you have ?

FIGURE 8-8 Utterances with Entites.

Figure 8-8 shows several examples of things a user might say to a chatbot if their intent was to search for wine. Notice how assigning the *WineType* entitity shows in the list as *[$WineType]* and *Rating* appears as the *[$number]* entity. There isn't an entity called *Rating* – it's just that the meaning of the *number* entity is that it will be used in the chatbot for *Rating*.

What you might notice is that the prebuilt entities, *number* in this example, don't appear in the context list when defining an utterance. That's because the current LUIS interface automatically recognizes prebuilt entities and replaces them with the placeholder for you, which is *[$number]* in this example. At this point, you're ready to test and finish training the model.

Training and Testing a Model

As you've seen earlier, a LUIS model consists of intents, entities, and utterances. LUIS uses machine learning to create the model, which means you must train it. To get started, click the Train And Test menu item, which takes you to the Test Your Application page, shown in Figure 8-9.

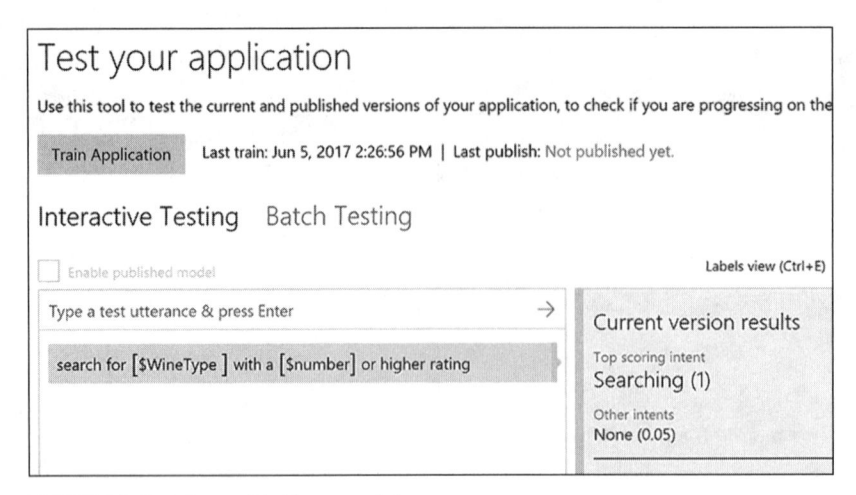

FIGURE 8-9 Training and testing a model.

To train, click the *Train Application* button, shown in Figure 8-9. Until you train the model, any changes since the last training (or new model creation) won't be available to applications.

This *Train and Test* page lets you test too. As the text box in Figure 8-9 suggests, *Type A Test Utterance & Press Enter*. You can see how **search for champaigne with a 90 or higher rating** has been typed. The LUIS UI normalized that utterance, added entity placeholders, and scored the results.

> **Tip** Notice that the test utterance in Figure 8-9 doesn't match any of the example utterances from Figure 8-8. This is important because it demonstrates how the trainined model generalizes enough to recognize many different ways to say the same thing. With only a few utterances, you have a working model that's sufficient for building code to support that intent without a lot of initial effort.

The score on the right side, under *Current Version Results*, shows that the utterance matched the *Searching* intent, along with a parenthesized probability score. It also shows losing intents and their score, for example, the *None* intent lost with a probability score of *0.05*.

Publishing a Model

Before using the LUIS model, you must publish it. To do so, click the *Publish* menu item and click the *Publish* button. In the resulting list, you'll see an *Endpoint url* with the following format:

```
https://westus.api.cognitive.microsoft.com/luis/v2.0/apps/<model-id>?subscription-key=<subscription-key>&verbose=true&timezoneOffset=0&q=
```

Copy and paste the contents of where you see *<model-id>* and *<subscription-key>*, because you'll need them to tell the chatbot how to communicate with this model, which you'll learn about next.

Using LUIS in Your Chatbots

Now that you have a LUIS model, it's time to write code to use it. In this section you'll learn how to create a *LuisDialog<T>* that knows how to handle user text and communicate with LUIS. Then you'll see how to create methods that map to intents. Inside those intent handling methods, you'll write code to read entities, extracting more details about what the user wants to accomplish. Let's start by reviewing the example program for this chapter, *WineBotLuis*.

Introducing *WineBotLuis*

WineBotLuis is an adaptation of the *WineBot* program, started in Chapter 5 and continued for each chapter thereafter. It uses the *WineApi* library, created in Chapter 5, to communicate with the Wine.com API. Remember that you'll need an API key from Wine.com for this program to work. Listing 8-1 shows the *WineBotDialog* class, which is part of the *WineBotLuis* project in the accompanying source code.

LISTING 8-1 A *LuisDialog<T>*-Derived Type for Handling Natural Language Input

```csharp
using System;
using System.Threading.Tasks;
using Microsoft.Bot.Builder.Dialogs;
using Microsoft.Bot.Builder.Luis.Models;
using Microsoft.Bot.Builder.Luis;
using WineBotLib;
using System.Collections.Generic;
using System.Text.RegularExpressions;
using System.Linq;
using Microsoft.Bot.Connector;

namespace WineBotLuis.Dialogs
{
    [LuisModel(
        modelID: "<model-id>",
        subscriptionKey: "<subscription-key>")]
    [Serializable]
    public class WineBotDialog : LuisDialog<object>
    {
        [LuisIntent("")]
        public async Task NoneIntent(IDialogContext context, LuisResult result)
        {
            string message = @"
Sorry, I didn't get that. 
Here are a couple examples that I can recognize: 
'What type of red wine do you have with a rating of 70?' or 
'Please search for champaigne.'";

            await context.PostAsync(message);
            context.Wait(MessageReceived);
        }
```

```csharp
[LuisIntent("Searching")]
public async Task SearchingIntent(IDialogContext context,
    IAwaitable<IMessageActivity> activity, LuisResult result)
{
    if (!result.Entities.Any())
        await NoneIntent(context, result);

    (int wineCategory, int rating) = ExtractEntities(result);

    var wines = await new WineApi().SearchAsync(
        wineCategory, rating, inStock: true, searchTerms: string.Empty);
    string message;

    if (wines.Any())
        message = "Here are the top matching wines: " +
                    string.Join(", ", wines.Select(w => w.Name));
    else
        message = "Sorry, No wines found matching your criteria.";

    await context.PostAsync(message);

    context.Wait(MessageReceived);
}

(int wineCategory, int rating) ExtractEntities(LuisResult result)
{
    const string RatingEntity = "builtin.number";
    const string WineTypeEntity = "WineType";

    int rating = 1;
    result.TryFindEntity(RatingEntity, out EntityRecommendation ratingEntityRec);
    if (ratingEntityRec?.Resolution != null)
        int.TryParse(ratingEntityRec.Resolution["value"] as string, out rating);

    int wineCategory = 0;
    result.TryFindEntity(WineTypeEntity, out EntityRecommendation wineTypeEntityRec);

    if (wineTypeEntityRec != null)
    {
        string wineType = wineTypeEntityRec.Entity;

        wineCategory =
            (from wine in WineTypeTable.Keys
             let matches = new Regex(WineTypeTable[wine]).Match(wineType)
             where matches.Success
             select (int)wine)
            .FirstOrDefault();
    }

    return (wineCategory, rating);
}

Dictionary<WineType, string> WineTypeTable =
    new Dictionary<WineType, string>
    {
```

```
                [WineType.ChampagneAndSparkling] = "champaign and
sparkling|champaign|sparkling",
                [WineType.DessertSherryAndPort] = "dessert sherry and
port|desert|sherry|port",
                [WineType.RedWine] = "red wine|red|reds|cabernet|merlot",
                [WineType.RoseWine] = "rose wine|rose",
                [WineType.Sake] = "sake",
                [WineType.WhiteWine] = "white wine|white|whites|chardonnay"
            };
        }
    }
```

A chatbot that uses LUIS to translate user text can derive from *LuisDialog<T>*, like *WineBotDialog* in Listing 8-1. *WineBotDialog* has a *Serializable* attribute, which the Bot Framework requires for all dialogs needing to persist state in the Bot State Service . The *LuisModel* attribute specifies the model id and subscription key that you learned about in the previous, Publishing a Model, section. Replace *<model-id>* and *<subscription-key>* with the values from the LUIS *Endpoint url*. The following sections explain each of the parts of the *WineBotDialog*.

Adding Intents

As described in Figure 8-1, *WineBotDialog* forwards the user input, utterance text, to LUIS for translation. *LuisDialog<T>*, *WineBotDialog's* base class, handles that message and takes care of communicating with LUIS so you don't have to write that code. When the response comes back, *LuisDialog<T>* evaluates the JSON, picks the intent with the highest score and calls the derived class (*WineBotDialog*) method that can handle that intent. *LuisDialog<T>* uses reflection to examine *LuisIntent* attributes and invokes the matching method.

Whenever LUIS can't map an utterance to an intent, it picks the *None* intent. The first method in *WineBotDialog* handles the *None* intent, repeated here for convenience:

```
        [LuisIntent("")]
        public async Task NoneIntent(IDialogContext context, LuisResult result)
        {
            string message = @"
Sorry, I didn't get that.
Here are a couple examples that I can recognize:
'What type of red wine do you have with a rating of 70?' or
'Please search for champaigne.'";

            await context.PostAsync(message);
            context.Wait(MessageReceived);
        }
```

For the *None* intent, give the *LuisIntent* attribute a blank string. Intent handler methods have two or three parameters. *IDialogContext* and *LuisResult* for *context* and *result* are required. Optionally, you can also add a second *Awaitable<T>* parameter, as shown for the *SearchingIntent* method of Listing 8-1. The *context* parameter is the same *IDialogContext* type described in depth in Chapter 5. *LuisResult* is a special type for holding entity information. These examples use a suggested Bot Framework convention of adding the *Intent* suffix to the method name.

This implementation of the *None* intent simply explains that it doesn't understand and gives a couple suggestions on what the user should do. After posting the message to the user, and before completing the method, call *context.Wait(MessageReceived)*. You might have noticed that *WineBotDialog* doesn't have a *MessageReceived* method, but it does inherit *MessageReceived* from *LuisDialog<T>*.

> **Tip** Whenever handling situations for when the chatbot doesn't understand the user, the response should give the user some idea on what the chatbot expects. Over time, you'll want to write logic that adapts to the user and gets better at understanding them, but it's still good to try to help them out when something isn't working.

That was the method for the *None* intent and there's also the *SearchingIntent* method. The *Searching* intent has the same *LuisIntent* attribute, but with the *Searching* string and the same parameters. The next section discusses how the *Searching* intent method handles entities.

> **Tip** You could declare multiple *LuisModel* attributes, each with a separate model. This can be useful if you encounter limitations in the number of intents your chatbot needs or if you prefer to organize via separate models. Just make sure your intent names are unique, possibly using a namespace-like approach like *pseudonamespace.myspecificintent*. LUIS ranks intent matches by probability and *LuisDialog* selects the intent handler with the highest probability. This works with multiple models because intents that don't match should have lower probabilities.

Handling Entities

The *Searching* intent method, *SearchingIntent*, handles the *LuisResult* parameter, *result*, to extract parameters it needs for the WineApi. Here's *SearchingIntent*, repeated below for convenience:

```
[LuisIntent("Searching")]
public async Task SearchingIntent(IDialogContext context, IAwaitable<IMessageActivity> activity, LuisResult result)
{
    if (!result.Entities.Any())
        await NoneIntent(context, result);

    (int wineCategory, int rating) = ExtractEntities(result);

    var wines = await new WineApi().SearchAsync(
        wineCategory, rating, inStock: true, searchTerms: string.Empty);
    string message;

    if (wines.Any())
        message = "Here are the top matching wines: " +
            string.Join(", ", wines.Select(w => w.Name));
```

```
        else
            message = "Sorry, No wines found matching your criteria.";

        await context.PostAsync(message);

        context.Wait(MessageReceived);
    }
```

LuisResult has an *Entites* property that is a *List<EntityRecommendation>*. If this list is empty, that means either the user didn't provide those values or LUIS is unable to recognize any entities. Though *SearchingIntent* calls *NoneIntent* in this case, it's possible that in the future you might want to write additional logic to see if it can figure out what is missing. Calling *NoneIntent* was a simple mitigation strategy, but the main point is that you need input validation because LUIS won't always recognize entities correctly. Remember, the user can literally say anything to the chatbot.

SearchingIntent gets *wineCategory* and *rating* values from the entities and uses *WineApi* to perform a wine search. The *ExtractEntities*, repeated below, reads a *LuisResult* instance and returns a *wineCategory* and *rating*:

```
(int wineCategory, int rating) ExtractEntities(LuisResult result)
{
    const string RatingEntity = "builtin.number";
    const string WineTypeEntity = "WineType";

    int rating = 1;
    result.TryFindEntity(RatingEntity, out EntityRecommendation ratingEntityRec);
    if (ratingEntityRec?.Resolution != null)
        int.TryParse(ratingEntityRec.Resolution["value"] as string, out rating);

    int wineCategory = 0;
    result.TryFindEntity(WineTypeEntity, out EntityRecommendation wineTypeEntityRec);

    if (wineTypeEntityRec != null)
    {
        string wineType = wineTypeEntityRec.Entity;

        wineCategory =
            (from wine in WineTypeTable.Keys
             let matches = new Regex(WineTypeTable[wine]).Match(wineType)
             where matches.Success
             select (int)wine)
            .FirstOrDefault();
    }

    return (wineCategory, rating);
}
```

The two *const string* values at the top of the *ExtractEntities* method map to the entity names in the LUIS model: *number* and *WineType*. Those are the first parameters to *TryFindEntity*, a *LuisResult* method for extracting an *EntityRecommendation* from the results.

> **Tip** Notice that the *RatingEntity* const in *ExtractEntities*, Listing 8-1, has a "builtin." Prefix, as in *builtin.number*. This differs from the *number* in Figure 8-6 and the *[$number]* from Figure 8-8. All prebuilt entities, from the same dialog in Figure 8-6, have this prefix. Additionally, all prebuilt/builtin entities and intents will have some form of prefix that you'll need to ensure are added to code properly. For entities, one way to double-check the spelling is to set a breakpoint in an intent handling method and use the Visual Studio debugger to drill into the *LuisResult* parameter instance and manually inspect entities.

The *out* parameter to *TryFindEntity* is an *EntityRecommendation* and you'll need to check that for *null* in case the entity isn't present. *EntityRecommendation* has two properties for results: *Entity* and *Resolution*. Whenever you've defined your own custom entity, you can read the *Entity* property, as with the *WineType* entity. When you use a prebuilt entity, it's best to use the *Resolution* property. That's because while you can read the value from the *Entity* property, it will be the string representing what the user typed in. In the case of *builtin.number*, a user could spell out *50* as *fifty* and it would be recognized. That means trying to parse *Entity* won't work on the string respresentation, but will work if the user typed the numeric representation, which could be confusing. LUIS provides the numeric representation of the string in *Resolution["value"]*, which will always be a parsable number.

The next dilemma is how to get an *int* representation of whatever word the user typed in for the *WineType* entity. Essentially, we need to figure out how to map the user's input to the *WineType* enum, shown here:

```
public enum WineType
{
    None = 0,
    RedWine = 124,
    WhiteWine = 125,
    ChampagneAndSparkling = 123,
    RoseWine = 126,
    DessertSherryAndPort = 128,
    Sake = 134
}
```

WineType assigns the *wineCategory* id to a matching member and those IDs match a category in the Wine.com API. The problem is that the user can type anything, such as *red*, *cabernet*, *rose*, and more. This is where a good regular expression skill set can help. One of many potential techniques is to use a dictionary table that maps regular expressions to the proper category, from the *WineTypeTable* from Listing 8-1, shown here:

```
Dictionary<WineType, string> WineTypeTable =
    new Dictionary<WineType, string>
    {
        [WineType.ChampagneAndSparkling] = "champaign and sparkling|champaign|sparkling",
        [WineType.DessertSherryAndPort] = "dessert sherry and port|desert|sherry|port",
        [WineType.RedWine] = "red wine|red|reds|cabernet|merlot",
```

```
            [WineType.RoseWine] = "rose wine|rose",
            [WineType.Sake] = "sake",
            [WineType.WhiteWine] = "white wine|white|whites|chardonnay"
        };
```

The keys in *WineTypeTable* are *WineType* members and values are regular expressions, using the *or* operator to match potential options. Clearly, this list and regular expressions are incomplete, but demonstrate the approach.

ExtractEntites has a LINQ expression that does this mapping for us, shown here:

```
wineCategory =
    (from wine in WineTypeTable.Keys
     let matches = new Regex(WineTypeTable[wine]).Match(wineType)
     where matches.Success
     select (int)wine)
    .FirstOrDefault();
```

The *wineCategory* variable is an *int*, to hold the one of the values of *WineType* that maps to an ID in the Wine.com API. The *from* clause reads each *WineTypeTable* key, *wine*, which is a member of the *WineType* enum. The *let* clause uses the *Regex* class to match the *wineType* entity value with the regular expression value returned by *WineTypeTable[wine]*. If the regular expression contains a value matching *wineType*, *matches* will be *true* and the *select* clause casts the *WineType* key to its underlying *int* representation. Depending on how your regular expressions matched and the number of matches, you could potentially want to write more logic to score and find the winning *WineType*. However, this example takes a first match wins approach with the *FirstOrDefault* clause. When the query finds a match, *wineCategory* will have an *int* representation for a category to search for, otherwise it will be the default value of *0*, which won't return any results from *WineApi*.

There is no one-size-fits-all solution for recognizing user input. While this technique might not be perfect for every situation, and you might need to code more of your own logic to interpret user input, it can move you far along the road in many scenarios. Figure 8-10 shows how *WineBotLuis* works, translating conversational text into actionable logic.

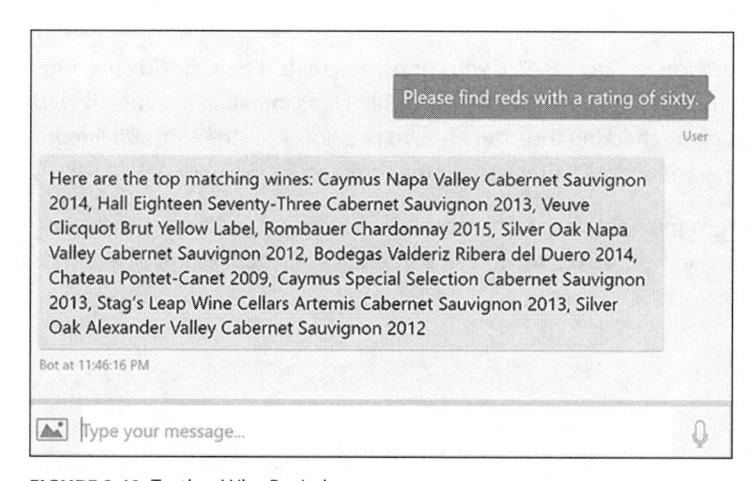

FIGURE 8-10 Testing WineBotLuis.

Now you have a working chatbot that understands natural language input. The next section explains how to make it better.

Continuous LUIS Model Improvement

Earlier sections in this chapter explain how to create a LUIS model. It was a minimal model and good to start off with. Though this is quick and powerful, you'll want to make it more robust and resilient to the unpredictability of user input. Fortunately, LUIS has additional tools to improve models.

To improve the model, visit the *https://luis.ai* Web page, navigate to the *WineBot Searching* intent, and click the *Suggested Utterances* tab, as shown in Figure 8-11.

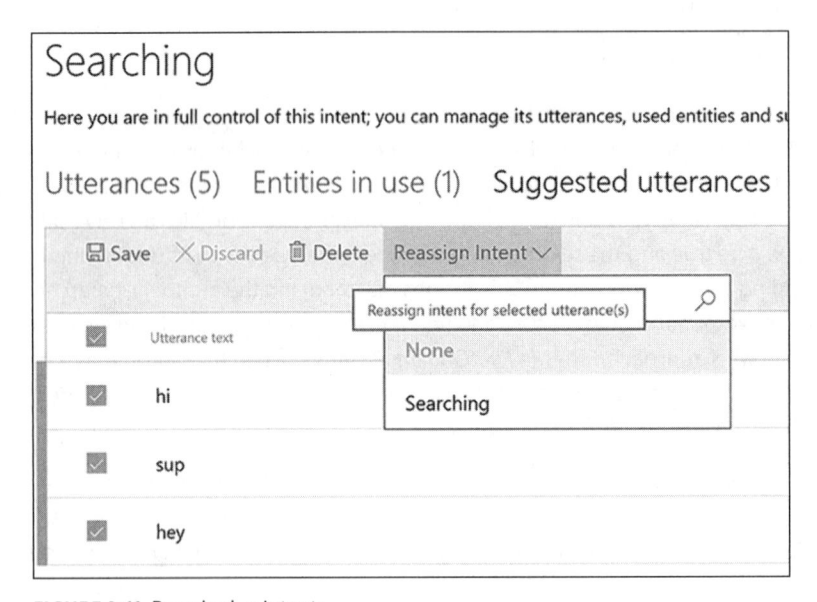

FIGURE 8-11 Reassigning Intents.

Notice the *Suggested Utterances* in Figure 8-11. If you choose to create a *Hello* or *Greeting* intent, these items might belong there, but they don't belong in the *Searching* intent. In this case, they should be reassigned. You can reassign by checking the utterances to reassign, click the *Reassign Intent* menu, and click the intent to move each item to. This reclassifies the utterances properly.

Sometimes LUIS doesn't recognize entities or mis-recognizes enties and you'll need to fix those problems too. The *None* intent, in Figure 8-12, shows some utterances that LUIS should have classified as *Searching* intents.

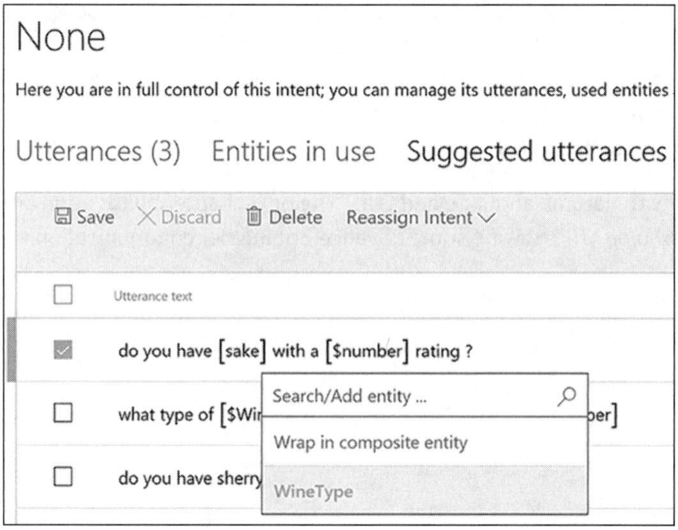

FIGURE 8-12 Fixing Entities.

Figure 8-12 shows how some of the utterances that clearly belong to the *Searching* intent appear in the *None* intent. In this case, LUIS didn't recognize *sake* as a *WineType*, and you need to click on *sake* and assign the *WineType* entity to it. After fixing entities, reassign the utterance to the proper intent, as explained previously.

The *Suggested Utterances* tab appears for all intents and entities and it's useful to re-visit them to see new utterances and fix them to improve the model. Additionally, you can re-visit the *Utterances* tab for any intent and add new utterances as you think about them. Once you've worked through suggestions and added more of your own utterances, test and do more training, as explained in previous sections. Finally, publish the newly trained model, making it immediately available for the chatbot.

Remember that working on chatbots is an iterative process. Working with LUIS models is the same and by periodically revisiting the model and improving, you're also improving the user experience.

Summary

In this chapter, you learned how to build a LUIS model and build chatbot code to handle plain language text from the user. To build a model, you need to create intents and entities, where intents are what the user wants to accomplish and entities are facts to extract from a user utterance. You saw how to enter utterances. On each utterance, you can label words that need to be recognized as entities.

Once you have intents, entities, and utterances defined, it's time to test, train, and publish the model. Training creates a machine learning model that recognizes generalized utterances from the user. The testing interface lets you try different utterances to see if your model is recognizing user input correctly. After training, visit the *Publish* tab and publish the model so that it can be used by your chatbot.

After creating the LUIS model, you learned how to write code to handle the model. You learned about the *LuisIntent* attribute, letting code know how to communicate with the LUIS model. Deriving from *LuisDialog<T>* handles the chatbot communication with LUIS automatically. Inside of the *LuisDialog<T>*-derived type, *WineBot*, you saw how to define methods to handle intents. You also learned how to process entities and write logic to handle the user's request.

Now you know how to work with natural language and LUIS. The next chapter builds upon all of the different dialogs, including *LuisDialog<T>*, showing some advance Bot Builder communication support.

CHAPTER 9

Managing Advanced Conversation

Earlier chapters in this part of the book detailed the inner workings of *IDialog<T>*, *FormFlow*, and *LuisDialog<T>*. Individually, these dialog types are powerful and allow for useful chatbots–with chatbots, simplicity is often the essence of success. Yet, there will be times when your requirements are sophisticated and require more advanced communication techniques. This chapter introduces several techniques for moving beyond the times you might feel limited by the functionality of a single dialog type.

An important concept in navigation is the dialog stack and we'll start off by explaining how this works, moving from one dialog to another. Because real life doesn't follow a script, your chatbot shouldn't either and you'll learn how to accept out-of-band information that you otherwise wouldn't be able to handle. You'll learn about chaining in this chapter, which supports complex dialog communication and navigation. You'll also learn how to format message text, thus improving the user experience.

Managing the Dialog Stack

All of the chatbots developed in previous chapters were simple, in that the developer needed to only code a single dialog type instance. The chatbot logic was to hand off control to a dialog, which would do the work and the underlying implementation of that wasn't a design factor. When working with multiple dialogs, however, this handoff logic between dialogs becomes more important, so you need to know about the dialog stack.

What is the Dialog Stack?

Before discussing the dialog stack, let's discuss stacks in general. A common metaphor is a stack of dishes. What makes this stack special is how people place dishes on and off the stack. When putting dishes in the cupboard, people place each dish on top of the other, in a stack. When retrieving dishes, they pull the dish from the top of the stack. There's a pattern here where the last dish placed in the stack will be the first dish to be removed.

This pattern is a primary consideration for computer stacks that is called Last-In First Out (LIFO). In computing, software is based on stacks. When a program starts, it sets the entry point to the first item in the stack. As the program calls methods, those methods get pushed to the top of the stack, so you can imagine methods calling methods as the stack grows. When those methods return, they pop the old method off the stack, which is how the program knows how to return to its caller.

The dialog stack is based on this same concept. When the Bot Framework calls a dialog, that dialog is pushed onto the dialog stack. Then, when the current dialog is done, it returns and is popped from the stack, letting its caller resume.

Designing Bot Framework dialog navigation around a stack is a well-worn path that's been proven for other technologies. While stack-based navigation handles a lot of common scenarios, you still aren't stuck because the Bot Framework has support for *scorables*, which is a way to break out of the single path of a stack to handle random paths that accommodate the unpredictability of human conversation and you'll see how that works later in this chapter.

> **Note** If you've ever written a Windows Phone App, you might already be familiar with an application stack. Essentially, opening a new page puts that page at the top of the stack. Tapping the back button pops that page from the application stack, bringing the user either back to the previous page or closing the app.

The next section shows how to build a chatbot that navigates between dialogs, using the dialog stack.

Navigating to Other Dialogs

Using the dialog stack, you can build chatbots that navigate from one dialog to another and back. The example here builds on on the WineBot chatbots from earlier chapters in this part of the book. This time, searching for wine is only a part of what a user can do and you'll see another feature for managing a profile. Listings 9-1 and 9-2, from the *WineBotDialogStack* project in the accompanying source code, show how to manage navigation between multiple dialogs with the dialog stack.

Listing 9-1 is the typical *Post* method, invoking *SendAsync* to call the first dialog the user interacts with, *RootDialog*. Behind the scenes, *SendAsync* pushes *RootDialog* on the stack, making it the current dialog.

LISTING 9-1 Using the Dialog Stack to Navigate Between Dialogs – *MessagesController.cs*

```
using System.Net;
using System.Net.Http;
using System.Threading.Tasks;
using System.Web.Http;
using Microsoft.Bot.Builder.Dialogs;
using Microsoft.Bot.Connector;
using System;

namespace WineBotDialogStack
{
    [BotAuthentication]
```

```
    public class MessagesController : ApiController
    {
        public async Task<HttpResponseMessage> Post([FromBody]Activity activity)
        {
            if (activity.Type == ActivityTypes.Message)
            {
                try
                {
                    await Conversation.SendAsync(activity, () => new Dialogs.RootDialog());
                }
                catch (InvalidOperationException ex)
                {
                    var client = new ConnectorClient(new Uri(activity.ServiceUrl));
                    var reply = activity.CreateReply($"Reset Message: {ex.Message}");
                    client.Conversations.ReplyToActivity(reply);
                }
            }

            var response = Request.CreateResponse(HttpStatusCode.OK);
            return response;
        }
    }
}
```

RootDialog, from Listing 9-2 is an *IDialog<T>*, which you've seen in previous chapters. Its *StartAsync* sets *MessageReceivedAsync* as the next method to handle input from the user. After *SendAsync* calls *RootDialog*, it sends the *IMessageActivity* to *RootDialog*, which forwards the call to *MessageReceivedAsync*.

LISTING 9-2 Using the Dialog Stack to Navigate Between Dialogs – *RootDialog.cs*

```
using System;
using System.Threading.Tasks;
using Microsoft.Bot.Builder.Dialogs;
using Microsoft.Bot.Builder.FormFlow;
using WineBotLib;
using System.Linq;

namespace WineBotDialogStack.Dialogs
{
    [Serializable]
    public class RootDialog : IDialog<object>
    {
        public Task StartAsync(IDialogContext context)
        {
            context.Wait(MessageReceivedAsync);

            return Task.CompletedTask;
        }

        Task MessageReceivedAsync(IDialogContext context, IAwaitable<object> result)
        {
```

```csharp
            string prompt = "What would you like to do?";
            var options = new[]
            {
                "Search Wine",
                "Manage Profile"
            };

            PromptDialog.Choice(context, ResumeAfterChoiceAsync, options, prompt);

            return Task.CompletedTask;
        }

        async Task ResumeAfterChoiceAsync(IDialogContext context, IAwaitable<string> result)
        {
            string choice = await result;

            if (choice.StartsWith("Search"))
                await context.Forward(
                    FormDialog.FromForm(new WineForm().BuildForm),
                    ResumeAfterWineSearchAsync,
                    context.Activity.AsMessageActivity());
            if (choice.StartsWith("Manage"))
                context.Call(new ProfileDialog(), ResumeAfterProfileAsync);
            else
                await context.PostAsync($"'{choice}' isn't implemented.");
        }

        async Task ResumeAfterWineSearchAsync(
            IDialogContext context, IAwaitable<WineForm> result)
        {
            WineForm wineResults = await result;

            List[] wines =
                await new WineApi().SearchAsync(
                    (int)wineResults.WineType,
                    (int)wineResults.Rating,
                    wineResults.InStock == StockingType.InStock,
                    "");

            string message;

            if (wines.Any())
                message = "Here are the top matching wines: " +
                            string.Join(", ", wines.Select(w => w.Name));
            else
                message = "Sorry, No wines found matching your criteria.";

            await context.PostAsync(message);

            context.Wait(MessageReceivedAsync);
        }

        async Task ResumeAfterProfileAsync(IDialogContext context, IAwaitable<string> result)
```

```
            {
                try
                {
                    string email = await result;

                    await context.PostAsync($"Your profile email is now {email}");
                }
                catch (ArgumentException ex)
                {
                    await context.PostAsync($"Fail Message: {ex.Message}");
                }

                context.Wait(MessageReceivedAsync);
            }
        }
    }
```

Inside of *MessageReceivedAsync*, notice the call to *PromptDialog.Choice*. While you've seen this in Chapter 5, Building Dialogs, it's important to point out the *PromptDialog* sets the *ResumeAfterChoiceAsync* method as the next method in *RootDialog* to run and then pushes itself as a new dialog on the stack. When the user responds, the *PromptDialog.Choice* dialog pops from the stack and returns control to *RootDialog*, the previous dialog on the stack, to resume at *ResumeAfterChoiceAsync*.

The following sections detail more options for the dialog stack and navigation, starting where we left off here at *ResumeAfterChoiceAsync*.

Navigating via *Forward*

One of the ways you can navigate to another dialog is by forwarding the *IMessageActivity* so the new dialog can handle the user's input. Listing 9-2 shows how to use the *IDialogContext*'s *Forward* method, repeated next, to accomplish this:

```
await context.Forward(
    FormDialog.FromForm(new WineForm().BuildForm),
    ResumeAfterWineSearchAsync,
    context.Activity.AsMessageActivity());
```

This example uses three parameters from forward, taking the default for the fourth, which is an async *CancellationToken*. The first parameter is an *IDialog<T>* instance and you can see that it's using FormFlow. The second parameter refers to the method to return to when the *WineForm* dialog returns. The third parameter is the activity containing the user's method–it uses the *AsMessageActivity* method to convert *Activity* and pass an *IMessageActivity* instance.

Forward pushes the new *WineForm* dialog on the stack and starts it, which passes the user's *IMessageActivity*. Listing 9-3 shows the *WineForm* dialog.

LISTING 9-3 Forwarding *IMessageActivity* to a Dialog – *WineForm.cs*

```csharp
using Microsoft.Bot.Builder.Dialogs;
using Microsoft.Bot.Builder.FormFlow;
using System;
using System.Linq;
using System.Threading.Tasks;
using WineBotLib;

namespace WineBotDialogStack.Dialogs
{
    [Serializable]
    public class WineForm
    {
        public WineType WineType { get; set; }
        public RatingType Rating { get; set; }
        public StockingType InStock { get; set; }

        public IForm<WineForm> BuildForm()
        {
            return new FormBuilder<WineForm>()
                .Message(
                    "I have a few questions on your wine search. " +
                    "You can type \"help\" at any time for more info.")
                .Build();
        }
    }
}
```

WineForm is intentionally implemented minimally to demonstrate the behavior of calling a Form-Flow form. When complete, *WineForm* holds all of the user's values in properties, which the caller can access, as shown in *ResumeAfterWineSearchAsync*, repeated next from Listing 9-2:

```csharp
        async Task ResumeAfterWineSearchAsync(IDialogContext context, IAwaitable<WineForm> result)
        {
            WineForm wineResults = await result;

            List[] wines =
                await new WineApi().SearchAsync(
                    (int)wineResults.WineType,
                    (int)wineResults.Rating,
                    wineResults.InStock == StockingType.InStock,
                    "");

            string message;

            if (wines.Any())
                message = "Here are the top matching wines: " +
                          string.Join(", ", wines.Select(w => w.Name));
            else
                message = "Sorry, No wines found matching your criteria.";
```

```
        await context.PostAsync(message);

        context.Wait(MessageReceivedAsync);
}
```

After *WineForm* completes and returns, it pops from the stack and resumes on *RootDialog.ResumeAfterWineSearchAsync*. The awaited *result* parameter is the instance of *WineForm* that is now done and has properties holding values from the user's responses. As seen in previous chapters, the *SearchAsync* call returns results based on the *WineForm* instance, *wineResults*, properties and responds to the user.

Navigating via *Call*

Sometimes you don't need to send the *IMessageActivity* to the called dialog, as demonstrated in the previous discussion on *Forward*. Maybe you just need to start a dialog and let it interact with the user, regardless of what the user's initial communication was. In that case, you can use the *IDialogContext's Call* method, repeated here from the *ResumeAfterChoiceAsync* method in Listing 9-2:

```
context.Call(new ProfileDialog(), ResumeAfterProfileAsync);
```

Call has two parameters: the new dialog and a resume method. The new dialog in this example is *ProfileDialog*. Similar to how resume methods in *Forward* and *PromptDialog* work, the second parameter refers to *ResumeAfterProfileAsync* as the method to call after *ProfileDialog*, shown in Listing 9-4, completes.

LISTING 9-4 Calling a Dialog – *ProfileDialog.cs*

```
using Microsoft.Bot.Builder.Dialogs;
using System;
using System.Threading.Tasks;

namespace WineBotDialogStack.Dialogs
{
    [Serializable]
    public class ProfileDialog : IDialog<string>
    {
        public Task StartAsync(IDialogContext context)
        {
            string prompt = "What would you like to do?";
            var options = new[]
            {
                "Change Email",
                "Reset",
                "Fail",
            };

            PromptDialog.Choice(context, MessageReceivedAsync, options, prompt);
```

```
            return Task.CompletedTask;
    }

    async Task MessageReceivedAsync(IDialogContext context, IAwaitable<string> result)
    {
        string choice = await result;

        switch (choice)
        {
            case "Change Email":
                string prompt = "What is your email address?";
                PromptDialog.Text(context, ResumeAfterEmailAsync, prompt);
                break;
            case "Fail":
                context.Fail(new ArgumentException("Testing Fail."));
                break;
            case "Reset":
                context.Reset();
                break;
            default:
                await context.PostAsync($"'{choice}' isn't implemented.");
                break;
        }
    }

    async Task ResumeAfterEmailAsync(IDialogContext context, IAwaitable<string> result)
    {
        string email = await result;
        context.Done(email);
    }
}
}
```

As with all *IDialog<T>* implementations, *StartAsync* is the entry point. This example is different from previous *StartAsync* implementations because instead of calling *Wait* on *MessageReceivedAsync*, it calls *PromptDialog.Choice* to give the user a menu for what to do next. This was necessary because after calling *ProfileDialog*, *StartAsync* runs and then waits until the next user input. The user needs some indication of what they should do. This differs from *SendAsync* and *Forward*, both of which not only call the dialog, but subsequently send the user's *IMessageActivity*, which then invokes the next method, which would have been *MessageReceivedAsync* if *StartAsync* called the *Wait* method. In this case, *Forward* doesn't make sense and *Call* is the more logical choice.

After the user responds to the *PromptDialog.Choice* in *StartAsync*, the dialog resumes on the *MessageReceivedAsync* method. The *switch* statement handles all three choices, discussed next.

Finishing a Dialog

The choices handled in the *switch* statement in *MessageReceivedAsync*, from Listing 9-4, represent the three ways to handle finishing a dialog: returning results, resetting the stack, or failing. The next three sections discuss these options.

The *Done* Method

The *Change Email* case shows how to handle the *Done* method. It uses *PromptDialog.Text* to get the user's email:

```
case "Change Email":
    string prompt = "What is your email address?";
    PromptDialog.Text(context, ResumeAfterEmailAsync, prompt);
    break;
```

PromptDialog.Text specifies *ResumeAfterEmailAsync* to handle the user's response:

```
async Task ResumeAfterEmailAsync(IDialogContext context, IAwaitable<string> result)
{
    string email = await result;
    context.Done(email);
}
```

Notice the call to *IDialogContext's Done* method. This is what transfers control back to the calling dialog. The *email* parameter is type *string*, corresponding with the fact that *PromptDialog* implements *IDialog<string>*. The *IDialog<T>* type parameter *T* is the return type of the dialog. Earlier examples just set this to *object* because they didn't return any values. Since this example needs to return a *string* through its *Done* method, the *IDialog<T>* type must also be *string*. Both the *T* type in *IDialog<T>* and the type returned by *Done* must be the same, or at least assignable.

The *Done* method pops the current dialog from the dialog stack and returns its argument to the caller. The *RootDialog*, from Listing 9-2 has a *ResumeFromProfileAsync* method that handles the return result from the call to *ProfileDialog*, repeated here:

```
async Task ResumeAfterProfileAsync(IDialogContext context, IAwaitable<string> result)
{
    try
    {
        string email = await result;

        await context.PostAsync($"Your profile email is now {email}");
    }
    catch (ArgumentException ex)
    {
        await context.PostAsync($"Fail Message: {ex.Message}");
    }

    context.Wait(MessageReceivedAsync);
}
```

ResumeAfterProfileAsync awaits the result parameter, which is an *IAwaitable<T>*, where *T* must be the same or an assignable type. There's also a *try/catch* handler for the case when a called dialog fails, discussed next.

The *Fail* Method

The *Fail* method is a way for a chatbot to indicate that it is unable to complete the dialog. The *Fail* case, from *MessageReceivedAsync* in Listing 9-4, calls the *Fail* method:

```
case "Fail":
    context.Fail(new ArgumentException("Testing Fail."));
    break;
```

The *Fail* method argument is an *Exception* type. This case uses an *ArgumentException*, but you can choose whatever exception you feel is appropriate. When the *Fail* method executes, it takes care of internal Bot Builder record keeping logic and throws the exception parameter for the resume method of the calling dialog.

The *Fail* method pops the current dialog from the dialog stack and throws its exception argument to the call chain of its caller's resume method, which is the *ResumeAfterProfileAsync* method from the *RootDialog* in Listing 9-2:

```
async Task ResumeAfterProfileAsync(IDialogContext context, IAwaitable<string> result)
{
    try
    {
        string email = await result;

        await context.PostAsync($"Your profile email is now {email}");
    }
    catch (ArgumentException ex)
    {
        await context.PostAsync($"Fail Message: {ex.Message}");
    }

    context.Wait(MessageReceivedAsync);
}
```

In the resume method, wrap the await call to the result parameter in a *try/catch* block, where the *catch* block type is the same type as the *Fail* method parameter. In *ResumeAfterProfileAsync*, the *catch* type is *ArgumentException*, matching the *ArgumentException* argument to the *Fail* method.

The *Reset* Method

When you want to let a user cancel the current dialog stack and start over, you can use the *Reset* method. The *Reset* method unwinds the entire stack. The *Reset* case from the *MessageReceivedAsync* method in Listing 9-4 calls the *Reset* method, repeated here:

```
            case "Reset":
                context.Reset();
                break;
```

Calling *Reset*, like the code above, pops all of the dialogs from the stack and throws an *InvalidOperationException* to the beginning of the call chain. The *Post* method, repeated below from Listing 9-1, shows how to handle a *Reset*:

```
public async Task<HttpResponseMessage> Post([FromBody]Activity activity)
{
    if (activity.Type == ActivityTypes.Message)
    {
        try
        {
            await Conversation.SendAsync(activity, () => new Dialogs.RootDialog());
        }
        catch (InvalidOperationException ex)
        {
            var client = new ConnectorClient(new Uri(activity.ServiceUrl));
            var reply = activity.CreateReply($"Reset Message: {ex.Message}");
            client.Conversations.ReplyToActivity(reply);
        }
    }

    var response = Request.CreateResponse(HttpStatusCode.OK);
    return response;
}
```

This code wraps *SendAsync* in a *try/catch* block because that's where the *InvalidOperationException* propagates. The code uses the *ConnectorClient*, which is discussed in Chapter 2, Setting Up a Project, to communicate with Bot Connector, sending an information message back to the user.

So far, you've seen navigation between dialogs based on C# conditional logic syntax. However, Bot Builder has a much more sophisticated set of tools to manage conversations, which you'll learn about next.

Managing Conversations with Chaining

The ability to call and return values from dialogs, managing the dialog stack is very useful for managing conversations. You might like that and it could be all that's necessary for a particular chatbot. However, there's an additional set of chaining tools for even more diverse conversational patterns. This section discusses the *Chain* class shows how to use several of its methods to accomplish relatively sophisticated navigation tasks.

This section starts with the *WineBotChain* program and then breaks down several parts of that program into more manageable pieces. What you should see is how *Chain* allows creating very complex conversation patters. Yet, through the breakdown, you'll see that it's not hard to create conversations that might even be simpler.

> **Note** Whether you use *Call/Forward/Done* or *Chain*–it doesn't matter because you can achieve the same goal with either technical approach. This choice is a matter of opinion and style for you and/or your team. Seeing how complex chains can be might lead some to decide they don't want to use them at all. Yet others might find beauty in the ability to combine navigation logic in a smaller space, rather than having separate dialogs everywhere. My opinion is that neither extreme is optimal and somewhere in-between might lead to better designs.

The *WineBotChain* Program

WineBotChain demonstrates several features of using the *Chain* class and its members. It has several layers of dialogs, essentially pulling a lot of navigation code into one place. Because of the versatility of *Chain*, this isn't an exhaustive set of examples, but a set of techniques you might use to think about how to use. Listing 9-5 shows the *RootDialog* for *WineBotChain*.

LISTING 9-5 Using the Chain Class – *RootDialog.cs*

```
using Microsoft.Bot.Builder.Dialogs;
using Microsoft.Bot.Builder.FormFlow;
using System;
using System.Linq;
using System.Text.RegularExpressions;
using System.Threading.Tasks;
using WineBotLib;
using static Microsoft.Bot.Builder.Dialogs.Chain;

namespace WineBotChain.Dialogs
{
    [Serializable]
    public class RootDialog : IDialog<object>
    {
        public Task StartAsync(IDialogContext context)
        {
            context.Wait(MessageReceivedAsync);

            return Task.CompletedTask;
        }

        Task MessageReceivedAsync(IDialogContext context, IAwaitable<object> result)
        {
            string prompt = "Which chain demo?";
            var options = new[]
            {
                "From",
                "LINQ",
                "Loop",
                "Switch"
            };

            PromptDialog.Choice(context, ResumeAfterChoiceAsync, options, prompt);
```

214 PART II Bot Builder

```csharp
            return Task.CompletedTask;
    }

    async Task ResumeAfterChoiceAsync(IDialogContext context, IAwaitable<string> result)
    {
        string choice = await result;

        switch (choice)
        {
            case "From":
                await DoChainFromAsync(context);
                break;
            case "LINQ":
                await DoChainLinqAsync(context);
                break;
            case "Loop":
                await DoChainLoopAsync(context);
                break;
            case "Switch":
                DoChainSwitch(context);
                break;
            default:
                await context.PostAsync($"'{choice}' isn't implemented.");
                break;
        }
    }

    async Task<string> ProcessWineResultsAsync(WineForm wineResult)
    {
        List[] wines =
            await new WineApi().SearchAsync(
                (int)wineResult.WineType,
                (int)wineResult.Rating,
                wineResult.InStock == StockingType.InStock,
                "");

        string message;

        if (wines.Any())
            message = "Here are the top matching wines: " +
                      string.Join(", ", wines.Select(w => w.Name));
        else
            message = "Sorry, No wines found matching your criteria.";

        return message;
    }

    async Task ResumeAfterWineFormAsync(
        IDialogContext context, IAwaitable<WineForm> result)
    {
        WineForm wineResult = await result;

        string message = await ProcessWineResultsAsync(wineResult);

        await context.PostAsync(message);
```

```csharp
        context.Wait(MessageReceivedAsync);
    }

    async Task DoChainFromAsync(IDialogContext context)
    {
        IDialog<WineForm> chain =
            Chain.From(() => FormDialog.FromForm<WineForm>(new WineForm().BuildForm));

        await context.Forward(
            chain,
            ResumeAfterWineFormAsync,
            context.Activity.AsMessageActivity());
    }

    async Task DoChainLinqAsync(IDialogContext context)
    {
        var chain =
            from wineForm in FormDialog.FromForm(new WineForm().BuildForm)
            from searchTerm in new PromptDialog.PromptString(
                "Search Terms?", "Search Terms?", 1)
            where wineForm.WineType.ToString().Contains("Wine")
            select Task.Run(() => ProcessWineResultsAsync(wineForm)).Result;

        await context.Forward(
            chain,
            ResumeAfterChainLinqAsync,
            context.Activity.AsMessageActivity());
    }

    async Task ResumeAfterChainLinqAsync(
        IDialogContext context, IAwaitable<string> result)
    {
        try
        {
            string response = await result;
            await context.PostAsync(response);
            context.Wait(MessageReceivedAsync);
        }
        catch (WhereCanceledException wce)
        {
            await context.PostAsync($"Where cancelled: {wce.Message}");
        }
    }

    async Task DoChainLoopAsync(IDialogContext context)
    {
        IDialog<WineForm> chain =
            Chain.From(() => FormDialog.FromForm(
                new WineForm().BuildForm, FormOptions.PromptInStart))
                .Do(async (ctx, result) =>
                {
                    try
                    {
                        WineForm wineResult = await result;
                        string message = await ProcessWineResultsAsync(wineResult);
```

```csharp
                            await ctx.PostAsync(message);
                        }
                        catch (FormCanceledException fce)
                        {
                            await ctx.PostAsync($"Cancelled: {fce.Message}");
                        }
                    })
                    .Loop();

    await context.Forward(
        chain,
        ResumeAfterWineFormAsync,
        context.Activity.AsMessageActivity());
}

void DoChainSwitch(IDialogContext context)
{
    string prompt = "What would you like to do?";
    var options = new[]
    {
        "Search Wine",
        "Manage Profile"
    };

    PromptDialog.Choice(context, ResumeAfterMenuAsync, options, prompt);
}

async Task ResumeAfterMenuAsync(IDialogContext context, IAwaitable<string> result)
{
    IDialog<string> chain =
        Chain
            .PostToChain()
            .Select(msg => msg.Text)
            .Switch(
                new RegexCase<IDialog<string>>(
                    new Regex("^Search", RegexOptions.IgnoreCase),
                    (reContext, choice) =>
                    {
                        return DoSearchCase();
                    }),
                new Case<string, IDialog<string>>(choice => choice.Contains("Manage"),
                    (manageContext, txt) =>
                    {
                        manageContext.PostAsync("What is your name?");
                        return DoManageCase();
                    }),
                new DefaultCase<string, IDialog<string>>(
                    (defaultCtx, txt) =>
                    {
                        return Chain.Return("Not Implemented.");
                    })
            )
            .Unwrap()
            .PostToUser();
```

```
            await context.Forward(
                chain,
                ResumeAfterSwitchAsync,
                context.Activity.AsMessageActivity());
        }

        IDialog<string> DoSearchCase()
        {
            return
                Chain
                    .From(() => FormDialog.FromForm(
                        new WineForm().BuildForm, FormOptions.PromptInStart))
                    .ContinueWith(async (ctx, res) =>
                    {
                        WineForm wineResult = await res;
                        string message = await ProcessWineResultsAsync(wineResult);
                        return Chain.Return(message);
                    });
        }

        IDialog<string> DoManageCase()
        {
            return
                Chain
                    .PostToChain()
                    .Select(msg => $"Hi {msg.Text}'! What is your email?")
                    .PostToUser()
                    .WaitToBot()
                    .Then(async (ctx, res) => (await res).Text)
                    .Select(msg => $"Thanks - your email, {msg}, is updated");
        }

        async Task ResumeAfterSwitchAsync(IDialogContext context, IAwaitable<string> result)
        {
            string message = await result;
            context.Done(message);
        }
    }
}
```

As demonstrated plenty of times before, the chatbot's *Post* method calls *SendAsync* on *RootDialog*, from Listing 9-5. *RootDialog* has the required *StartAsync* and typical *MessageReceivedAsync* for *IDialog<T>* types. *MessageReceivedAsync* uses a *PromptDialog.Choice* to ask the user what options they want: *From*, *LINQ*, *Loop*, or *Switch*. The continuation from *Choice*, *ResumeAfterChoiceAsync* runs the user's response through a *switch* statement to launch the proper method to handle the user's choice.

Each of the user's choices represent approaches to take when designing navigation with *Chain*. The *From* option specifies a dialog to run and *Loop* runs a dialog multiple times. The *LINQ* option does exactly what it says, allowing LINQ statements that use dialogs. *Switch* is a way to choose which dialog to run, based on user input. From *Switch*, we'll drill down some more to show new and different *Chain* methods. Let's start with *From*.

> **Tip** The *Chain* class has several methods, each supporting a fluent interface. This allows you to create conversations that navigate from one dialog or response handler to another dialog or response handler in a seamless way. No doubt this chaining behavior inspired the name of the class.

Chain.From

The *Chain.From* method supports launching a dialog as part of a chain. In Listing 9-5, the *DoChainFromAsync* method, repeated below, handles the choice when the user selects *From*:

```
async Task DoChainFromAsync(IDialogContext context)
{
    IDialog<WineForm> chain =
        Chain.From(() => FormDialog.FromForm<WineForm>(new WineForm().BuildForm));

    await context.Forward(
        chain,
        ResumeAfterWineFormAsync,
        context.Activity.AsMessageActivity());
}
```

DoChainFromAsync uses *Chain.From* to launch the FormFlow form, *WineForm*. *FormDialog. FromForm* returns *IFormDialog*, an *IDialog<T>*, that *From* requires as part of its *Func<IDialog<T>>* parameter type.

Behind the scenes, *From* takes care of the *Call* and *Done* for dialog stack management. The subsequent call to *Forward* starts the *Chain*, handling the response in *ResumeAfterWineFormAsync*, repeated here:

```
async Task ResumeAfterWineFormAsync(IDialogContext context, IAwaitable<WineForm> result)
{
    WineForm wineResult = await result;

    string message = await ProcessWineResultsAsync(wineResult);

    await context.PostAsync(message);

    context.Wait(MessageReceivedAsync);
}
```

Like a normal resume method, explained earlier in this chapter, this code processes the response from the dialog. This is the end of processing from the *Chain*, so the code calls *context.Wait* to set *MessageReceivedAsync* as the next method for *RootDialog* to send messages to.

Chain.Loop

When users select the *Loop* option, the *DoChainLoopAsync* method handles the request. The following code, repeated from Listing 9-5, shows how to loop, continually running the same dialog:

```
async Task DoChainLoopAsync(IDialogContext context)
{
    IDialog<WineForm> chain =
        Chain.From(() => FormDialog.FromForm(new WineForm().BuildForm, FormOptions.PromptInStart))
            .Do(async (ctx, result) =>
            {
                try
                {
                    WineForm wineResult = await result;
                    string message = await ProcessWineResultsAsync(wineResult);
                    await ctx.PostAsync(message);
                }
                catch (FormCanceledException fce)
                {
                    await ctx.PostAsync($"Cancelled: {fce.Message}");
                }
            })
            .Loop();

    await context.Forward(
        chain,
        ResumeAfterWineFormAsync,
        context.Activity.AsMessageActivity());
}
```

Similar to the previous section, this code calls *Chain.From* to launch the *WineForm* FormFlow dialog. The difference here is the call to *Loop*, causing the dialog to continually run. With a FormFlow dialog, a user can type *Quit* to exit the dialog, throwing a *FormCanceledException*. The *Do* method handles the response from the form, handling the *FormCanceledException* for when the user quits.

This example highlights the purpose of chaining, allowing navigation that launches a dialog, handles the results, and continues navigation operations until complete.

While the *Forward* starts the chain, the resume method, *ResumeAfterWineFormAsync*, won't be called in this example because the only way out was to *Quit*, with the *FormCanceledException*. Fortunately, *Chain* has another method, *While*, that will support looping while a condition is *true*.

Chain.Switch

Whenever a chatbot needs to implement *Chain* logic based on user input, it can use the *Switch* method. *Switch* has three ways to handle user input: regex, custom logic, or default. The following *DoChainSwitchAsync* method, repeated from Listing 9-5, shows each of the cases:

```
void DoChainSwitch(IDialogContext context)
{
    string prompt = "What would you like to do?";
```

```csharp
        var options = new[]
        {
            "Search Wine",
            "Manage Profile"
        };

        PromptDialog.Choice(context, ResumeAfterMenuAsync, options, prompt);
    }
```

DoChainSwitch uses *PromptDialog.Choice*, specifying *ResumeAfterMenuAsync*, repeated here:

```csharp
    async Task ResumeAfterMenuAsync(IDialogContext context, IAwaitable<string> result)
    {
        IDialog<string> chain =
            Chain
                .PostToChain()
                .Select(msg => msg.Text)
                .Switch(
                    new RegexCase<IDialog<string>>(new Regex("^Search", RegexOptions.IgnoreCase),
                        (reContext, choice) =>
                        {
                            return DoSearchCase();
                        }),
                    new Case<string, IDialog<string>>(choice => choice.Contains("Manage"),
                        (manageContext, txt) =>
                        {
                            manageContext.PostAsync("What is your name?");
                            return DoManageCase();
                        }),
                    new DefaultCase<string, IDialog<string>>(
                        (defaultCtx, txt) =>
                        {
                            return Chain.Return("Not Implemented.");
                        })
                )
                .Unwrap()
                .PostToUser();

        await context.Forward(
            chain,
            ResumeAfterSwitchAsync,
            context.Activity.AsMessageActivity());
    }
```

This *Chain* has several methods: *PostToChain*, *Select*, *Switch*, *Unwrap*, and *PostToUser*. Each of these methods, except for the first, process the output dialog of the antecedent (previous) method. We need to go step-by-step, following the processing to see how this works.

Looking past the *Chain*, the *context.Forward* call starts the chain. As you know, *Forward* passes its third parameter, an *IMessageActivity*, to the dialog and *Chain* is an *IDialog<T>*, receiving that *IMessageActivity* instance. The *PostToChain* method sends the *IMessageActivity*, received from the caller, to the next method in the chain, *Select*.

> **Tip** While *Chain*, in the *ResumeAfterMenuAsync* method, is a couple levels into the hierarchy, you could also pass a *Chain* to *SendAsync* at the chatbot's *Post* method, as the main dialog, which would pass the user's *IMessageActivity*, requiring a call to *PostToChain* to pass that *IMessageActivity* to the next method in the chain.

The *Select* method takes the instance for whatever type the previous method passed and allows you to build a new projection to pass to the next method. In this example, *Select* takes the *IMessageActivity*, passed from *PostToChain*, and builds a string type projection, which is the *Text* property of the *IMessageActivity*. *Select* passes this string to the *Switch* method.

Switch takes antecedent input and runs it through each of its cases. Once *Switch* matches a case, it passes the results of that case to the next method in the chain. Logically, this works much the same as a C# *switch* statement, except for the addition of the input and output value flows. *Switch* supports three types of cases: *RegexCase*, *Case*, and *DefaultCase*.

The *RegexCase* has two parameters–a *Regex* instance and a case handler. The first parameter is an instance of the .NET *Regex* class, which allows specifying a regular expression to match the input on. *RegexCase* uses the *Regex* instance to determine if the input matches the regular expression. In this example, if the input starts with *Search*, the case matches and executes the handler. The handler type is *ContextualSelector*, accepting *IBotContext* and *string* parameters and returning an *IDialog<T>*, where *T* is *string* in this example.

The second parameter to *Case* is also a *ContextualSelctor*, but the first parameter, condition, is *Func<T, bool>*, where *T* is *string* in this case. The *condition* evaluates the input and returns a *bool* to indicate if the condition matches. This example matches if the user input contains *Manage*.

When none of the cases match, the *Switch* method executes the *ContextSelector* for *DefaultCase*. The *DefaultCase* implementation is another *Chain* method, *Return*, which returns an *IDialog<T>*, where *T* is a *string* in this example. Both the *DoSearchCase* and *DoManageCase* also return *IDialog<string>*. I'll discuss these methods soon, but first explain why the next method, *UnWrap*, is required.

Switch passes *IDialog<T>* to the next method. What's important about this scenario is that *T* is *IDialog<string>* because that's what the *ContextSelector* of each *Switch* case returns. This means, *Switch* passes *IDialog<IDialog<string>>* to the next method. The code and dialogs called by a *Chain* method shouldn't have to understand whether they're being called by a *Chain*, *Call*, or *Forward*. Therefore, the *Chain.Unwrap* helps because it dereferences the inner *IDialog<T>* for you. Coming full circle, *Unwrap* accepts *IDialog<IDialog<string>>* from *Swich*, extracts the *IDialog<string>*, and passes the *IDialog<string>* to the next method, *PostToUser*.

The *PostToUser* method accepts an *IDialog<string>* from its antecedent and, as its name suggests, posts the *string* to the user. In this example, the strings are the responses from the *Switch* cases.

The nesting in this sample is extreme and might not be the way you would design your code, yet it demonstrates different ways to use chains. Let's have fun and explore a bit more. The next sections drill-down on the *DoSearchCase* and *DoManageCase* logic, from the *Switch* cases.

Chain.ContinueWith

The previous example, for *Chain.Loop*, handles the response from the the dialog with a *Do*, which is for side-effect operations with the result of the antecedent dialog. There are also a couple other ways to handle dialog responses, including the *ContinueWith*, which is part of the *DoSearchCase* method, repeated below.

```
IDialog<string> DoSearchCase()
{
    return
        Chain
            .From(() => FormDialog.FromForm(new WineForm().BuildForm, FormOptions.PromptInStart))
            .ContinueWith(async (ctx, res) =>
            {
                WineForm wineResult = await res;
                string message = await ProcessWineResultsAsync(wineResult);
                return Chain.Return(message);
            });
}
```

ContinueWith takes a parameter of type *Continuation* that passes an *IBotContext* and *IAwaitable<T>* to the handler, where *T* is *WineForm* in this example. The *Chain.Return* returns *IDialog<string>*—the *DoSearchCase* method return type for the *Switch* method.

The next example shows even more chatbot conversation customization with *Chain*.

An Assortment of Posting and Waiting *Chain methods*

Rather than calling a separate dialog, *Chain* methods can interact with the user directly, using several posting and waiting methods. Here's an example from *DoManageCase*:

```
IDialog<string> DoManageCase()
{
    return
        Chain
            .PostToChain()
            .Select(msg => $"Hi {msg.Text}'! What is your email?")
            .PostToUser()
            .WaitToBot()
            .Then(async (ctx, res) => (await res).Text)
            .Select(msg => $"Thanks - your email, {msg}, is updated");
}
```

This example starts with *PostToChain*, taking the input from the *Switch* case and passing it to the next method in the *Chain*. If you recall, the case handler from the *Switch* method called *manageContext.PostAsync("What is your name?")* before calling *DoManageCase*, setting the *Chain* in *DoManageCase* to the next dialog to handle the user's response.

Select receives the *IMessageActivity*, *msg*, from *PostToChain* and projects a new *string* for *PostToUser.PostToUser* sends the input *string* to the user.

At this point in time, we don't have a message from the user, so we must wait for the user's response, so the *Chain* calls *WaitToBot*, waiting for a user message to arrive for the chatbot.

You've seen *Do* and *ContinueWith* methods. Now the chatbot takes the user message that arrives after *WaitToBot* receives the user's response and passes that response to *Then*. The *Then* method is another type of handler that accepts a *Func<IBotContext, IAwaitable<T>, IDialog<T>>* where *T* is *IMessageActivity* for this example. Then processes the result, and passes that result to *Select*.

Using the *IMessageActivity* from *Then*, *Select* projects a new *string* and returns a new *IDialog<string>* that is sent back to the *Switch* case.

This example demonstrates how you might use *Chain* for a quick interaction with the user. While there's potential for this to be a complex set of interactions, you might consider it for simple interactions, rather than engage in the ceremony of a new *IDialog<T>*.

> ### *ContinueWith*, *Do*, or *Then* – Which to Choose?
>
> You've seen three *Chain* methods that implement lambdas: *ContinueWith*, *Do*, and *Then*. They're syntax makes them look similar. Sometimes one might work in place of another, but they each have different features that can make them unique in specific circumstances. Listing 9-5 offers some indication of how these methods are used and the following discussion highlights aspects that illuminate differences.
>
> *ContinueWith* lets you create a brand new *IDialog<T>* to return as the next dialog in the *Chain*. You can see in the *DoSearchCase* method, where *ContinueWith* uses *Chain.Return* to return a new *IDialog<T>*, where *T* is a new *string*.
>
> *Do* doesn't return any value. Its purpose is to perform some action, aka side-effect, and let the antecedent dialog continue to be the current dialog in the chain. In the *DoChainLoopAsync* method, *Do* just performs an action, sends a message to the user, and ends.
>
> The purpose of *Then* is to extract a value from an *IMessageActivity* and return a value to the chain. In *DoManageCase*, the *Then* lambda returns the *Text* property because that's all the next method needs. This is a useful way to project or transform any part of an *IMessageActivity* so the next method in the *Chain* doesn't have to.
>
> To summarize, here are a few quick rules to think about when choosing *ContinueWith*, *Do*, or *Then*:
>
> - *ContinueWith*: Return a new *IDialog<T>*.
> - *Do*: Perform a side-effect and stay on the antecedent (same) *IDialog<T>*.
> - *Then*: Project input into a new value for the next method in the *Chain*.

Because of how they're named, it's easy to accidentally use the wrong method, but we do have help from the compiler and IDE, giving us syntax errors because of type mismatches.

LINQ to Dialog

Previous sections show how you can use *Chain* in many ways. Another way to use *Chain* is via LINQ statements. In LINQ to Dialog, the data source is *IDialog<T>*. You can do *from*, *let*, *select*, and *where* clauses. While you can use the fluent syntax, such as earlier examples that used *Select*, you can also use LINQ query syntax, as demonstrated in the *DoChainLinqAsync*:

```
async Task DoChainLinqAsync(IDialogContext context)
{
    var chain =
        from wineForm in FormDialog.FromForm(new WineForm().BuildForm)
        from searchTerm in new PromptDialog.PromptString("Search Terms?", "Search Terms?", 1)
        where wineForm.WineType.ToString().Contains("Wine")
        select Task.Run(() => ProcessWineResultsAsync(wineForm)).Result;

    await context.Forward(
        chain,
        ResumeAfterChainLinqAsync,
        context.Activity.AsMessageActivity());
}
```

The two *from* clauses demonstrate how to perform a select-many. This example runs a sequence of dialogs, where *WineForm* runs first, returns a FormFlow dialog result in *wineForm* and then *PromptDialog.PromptString* runs, returning the dialog result value in *searchTerm*.

If all goes well, the select executes *ProcessWineResultsAsync* and returns the results of the chain as a *string*. However, notice that the *where* clause filters on whether *wineForm.WineType* has a value containing *Wine*. As long as the *where* clause is *true* all is good. If the user selects *sake* for *WineType*, the *where* clause is *false*, resulting in the *Chain* throwing a *WhereCanceledException*. This might be unexpected for developers looking at LINQ data implementations that filter on the *where* clause, but this is LINQ to Dialog and it indicates that the *Chain* is unable to complete its intended purpose, resulting in an exception.

Another gotcha that might not seem obvious is that you can't wrap the code in *DoChainLinqAsync* in a *try/catch* to handle the *WhereCanceledException*. Remember, we're working with async code, and the code awaits *Forward*, which starts the *Chain* and specifies the *ResumeAfterChainLinqAsync* method, repeated here:

```
async Task ResumeAfterChainLinqAsync(IDialogContext context, IAwaitable<string> result)
{
    try
    {
        string response = await result;
        await context.PostAsync(response);
        context.Wait(MessageReceivedAsync);
    }
    catch (WhereCanceledException wce)
    {
        await context.PostAsync($"Where cancelled: {wce.Message}");
    }
}
```

The *ResumeAfterChainLinqAsync* result parameter is *IAwaitable<string>*, like all callbacks for *Call* and *Forward*. Here, there is a *try/catch* block and the statement of interest is *await result*. Exceptions from the *Chain* can't be handled until the code awaits the *result*, which is what *ResumeAfterChainLinqAsync* does.

Now you've seen several ways to use *Chain*. There are myriad options and the Bot Framework will no doubt be adding and enhancing features. You can take what is there, in the *Chain* feature set, and build on it to see which features work for you, your team, and project.

Handling Interruptions with *IScorable*

A lot of chatbot work goes into designing conversation flow. The easy part is what is often termed as the happy path, the conversation patterns that work perfectly as long as the user follows the script. The remaining work comes from handling the conversation paths that might not go as planned, such as that which happens in real life. An example might be a user booking an airline ticket and somewhere in the conversation asks how the weather is at the destination. Another example might be if a user is in the midst of answering questions for a registration form for a service and asks what the cancellation policy is.

Because users can say anything to a chatbot at any time, you need a tool to handle out-of-band communication, which is where *IScorable* comes in. An *IScorable* is a type that listens for incoming messages, evaluates whether it should handle the message, and votes to be the handler. Bot Builder picks the *IScorable* with the highest score and allows it to handle the message. Whenever an *IScorable* type handles a message, that overrides any handling on the dialog stack, allowing the winning *IScorable* to handle the user's message and reply as appropriate.

The example in this section is for a situation where a user might ask for help. This is built into Form-Flow, but other dialog types don't have help unless they've been coded to handle it. In this scenario an *IScorable* could be useful, intercepting a help command to assist the user when their command would otherwise be interpreted as input to the current dialog at the top of the stack. Listing 9-6, from the *ScorableHelp* project in the accompanying source code, shows the *HelpScorable* that allows this.

LISTING 9-6 Creating an *IScorable* – HelpScorable.cs

```
using Microsoft.Bot.Builder.Dialogs;
using Microsoft.Bot.Builder.Dialogs.Internals;
using Microsoft.Bot.Builder.Scorables.Internals;
using Microsoft.Bot.Connector;
using System.Text.RegularExpressions;
using System.Threading;
using System.Threading.Tasks;

namespace ScorableHelp.Dialogs
{
    public class HelpScorable : ScorableBase<IActivity, string, double>
    {
```

```csharp
    readonly IBotToUser botToUser;

    public HelpScorable(IBotToUser botToUser)
    {
        this.botToUser = botToUser;
    }

    protected override async Task<string> PrepareAsync(
        IActivity activity, CancellationToken token)
    {
        var text = (activity as IMessageActivity)?.Text ?? "";
        var regex = new Regex("/help", RegexOptions.IgnoreCase);
        var match = regex.Match(text);

        return match.Success ? match.Groups[0].Value : null;
    }

    protected override bool HasScore(IActivity item, string state)
    {
        return state != null;
    }

    protected override double GetScore(IActivity item, string state)
    {
        return 1.0;
    }

    protected override async Task PostAsync(
        IActivity item, string state, CancellationToken token)
    {
        await botToUser.PostAsync("How may I help?", cancellationToken: token);
    }

    protected override Task DoneAsync(IActivity item, string state, CancellationToken token)
    {
        return Task.CompletedTask;
    }
}
```

HelpScorable derives from the abstract class *ScorableBase*, which derives from *IScorable*. Each of the methods override abstract methods from *ScorableBase*. The *ScorableBase* class helps by supporting strongly typed inheritance. Here's a part of the *ScorableBase* class with all members, except the *GetScore* methods omitted for brevity:

```csharp
public abstract class ScorableBase<Item, State, Score> : IScorable<Item, Score>
{
    protected abstract Score GetScore(Item item, State state);

    [DebuggerStepThrough]
    Score IScorable<Item, Score>.GetScore(Item item, object opaque)
    {
        var state = (State)opaque;
```

```
            if (!HasScore(item, state))
            {
                throw new InvalidOperationException();
            }

            return this.GetScore(item, state);
        }
    }
```

Notice that *ScorableBase* has three type parameters, but *IScorable* only has two. *Item* is the type of parameter that the class evaluates, *State* is the type for the value allowing the code to hold any information it needs to make a decision throughout the scorable process, and *Score* is the type of the result that it the output of the class. The goal of the *IScorable* is to evaluate *Item* and store shared information for all methods in *State*, allowing a final *Score* to be found. In Listing 9-5, *Item* is *IActivity*, *State* is *string*, and *Score* is *double*.

The two *GetScore* overloads in *ScorableBase* demonstrate how it enables a strongly typed child class. Bot Builder uses the interface type to call the explicitly implemented *IScorable<Item, Score>.GetScore* method, which has some checks on whether a score has been assigned. We'll discuss the *HelpScorable* implementation of this in a following paragraph. What you should get out of this is that even though the *opaque* parameter is type *object*, *IScorable<Item, Score>.GetScore* delegates the call to the abstract *GetScore*, calling child class methods, like those in *HelpScorable*, with strongly typed parameters. Any class that implements *IScorable* directly would be weakly typed. *ScorableBase* is more convenient.

The *HelpScorable* constructor takes an *IBotToUser* parameter and saves the reference to use in the *PostAsync* method to respond to the user. Bot Builder calls each of the methods in the order shown in Listing 9-6, with a caveat that is explained shortly.

PrepareAsync is where the *IScorable* evaluates user input, via the *IActivity* parameter. This example uses a regular expression to see if the user typed **/help**. Notice the forward slash, this might be a useful convention to differentiate commands from normal conversation input. In fact, the Bot Framework has an *IScorable*, named */deleteprofile*, for deleting user information that also uses the forward slash convention. Just type **/deleteprofile** and it will clear out all user data in Bot State. *PrepareAsync* returns a string that represents the result of the evaluation and is passed as the state parameter to *HasScore*.

Bot Builder could potentially be evaluating multiple *IScorable* instances. In this example, we know that it's evaluating both */deleteprofile* and */help*. To minimize the list, each *IScorable* can indicate that it doesn't want to vote to handle the user's input, which is the purpose of the *HasScore* method. The *HelpScorable* example checks *status* for *null* as its test because *PrepareAsync* either returns a matched string or *null* if the user typed something other than */help*. If an *IScorable* wants to vote, it returns *true*. When an *IScorable* returns *false*, it no longer participates in evaluation and Bot Builder doesn't call any more of its methods.

GetScore returns its score to compete to handle the message. In this example, the score is *1.0,* which is consistent with the implementation of */deleteprofile*. The return value is a double and represents the level of certainty, confidence, or probability that this *IScorable* can properly handle the user's input. You have the freedom to set the return values to accommodate which *IScorable* to win in your design.

It's possible that multiple *IScorables* will return a score, resulting in competition to handle a single activity. When this happens, Bot Builder takes the *IScorable* with the highest score.

If an *IScorable* wins with the highest score, Bot Builder calls its *PostAsync* method. The purpose of the *PostAsync* method is to respond to user input. Doing so sooner might be a waste of processing if this *IScorable* didn't win. This example posts a simple message to the user, but could also launch another dialog, which would become the current dialog on the dialog stack.

The *DoneAsync* method performs clean-up work, if any, and is the last method called.

You must write code to register an *IScorable* so that Bot Builder can find it. List 9-7 shows how to do this.

LISTING 9-7 Registering an *IScorable – Global.asax.cs*

```
using Autofac;
using Microsoft.Bot.Builder.Dialogs;
using Microsoft.Bot.Builder.Scorables;
using Microsoft.Bot.Connector;
using ScorableHelp.Dialogs;
using System.Web.Http;

namespace ScorableHelp
{
    public class WebApiApplication : System.Web.HttpApplication
    {
        protected void Application_Start()
        {
            GlobalConfiguration.Configure(WebApiConfig.Register);

            Conversation.UpdateContainer(builder =>
            {
                builder.RegisterType<HelpScorable>()
                    .As<IScorable<IActivity, double>>()
                    .InstancePerLifetimeScope();
            });
        }
    }
}
```

Bot Framework uses an Inversion of Control (IoC) container named *AutoFac*. This is an open source tool, supporting unit testing and the loose coupling of code. It's also a way to plug in types for various purposes with Bot Builder. In this example, the *builder.RegisterType* allows Bot Builder to know that it should allow *HelpScorable* to compete to handle user messages, along with other registered *IScorable* types.

The *Conversation.UpdateContainer* method is a Bot Builder type, exposing a *ContainerBuilder* instance, *builder*, that Bot Builder uses to register its types. You'll see many examples that instantiate a *ContainerBuilder* and then call *Update* on the instance after registering types. However, AutoFac deprecated *Update* and that causes compiler warnings. Using *Conversation.UpdateContainer* fixes that problem.

That completes the discussion for *IScorables*, which is another way to respond to the user. On responses, some of the messages to the user have been quite plain and the next section shows how to improve the appearance of those messages.

Formatting Text Output

All of the output so far has been plain text. In the case of the response from the *WineBot* search, the response appears as a comma-separated list and that isn't easy to read. In this section, you'll learn how to format user output so it's more readable. Listing 9-8, from the *WineBotFormatted* project in the accompanying source code, shows how to format text in *Markdown* format.

LISTING 9-8 Formatting Text – *WineForm.cs*

```
using Microsoft.Bot.Builder.FormFlow;
using System;
using System.Collections.Generic;
using System.Linq;
using System.Web;
using WineBotLib;
using Microsoft.Bot.Builder.Dialogs;
using System.Threading.Tasks;
using System.Text;
using Microsoft.Bot.Connector;

namespace WineBotFormatted.Dialogs
{
    [Serializable]
    public class WineForm
    {
        public WineType WineType { get; set; }
        public RatingType Rating { get; set; }
        public StockingType InStock { get; set; }

        public IForm<WineForm> BuildForm()
        {
            return new FormBuilder<WineForm>()
                .Message(
                    "I have a few questions on your wine search. " +
                    "You can type \"help\" at any time for more info.")
                .OnCompletion(WineFormCompletedAsync)
                .Build();
        }

        async Task WineFormCompletedAsync(IDialogContext context, WineForm wineResults)
```

```csharp
{
    List[] wines =
        await new WineApi().SearchAsync(
            (int)wineResults.WineType,
            (int)wineResults.Rating,
            wineResults.InStock == StockingType.InStock,
            "");

    var message = new StringBuilder();

    if (wines.Any())
    {
        message.AppendLine("# Top Matching Wines ");

        foreach (var wine in wines)
            message.AppendLine($"* {wine.Name}");
    }
    else
    {
        message.Append("_Sorry, No wines found matching your criteria._");
    }

    //var reply = (context.Activity as Activity).CreateReply(message.ToString());
    //reply.TextFormat = "plain";
    //await context.PostAsync(reply);

    await context.PostAsync(message.ToString());
    }
  }
}
```

The *WineFormCompletedAsync* method shows how to format text. It's using *Markdown*, which is a popular form of text formatting that you can learn more about with a Web search. This example uses the *h1* symbol, which is a hash mark to create a heading - # *Top Matching Wines*. Each wine appears in a bulleted list, indicated by the star prefix - * *{wine.Name}*. Figure 9-1 shows what this looks like.

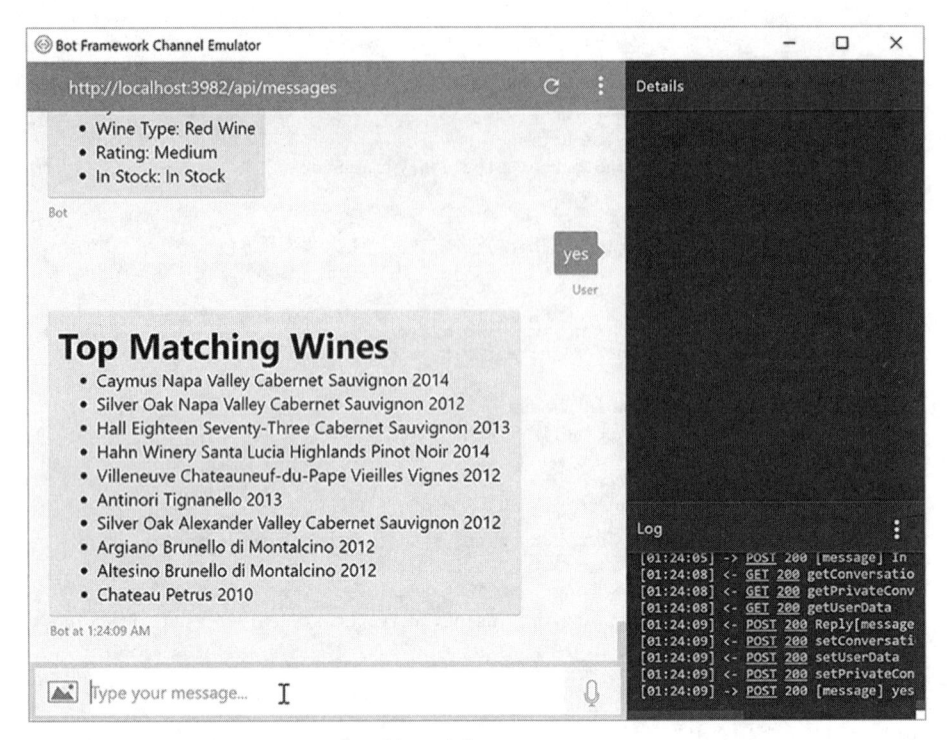

FIGURE 9-1 Formatting message text with markdown.

As Figure 9-1 shows, the wine list is much more readable than the previous comma-separated version.

The default format of text is *markdown*, which might be surprising because characters like # and * can alter the appearance of the output if it wasn't intended to be interpreted as markup. This assumes that a channel will try to translate the markdown to HTML, but on channels that are text only, such as SMS, you'll see all the markdown characters as normal. This highlights the benefits of markdown in that it is very easy to read whether translated to HTML or not. If you want to have plain text output, you can set the *TextFormat* property of the *reply* message to *plain*. The following commented code, repeated from Listing 9-8 shows how to set the format to plain text:

```
//var reply = (context.Activity as Activity).CreateReply(message.ToString());
//reply.TextFormat = "plain";
//await context.PostAsync(reply);
```

The *Activity* type has a *CreateReply* method that takes an input *Activity* and converts it to a response *Activity*, essentially switching the *Reciptient* and *From* properties behind the scenes. Once you have a new activity, set the *TextFormat* property to *plain*.

> **Note** *TextFormat* also accepts a value of *xml*. However, that only works for Skype. To use it, you can decorate text with HTML markup. This might be useful for porting an older Skype chatbot that used HTML formatting, but isn't practical for a chatbot that targets multiple platforms.

Summary

In this chapter, you learned various forms of advanced conversation. The chapter started with a discussion of the dialog stack, covered chainging, discussed *IScorables*, and showed how to format text.

You learned how the underlying navigation of dialogs are managed via stack, with the current dialog at the top of the stack. This continued to examples of using various methods to navigate from one dialog to another. You learned how dialogs can return results to a calling dialog that could process those results.

The *Chain* methods are very extensive and allow creating dialogs as well as other logic. You can manage navigation with chains via loops and case statements. There are ways to add continuations to handle the results of a dialog. You can even use LINQ to build dialog chains. There were several examples, showing how *Chain* could be used in a program and you're encouraged to experiment more to get a feel for the possibilities.

You learned how to create an *IScorable*, which allows a chatbot to handle user interactions outside the normal flow of the current conversation. *IScorables* also need to be registered via the AutoFac IoC container that Bot Builder uses.

Finally, you learned how to use markdown to format text for output. This can offer a better user experience than plain text. In the next chapter, you'll learn how to improve the user experience even more with cards.

PART III

Channels and GUI

CHAPTER 10	Attaching Cards................................237
CHAPTER 11	Configuring Channels 279
CHAPTER 12	Creating Email, SMS, and Web Chatbots......... 291

Besides a few buttons, most of the chatbots in earlier parts of this book were largely text. This makes sense because of the conversational nature of chatbots. However, there's a case to be made for some graphical user interface (GUI) interaction and now you'll see how that's possible. Since a large part of this book is about creating multi-channel chatbots, you'll also learn how top configure and use those channels.

Chapter 10 , *Attaching Cards* focuses heavily on adding GUI to a chatbot. The chapter is a hat tip to the underlying technology for how to add GUI to a message to a user, via attachments. The next two chapters explain how to configure channels. Chapter 11, *Configuring Channels* covers the general topic of channel configuration, including a few typical channels that you might want to surface a chatbot on. Chapter 12, *Creating Email, SMS, and Web Chatbots*, shows how to configure channels that don't fit into the traditional messaging app channel.

After completing this part of the book, you'll know what GUI elements are available to enhance interaction and offer multimedia to users. You'll also know how to configure both messaging app and other types of channels.

235

CHAPTER 10

Attaching Cards

Previous chapters displayed plenty of text for user interaction. That's great because it's where chatbots excel. However, sometimes text isn't enough to illustrate a concept. Sometimes an image can enhance communication. Other times, you need to share rich binary data, like audio and video. This chapter shows how to display rich graphical user interface (GUI) elements with cards.

This chapter starts out with an overview of the demo program, Music Chatbot. You'll start learning about some building-block features: Attachment and Suggested Actions. A Suggested Action is a button that a user can select for a specified action. Attachments allow sharing files between user and chatbot. Attachments are also instrumental in allowing chatbots to show cards to users. You'll learn how to build several types of cards, like Hero and Audio. We'll wrap up the chapter with a section on Adaptive cards that allow a more flexible cross-platform interface to a chatbot.

Music Chatbot Overview

This chapter shifts gears by presenting a new demo, named Music Chatbot with much more functionality than you've seen earlier. This demo gives you more conceptual information to consider when designing a chatbot. You'll see several listings throughout the chapter and they are all part of the same project, for Music Chatbot, available in the accompanying source code.

Before building a chatbot, you'll want to design the conversation paths. The sections that follow describe from beginning to end what a conversation with Music Chatbot might be. I've used Visio and the diagrams mostly resemble UML flowcharts. There's a growing list of 3rd party tools to help design chatbot conversations and you should pick what works best for you as long as it helps. I'll explain the design here, implement each in code, and refer back to each diagram throughout the chapter to associate the implementation with the design.

> **Note** These diagrams largely describe what is often referred to as the Happy Path, where conversation takes place in a problem-free environment. In real life, users will say anything and everything to a chatbot, deviating from the ideal of the Happy Path. For these reasons, you'll logically need to design responses to handle unexpected input or grow team conventions on how to handle unexpected scenarios during development.

The Groove API

Music Chatbot uses the Microsoft Groove API to access content. To run the accompanying code for this chapter, there are a few things you'll need to do:

1. Register Music Chatbot with the Bot Framework, as described in Chapter 2, Setting Up a Project.
2. Visit *https://docs.microsoft.com/en-us/groove/api-overview* and register with the Groove API.
3. Update Web.config with credentials obtained during Step #1 – necessary to get security tokens with the Groove API.

> **Tip** Remember to add credentials when connecting with the Bot Emulator.

Music Chatbot contains its own *GrooveService* class, Listing 10-1, for handling communication with the Groove API.

LISTING 10-1 The Music Chatbot *GrooveService*

```csharp
using MusicChatbot.Models;
using Newtonsoft.Json;
using System;
using System.Collections.Generic;
using System.Configuration;
using System.IO;
using System.Linq;
using System.Net;
using System.Text;
using System.Text.RegularExpressions;

namespace MusicChatbot.Services
{
    public class GrooveService
    {
        const string BaseUrl = "https://music.xboxlive.com";

        public string GetToken()
        {
            string service = "https://login.live.com/accesstoken.srf";

            string clientId = ConfigurationManager.AppSettings["MicrosoftAppId"];
            string clientSecret = ConfigurationManager.AppSettings["MicrosoftAppPassword"];
            string clientSecretEnc = System.Uri.EscapeDataString(clientSecret);

            string scope = "app.music.xboxlive.com";
            string scopeEnc = System.Uri.EscapeDataString(scope);

            string grantType = "client_credentials";

            string postData =
                $"client_id={clientId}&client_secret={clientSecretEnc}" +
```

```csharp
            $"&scope={scopeEnc}&grant_type={grantType}";

        string responseString = SendRequest("POST", service, postData);

        string token = ExtractTokenFromJson(responseString);
        return token;
    }

    string ExtractTokenFromJson(string json)
    {
        Match match = Regex.Match(
            json, ".*\"access_token\":\"(?<token>.*?)\".*", RegexOptions.IgnoreCase);

        string token = null;
        if (match.Success)
            token = match.Groups["token"].Value;

        return token;
    }

    string SendRequest(string method, string service, string postData)
    {
        HttpWebRequest request = (HttpWebRequest)WebRequest.Create(service);

        UTF8Encoding encoding = new UTF8Encoding();
        byte[] data = encoding.GetBytes(postData);

        request.Method = method;
        request.ContentType = "application/x-www-form-urlencoded";
        request.ContentLength = data.Length;

        using (Stream stream = request.GetRequestStream())
            stream.Write(data, 0, data.Length);

        string responseString = null;
        using (HttpWebResponse response = (HttpWebResponse)request.GetResponse())
            responseString = new StreamReader(response.GetResponseStream()).ReadToEnd();

        return responseString;
    }

    public List<string> GetGenres()
    {
        string token = GetToken();

        HttpWebRequest request = (HttpWebRequest)WebRequest.Create(
            $"{BaseUrl}/3/content/music/catalog/genres");
        request.Method = WebRequestMethods.Http.Get;
        request.Accept = "application/json";
        request.Headers["Authorization"] = "Bearer " + token;

        string responseJson;
        using (var response = (HttpWebResponse)request.GetResponse())
        using (var sr = new StreamReader(response.GetResponseStream()))
            responseJson = sr.ReadToEnd();
```

```csharp
        Genres genres = JsonConvert.DeserializeObject<Genres>(responseJson);
        var genreList =
            (from genre in genres.CatalogGenres
             where genre.ParentName == null
             select genre.Name)
            .ToList();

        return genreList;
    }

    public List<Item> GetTracks(string genre)
    {
        string token = GetToken();

        HttpWebRequest request = (HttpWebRequest)WebRequest.Create(
            $"{BaseUrl}/1/content/music/catalog/tracks/browse?" +
            $"genre={genre}&maxItems=5&extra=Tracks");
        request.Method = WebRequestMethods.Http.Get;
        request.Accept = "application/json";
        request.Headers["Authorization"] = "Bearer " + token;

        string responseJson;
        using (var response = (HttpWebResponse)request.GetResponse())
        using (var sr = new StreamReader(response.GetResponseStream()))
            responseJson = sr.ReadToEnd();

        TracksRoot tracks = JsonConvert.DeserializeObject<TracksRoot>(responseJson);
        var genreList =
            (from track in tracks.Tracks.Items
             select track)
            .Take(5)
            .ToList();

        return genreList;
    }

    public Preview GetPreview(string namespaceId)
    {
        string token = GetToken();

        string clientInstanceId = ConfigurationManager.AppSettings["MicrosoftAppId"];
        HttpWebRequest request = (HttpWebRequest)WebRequest.Create(
            $"{BaseUrl}/1/content/{namespaceId}/preview?" +
            $"clientInstanceId={clientInstanceId}");
        request.Method = WebRequestMethods.Http.Get;
        request.Accept = "application/json";
        request.Headers["Authorization"] = "Bearer " + token;

        string responseJson;
        using (var response = (HttpWebResponse)request.GetResponse())
        using (var sr = new StreamReader(response.GetResponseStream()))
            responseJson = sr.ReadToEnd();

        Preview preview = JsonConvert.DeserializeObject<Preview>(responseJson);
        return preview;
    }
```

```csharp
public string Search(SearchArguments args)
{
    string token = GetToken();

    HttpWebRequest request = (HttpWebRequest)WebRequest.Create(
        $"{BaseUrl}/1/content/music/search?q={Uri.EscapeDataString(args.Query)}" +
        $"&maxItems={args.MaxItems}&filters={args.Filters}&source={args.Source}");
    request.Method = WebRequestMethods.Http.Get;
    request.Accept = "application/json";
    request.Headers["Authorization"] = "Bearer " + token;

    string responseJson;
    using (var response = (HttpWebResponse)request.GetResponse())
    using (var sr = new StreamReader(response.GetResponseStream()))
        responseJson = sr.ReadToEnd();

    return responseJson;
    }
  }
}
```

The Groove API requires a token for each API request and *GetToken* shows how that works. To get a token, *GetToken* uses the *MicrosoftAppId* and *MicrosoftAppPassword* from the chatbot registration, performed in step #1. The following code shows where the *MicrosoftAppId* and *MicrosoftAppPassword* are in the *appSettings* section of *Web.config*.

```xml
<?xml version="1.0" encoding="utf-8"?>
<configuration>
  <appSettings>
    <add key="BotId" value="MusicChatbot" />
    <add key="MicrosoftAppId" value="Your App ID" />
    <add key="MicrosoftAppPassword" value="Your Password" />
  </appSettings>
  <connectionStrings>
    <add name="StorageConnectionString"
        connectionString="Your Azure storage connection string" />
  </connectionStrings>
  <system.web>
    <!-- Omitted for clarity -->
  </system.web>
  <system.webServer>
    <!-- Omitted for clarity -->
  </system.webServer>
  <runtime>
    <!-- Omitted for clarity -->
  </runtime>
</configuration>
```

The rest of the methods use *GetToken* to make Groove API calls. *GetGenres* gets a list of music categories that a user can choose. *GetTracks* gets a list of tracks for browsing and playing music tracks. The *GetPreview* method returns a type with the URL for a track's audio, so the user can listen to a track. *Search* performs the search based on criteria from the user. All of this code is based on demos from the Groove API documentation, which you can find at the link in step #2.

The *Item*, *Preview*, and *SearchArguments* are custom types, defined in the Music Chatbot code, holding properties with data the program needs and you'll see how those come into play throughout this chapter.

The next section describes the design of the dialogs that use *GrooveService*.

The Root Dialog

Music Chatbot starts with a root menu asking a user what they would like to do. Figure 10-1 explains that workflow.

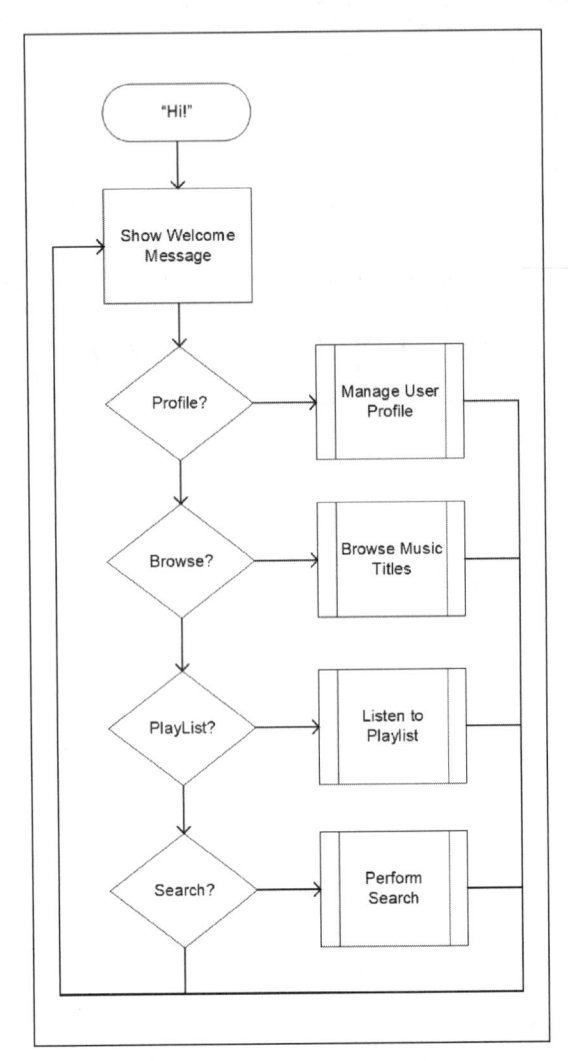

FIGURE 10-1 The Music Chatbot Root Dialog.

As shown in Figure 10-1, after the *Welcome* message, the user can select *Profile*, *Browse*, *Playlist* or *Search*. Each option launches a new dialog to fulfill the request.

The Profile Dialog

The profile dialog lets a user update or display their profile information. Figure 10-2 shows how that works.

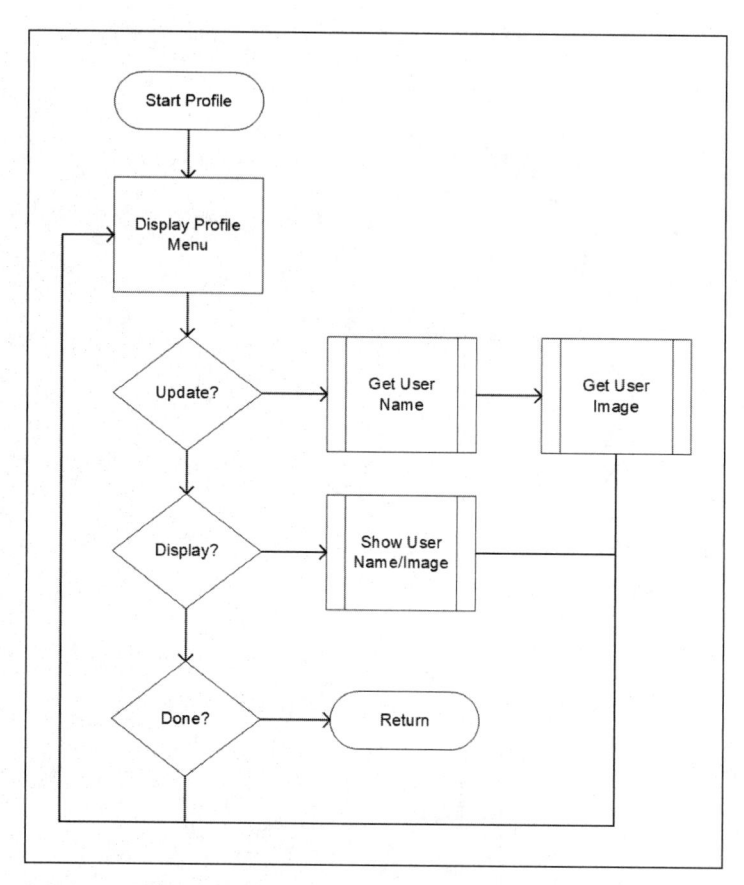

FIGURE 10-2 The Music Chatbot Profile Dialog.

The profile dialog in Figure 10-2 lets a user choose to *Update* or *Display* their profile information. *Update* branches into processes to request and save a users name and image. *Display* retrieves the user's name and image and displays the values. *Done* lets the user go back to the root dialog.

The Browse Dialog

When the user selects browse, they can see a list of tracks (songs) and optionally make a purchase. Figure 10-3 shows how browse dialog works.

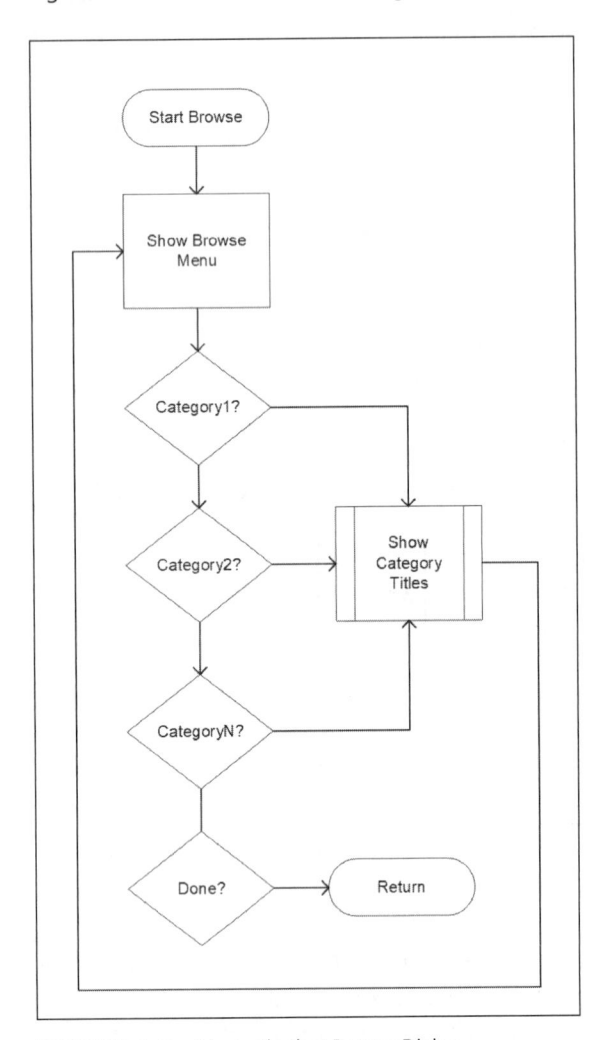

FIGURE 10-3 The Music Chatbot Browse Dialog.

Figure 10-3 shows how the browse dialog shows a menu. Music Chatbot does a direct request to the Groove service to get available categories, giving the user a dynamic list of categories to select. When a user picks a category, they'll see a set of tracks in that category and can choose to buy a track. Selecting *Done* takes the user back to the root dialog.

The Playlist Dialog

Users can choose the playlist to listen to tracks. Figure 10-4 shows how playlist dialog works.

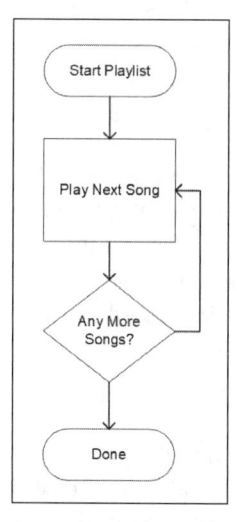

FIGURE 10-4 The Music Chatbot Playlist Dialog.

The playlist dialog in Figure 10-4 lets a user listen to a track. They can continue listing to tracks as long as they want. When they're done listening, they'll go back to the root dialog.

The Search Dialog

Search lets users search for various information, such as artists, tracks, and albums. Figure 10-5 shows how the search dialog works.

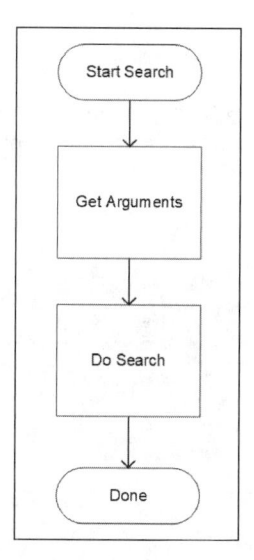

FIGURE 10-5 The Music Chatbot Search Dialog.

CHAPTER 10 Attaching Cards   245

Figure 10-5 shows how the search collects arguments and then does the search.

That was the Music Chatbot design and a description of the services for the data source, the Groove API. The rest of the chapter looks at the implementation of that design and how each dialog uses *GrooveService*.

Building Blocks

This section is called Building Blocks because each of these features are fundamental to the implementation of Bot Framework cards. Suggested Actions have *CardAction* types, which are the buttons in cards, and *Attachments* are the vehicle for adding cards to a chatbot response. This will become clearer as you learn more about cards in the coming section, but first we'll start with Suggested Actions.

Presenting Suggested Actions

A Suggested Action lets you add commands to a chatbot's message. These typically appear as buttons that the user can click or tap. There are various configurable actions, but this example receives the user's response as a normal *IMessageActivity*. Figure 10-6 shows what a message with Suggested Actions looks like.

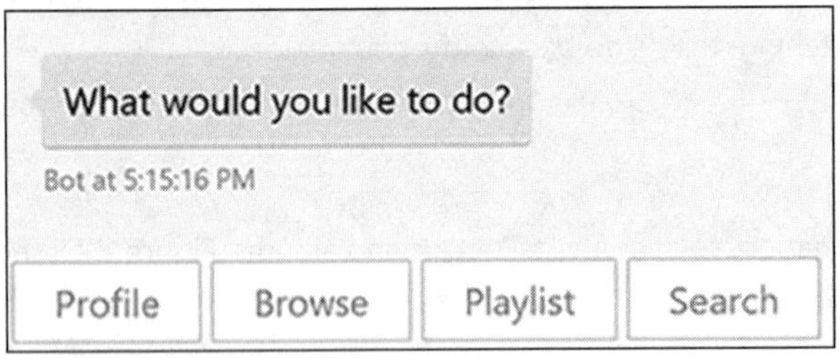

FIGURE 10-6 A message with Suggested Actions.

As shown in Figure 10-6, the suggested actions appear as a set of buttons for *Profile*, *Browse*, *Playlist*, and *Search*. Figure 10-6 is the user interface for the Root Dialog, diagramed in Figure 10-1 and implemented in Listings 10-2 and 10-3.

LISTING 10-2 The Music Chatbot *RootDialog*

```
using System;
using System.Threading.Tasks;
using Microsoft.Bot.Builder.Dialogs;
using Microsoft.Bot.Connector;
using MusicChatbot.Models;
using System.Linq;
using System.Collections.Generic;

namespace MusicChatbot.Dialogs
{
```

```csharp
[Serializable]
public class RootDialog : IDialog<object>
{
    public const string WelcomeMessage =
        "### Welcome to Music Chatbot!\n" +
        "Here are some of the things you can do:\n" +
        "* *Profile* to manage your profile information.\n" +
        "* *Browse* to find the music you like.\n" +
        "* *Playlist* for listening to favorite tunes.\n\n" +
        "Type \"Go\" to get started!";

    public Task StartAsync(IDialogContext context)
    {
        context.Wait(MessageReceivedAsync);
        return Task.CompletedTask;
    }

    async Task MessageReceivedAsync(IDialogContext context, IAwaitable<IMessageActivity> result)
    {
        IMessageActivity activity = await result;

        RootMenuItem choice;
        if (Enum.TryParse(activity.Text, out choice))
        {
            switch (choice)
            {
                case RootMenuItem.Profile:
                    await context.Forward(new ProfileDialog(), ResumeAfterDialogAsync, activity);
                    break;
                case RootMenuItem.Browse:
                    await context.Forward(new BrowseDialog(), ResumeAfterDialogAsync, activity);
                    break;
                case RootMenuItem.Playlist:
                    await context.Forward(new PlaylistDialog(), ResumeAfterDialogAsync, activity);
                    break;
                case RootMenuItem.Search:
                    await context.Forward(new SearchDialog(), ResumeAfterDialogAsync, activity);
                    break;
                default:
                    await context.PostAsync(WelcomeMessage);
                    context.Wait(MessageReceivedAsync);
                    break;
            }
        }
        else
        {
            await ShowMenuAsync(context);
        }
    }

    async Task ResumeAfterDialogAsync(IDialogContext context, IAwaitable<object> result)
```

```csharp
            {
                await ShowMenuAsync(context);
            }

            async Task ShowMenuAsync(IDialogContext context)
            {
                var options = Enum.GetValues(typeof(RootMenuItem)).Cast<RootMenuItem>().ToArray();

                var reply = (context.Activity as Activity).CreateReply("What would you like to do?");

                reply.SuggestedActions = new SuggestedActions
                {
                    To = new List<string> { context.Activity.From.Id },
                    Actions =
                        (from option in options
                         let text = option.ToString()
                         select new CardAction
                         {
                             Title = text,
                             Type = ActionTypes.ImBack,
                             Value = text
                         })
                        .ToList()
                };

                await
                    new ConnectorClient(new Uri(reply.ServiceUrl))
                        .Conversations
                        .SendToConversationAsync(reply);

                context.Wait(MessageReceivedAsync);
            }
        }
    }
```

LISTING 10-3 The Music Chatbot *RootMenuItem*

```csharp
namespace MusicChatbot.Models
{
    public enum RootMenuItem
    {
        Profile, Browse, Playlist, Search
    }
}
```

RootMenuItem in Listing 10-3 is an enum for what valid message choices could be. When *MessageReceivedAsync*, in Listing 10-2, can't match a *RootMenuItem* member, it calls the *ShowMenuAsync* method.

ShowMenuAsync uses the *CreateReply* method from the incoming *Activity* to create the new *Activity* that goes back to the user. *Activity* has a *SuggestedActions* property, which is type *SuggestedAction* that has two properties: *To* and *Actions*. The *To* property is optional and might be useful to specify which user can see the actions in a group chat scenario. *Actions* is a *List<CardAction>*, containing the options for the user. Each option is a *CardAction*, containing *Title*, *Type*, and *Value* properties, and *ShowMenuAsync* populates the *Actions* with a LINQ statement on the available options corresponding to *RootMenuItem* members. The *ActivityTypes.ImBack* causes the user's choice to come back to the dialog as a normal *IMessageActivity*. The following *ActivityTypes* class is a Bot Framework type showing what options work with the *CardAction Types* property:

```csharp
public class ActionTypes
{
    /// <summary>
    /// Client will open given url in the built-in browser.
    /// </summary>
    public const string OpenUrl = "openUrl";

    /// <summary>
    /// Client will post message to bot, so all other participants will see
    /// that was posted to the bot and who posted this.
    /// </summary>
    public const string ImBack = "imBack";

    /// <summary>
    /// Client will post message to bot privately, so other participants
    /// inside conversation will not see that was posted.
    /// </summary>
    public const string PostBack = "postBack";

    /// <summary>
    /// playback audio container referenced by url
    /// </summary>
    public const string PlayAudio = "playAudio";

    /// <summary>
    /// playback video container referenced by url
    /// </summary>
    public const string PlayVideo = "playVideo";

    /// <summary>
    /// show image referenced by url
    /// </summary>
    public const string ShowImage = "showImage";

    /// <summary>
    /// download file referenced by url
    /// </summary>
    public const string DownloadFile = "downloadFile";

    /// <summary>
    /// Signin button
    /// </summary>
    public const string Signin = "signin";
}
```

The difference between *ImBack* and *PostBack* is that all users (in a group conversation) can see *ImBack*, but *PostBack* is private to the user who chose it. *PostBack* can be useful for sending information, such as a JSON document, back and you don't want that to show in the chat window. This behavior can also vary by channel, so be sure to test to verify behaviors on all channels the chatbot surfaces on. Some of the *ActionTypes* members correspond to a card and you'll see how they work later in this chapter.

> **Tip** Notice how *ShowMenuAsync* is factored into it's own method. This lets *ResumeAfterDialogAsync* and the else clause in *MessageReceivedAsync* call *ShowMenuAsync* and always provide the user with some guidance. This might or might not make sense for your chatbot, but the concept of making sure a user doesn't get lost can help quality.

Cards use *CardAction* for their options, so you'll see it used as an integral part of cards. The next building block is *Attachments*.

Working with Attachments

Attachments allow chatbots and users to share content. When a user needs to send content to a chatbot, it will be in the form of a file, such as . an image to update their profile or a CSV file for data analysis. From the other direction, a chatbot might want to show the user an image. Listing 10-4 shows how Music Chatbot uses attachments in *ProfileDialog*.

LISTING 10-4 The Music Chatbot *ProfileDialog*

```
using Microsoft.Bot.Builder.Dialogs;
using Microsoft.Bot.Connector;
using MusicChatbot.Models;
using MusicChatbot.Services;
using Newtonsoft.Json.Linq;
using System;
using System.IO;
using System.Linq;
using System.Net.Http;
using System.Threading.Tasks;

namespace MusicChatbot.Dialogs
{
    [Serializable]
    public class ProfileDialog : IDialog<object>
    {
        public string Name { get; set; }
        public byte[] Image { get; set; }

        public Task StartAsync(IDialogContext context)
        {
            context.Wait(MessageReceivedAsync);
            return Task.CompletedTask;
        }
```

```csharp
Task MessageReceivedAsync(IDialogContext context, IAwaitable<object> result)
{
    ShowMainMenu(context);
    return Task.CompletedTask;
}

void ShowMainMenu(IDialogContext context)
{
    var options = Enum.GetValues(
        typeof(ProfileMenuItem)).Cast<ProfileMenuItem>().ToArray();
    PromptDialog.Choice(
        context, ResumeAfterChoiceAsync, options, "What would you like to do?");
}

async Task ResumeAfterChoiceAsync(
    IDialogContext context, IAwaitable<ProfileMenuItem> result)
{
    ProfileMenuItem choice = await result;

    switch (choice)
    {
        case ProfileMenuItem.Display:
            await DisplayAsync(context);
            break;
        case ProfileMenuItem.Update:
            await UpdateAsync(context);
            break;
        case ProfileMenuItem.Done:
        default:
            context.Done(this);
            break;
    }
}

Task UpdateAsync(IDialogContext context)
{
    PromptDialog.Text(context, ResumeAfterNameAsync, "What is your name?");
    return Task.CompletedTask;
}

async Task ResumeAfterNameAsync(IDialogContext context, IAwaitable<string> result)
{
    Name = await result;
    await context.PostAsync("Please upload your profile image.");
    context.Wait(UploadAsync);
}

async Task UploadAsync(IDialogContext context, IAwaitable<object> result)
{
    var activity = await result as Activity;

    if (activity.Attachments.Any())
    {
        Attachment userImage = activity.Attachments.First();
        Image = await new HttpClient().GetByteArrayAsync(userImage.ContentUrl);
```

```csharp
            StateClient stateClient = activity.GetStateClient();
            BotData userData = await stateClient.BotState.GetUserDataAsync
                (activity.ChannelId, activity.From.Id);
            userData.SetProperty(nameof(Name), Name);
            userData.SetProperty(nameof(Image), Image);
            await stateClient.BotState.SetUserDataAsync(
                activity.ChannelId, activity.From.Id, userData);
        }
        else
        {
            await context.PostAsync("Sorry, I didn't see an image in the attachment.");
        }

        ShowMainMenu(context);
    }

    async Task DisplayAsync(IDialogContext context)
    {
        Activity activity = context.Activity as Activity;

        StateClient stateClient = activity.GetStateClient();
        BotData userData =
            await stateClient.BotState.GetUserDataAsync(
                activity.ChannelId, activity.From.Id);

        if ((userData.Data as JObject)?.HasValues ?? false)
        {
            string name = userData.GetProperty<string>(nameof(Name));

            await context.PostAsync(name);

            byte[] image = userData.GetProperty<byte[]>(nameof(Image));

            var fileSvc = new FileService();
            string imageName = $"{context.Activity.From.Id}_Image.png";

            string imageFilePath = fileSvc.GetFilePath(imageName);
            File.WriteAllBytes(imageFilePath, image);

            string contentUrl = fileSvc.GetBinaryUrl(imageName);
            var agenda = new Attachment("image/png", contentUrl, imageName);
            Activity reply = activity.CreateReply();
            reply.Attachments.Add(agenda);

            await
                new ConnectorClient(new Uri(reply.ServiceUrl))
                    .Conversations
                    .SendToConversationAsync(reply);
        }
        else
        {
            await context.PostAsync("Profile not available. Please update first.");
        }
```

```
        ShowMainMenu(context);
    }
  }
}
```

ProfileDialog implements the Profile Dialog diagram in Figure 10-2. Music Chatbot uses the *IBotDataStore<BotData>* to store attachments. The first consideration with this approach is that *IBotDataStore<BotData>* uses the default Bot State Service, with a capacity of 32Kb, which doesn't have enough storage space for images.

> **Note** The Bot State Service isn't meant for production work and is there to only facilitate development.

The proper way to manage state data is to provide an *IBotDataStore<BotData>* implementation that uses your own data source. For Music Chatbot, the data source is Azure Table Storage and here are the steps for doing that:

1. Install the *Microsoft.Bot.Builder.Azure* NuGet package. This contains the *IBotDataStore<BotData>* implementation for multiple data sources, including Azure Table Storage. If you need more help with understanding Azure, the free Microsoft Press book, Fundamentals of Azure (*https://blogs.msdn.microsoft.com/microsoft_press/2016/09/01/free-ebook-microsoft-azure-essentials-fundamentals-of-azure-second-edition/*), is an excellent reference.

2. Create an Azure Storage account.

3. Create a connection string entry in *web.config*, as shown in Listing 10-5.

4. Register a *TableBotDataStore* as the *IBotDataStore<BotData>* implementation with the Autofac container in *Global.asax.cs*, shown in Listing 10-6.

LISTING 10-5 Table Storage connection string.

```
<connectionStrings>
  <add name="StorageConnectionString" connectionString="Your connection string" />
</connectionStrings>
```

LISTING 10-6 The Music Chatbot Global.asax

```
using Autofac;
using Microsoft.Bot.Builder.Azure;
using Microsoft.Bot.Builder.Dialogs;
using Microsoft.Bot.Builder.Dialogs.Internals;
using Microsoft.Bot.Connector;
using System.Configuration;
using System.Web.Http;
```

```csharp
namespace MusicChatbot
{
    public class WebApiApplication : System.Web.HttpApplication
    {
        protected void Application_Start()
        {
            GlobalConfiguration.Configure(WebApiConfig.Register);

            Conversation.UpdateContainer(builder =>
            {
                var store = new TableBotDataStore(
                    ConfigurationManager
                        .ConnectionStrings["StorageConnectionString"]
                        .ConnectionString);

                builder.Register(c => store)
                    .Keyed<IBotDataStore<BotData>>(AzureModule.Key_DataStore)
                    .AsSelf()
                    .SingleInstance();

                builder.Register(c => new CachingBotDataStore(store,
                    CachingBotDataStoreConsistencyPolicy
                    .ETagBasedConsistency))
                    .As<IBotDataStore<BotData>>()
                    .AsSelf()
                    .InstancePerLifetimeScope();
            });
        }
    }
}
```

As shown in Listing 10-6, the Bot Framework passes its *ContainerBuilder* as the *Conversation.UpdateContainer* lambda parameter. If you aren't familiar with Autofac, it's an open source Inversion of Control (IoC) container that the Bot Framework uses and you can visit *https://autofac.org/* for more information.

The following sections explain how uploading and displaying attachments works.

Accepting Attachments from Users

The Bot Emulator has a file upload button on the left of the text entry window and you can use that to upload an image to Music Chatbot. Music Chatbot asks the user for their name and then asks them to upload an image. The only image type Music Chatbot accepts is *.png*. The *UploadAsync* method, repeated below, handles the user's file:

```csharp
async Task UploadAsync(IDialogContext context, IAwaitable<object> result)
{
    var activity = await result as Activity;

    if (activity.Attachments.Any())
    {
        Attachment userImage = activity.Attachments.First();
        Image = await new HttpClient().GetByteArrayAsync(userImage.ContentUrl);
```

```
            StateClient stateClient = activity.GetStateClient();
            BotData userData = await stateClient.BotState.GetUserDataAsync(activity.
ChannelId, activity.From.Id);
            userData.SetProperty(nameof(Name), Name);
            userData.SetProperty(nameof(Image), Image);
            await stateClient.BotState.SetUserDataAsync(activity.ChannelId, activity.From.
Id, userData);
        }
        else
        {
            await context.PostAsync("Sorry, I didn't see an image in the attachment.");
        }

        ShowMainMenu(context);
    }
```

There's only one attachment, so *UploadAsync* reads the first item in the *Attachments* collection. When a user uploads an attachment, that file gets posted on the Bot Connector server for a short period of time, so you don't want to wait too long before getting it. The *Attachment* has a URL for where that file is located in the *ContentUrl* property. With that *ContentUrl*, the code can get a *byte[]*, which it places in the *Image* property. Then the chatbot saves the *Name* and *Image* in User State as covered in Chapter 3, Conversation Essentials.

Now that you know how to accept attachments from users, the next section explains how to send attachments to users.

Sending Attachments to Users

When a user wants to see their profile information, they choose the *Display* option from the *ProfileDialog* menu, which runs the *DisplayAsync* method below:

```
async Task DisplayAsync(IDialogContext context)
{
    Activity activity = context.Activity as Activity;

    StateClient stateClient = activity.GetStateClient();
    BotData userData =
        await stateClient.BotState.GetUserDataAsync(
            activity.ChannelId, activity.From.Id);

    if ((userData.Data as JObject)?.HasValues ?? false)
    {
        string name = userData.GetProperty<string>(nameof(Name));

        await context.PostAsync(name);

        byte[] image = userData.GetProperty<byte[]>(nameof(Image));

        var fileSvc = new FileService();
        string imageName = $"{context.Activity.From.Id}_Image.png";

        string imageFilePath = fileSvc.GetFilePath(imageName);
        File.WriteAllBytes(imageFilePath, image);
```

```csharp
            string contentUrl = fileSvc.GetBinaryUrl(imageName);
            var agenda = new Attachment("image/png", contentUrl, imageName);
            Activity reply = activity.CreateReply();
            reply.Attachments.Add(agenda);

            await
                new ConnectorClient(new Uri(reply.ServiceUrl))
                    .Conversations
                    .SendToConversationAsync(reply);
        }
        else
        {
            await context.PostAsync("Profile not available. Please update first.");
        }

        ShowMainMenu(context);
    }
```

User profile information is saved in User State by the *UploadAsync* method, so the *DisplayAsync* method must retrieve those values from User State.

Since *Attachments* passes files via *ContentUrl*, *DisplayAsync* saves the file to the local file system so it can create a URL. Another way to do this is to build a URL that points at a Web API endpoint that will return a file, but this was simple. Listing 10-7 shows the *FileService* class used to manage file names and URLs.

LISTING 10-7 The Music Chatbot *FileService*

```csharp
using System.Web;

namespace MusicChatbot.Services
{
    public class FileService
    {
        public string GetBinaryUrl(string fileName)
        {
            string absoluteUri =
                HttpContext.Current.Request.Url.AbsoluteUri +
                $"/Binaries/" + fileName;
            return absoluteUri.Replace("api/messages/", "");
        }

        public string GetFilePath(string fileName)
        {
            return HttpContext.Current.Server.MapPath("/Binaries/" + fileName);
        }
    }
}
```

Both *GetBinaryUrl* and *GetFilePath* use the ASP.NET *HttpContext* to build strings that use the *Binaries* folder, which is part of the MusicChatbot project in the downloadable source code.

After saving the file into the *Binaries* folder, the *DisplayAsync* method instantiates a new *Attachment* with the *contentUrl*, creates a new reply, and adds that *Attachment* to the *Attachments* collection. *DisplayAsync* calls *context.PostAsync* to display the name, but that won't work for attachments. To send the attachment with the image, the code instantiates a new *ConnectorClient* and then calls the the *Conversation* property's *SendToConversationAsync* method, covered in Chapter 3, to send the attachment to the user. Figure 10-7 shows what happens when the user chooses the *Display* option.

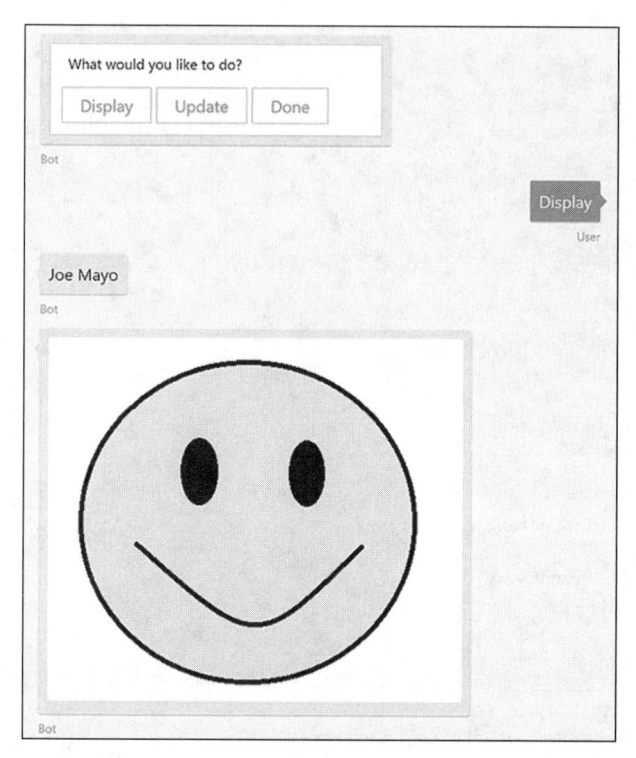

FIGURE 10-7 Displaying an attachment

Now you know how to receive and send attachments. Attachments are important for sending cards to users too, which you'll learn about next.

Displaying Cards

The Bot Framework has a growing list of cards. While each of these cards has specific purposes, their commonality is in presenting a GUI interface to the user. This section discusses some of the available cards and how they work.

Implementing *BrowseDialog*

The *BrowseDialog* implementation, in Listing 10-8, shows how to use Hero cards and Thumbnail cards. A Hero card lets you add a title, sub-title, a Hero image, and actions to a card. A Thumbnail card is similar to the Hero card, but a smaller thumbnail image. The *BrowseDialog* class implements the design in Figure 10-3.

LISTING 10-8 The Music Chatbot *BrowseDialog*

```csharp
using Microsoft.Bot.Builder.Dialogs;
using Microsoft.Bot.Connector;
using MusicChatbot.Models;
using MusicChatbot.Services;
using System;
using System.Collections.Generic;
using System.Linq;
using System.Threading.Tasks;

namespace MusicChatbot.Dialogs
{
    [Serializable]
    public class BrowseDialog : IDialog<object>
    {
        const string DoneCommand = "Done";

        public Task StartAsync(IDialogContext context)
        {
            context.Wait(MessageReceivedAsync);
            return Task.CompletedTask;
        }

        Task MessageReceivedAsync(IDialogContext context, IAwaitable<object> result)
        {
            List<string> genres = new GrooveService().GetGenres();
            genres.Add("Done");
            PromptDialog.Choice(context, ResumeAfterGenreAsync, genres, "Which music category?");
            return Task.CompletedTask;
        }

        async Task ResumeAfterGenreAsync(IDialogContext context, IAwaitable<string> result)
        {
            string genre = await result;

            if (genre == DoneCommand)
            {
                context.Done(this);
                return;
            }

            var reply = (context.Activity as Activity)
                .CreateReply($"## Browsing Top 5 Tracks in {genre} genre");
```

```csharp
        List<HeroCard> cards = GetHeroCardsForTracks(genre);
        cards.ForEach(card =>
            reply.Attachments.Add(card.ToAttachment()));

        ThumbnailCard doneCard = GetThumbnailCardForDone();
        reply.Attachments.Add(doneCard.ToAttachment());

        reply.AttachmentLayout = AttachmentLayoutTypes.Carousel;

        await
            new ConnectorClient(new Uri(reply.ServiceUrl))
                .Conversations
                .SendToConversationAsync(reply);

        context.Wait(MessageReceivedAsync);
    }

    List<HeroCard> GetHeroCardsForTracks(string genre)
    {
        List<Item> tracks = new GrooveService().GetTracks(genre);

        var cards =
            (from track in tracks
             let artists =
                 string.Join(", ",
                     from artist in track.Artists
                     select artist.Artist.Name)
             select new HeroCard
             {
                 Title = track.Name,
                 Subtitle = artists,
                 Images = new List<CardImage>
                 {
                     new CardImage
                     {
                         Alt = track.Name,
                         Tap = BuildBuyCardAction(track),
                         Url = track.ImageUrl
                     }
                 },
                 Buttons = new List<CardAction>
                 {
                     BuildBuyCardAction(track)
                 }
             })
            .ToList();
        return cards;
    }

    CardAction BuildBuyCardAction(Item track)
    {
        return new CardAction
        {
            Type = ActionTypes.OpenUrl,
            Title = "Buy",
            Value = track.Link
```

```csharp
            };
        }

        ThumbnailCard GetThumbnailCardForDone()
        {
            return new ThumbnailCard
            {
                Title = DoneCommand,
                Subtitle = "Click/Tap to exit",
                Images = new List<CardImage>
                {
                    new CardImage
                    {
                        Alt = "Smile",
                        Tap = BuildDoneCardAction(),
                        Url = new FileService().GetBinaryUrl("Smile.png")
                    }
                },
                Buttons = new List<CardAction>
                {
                    BuildDoneCardAction()
                }
            };
        }

        CardAction BuildDoneCardAction()
        {
            return new CardAction
            {
                Type = ActionTypes.PostBack,
                Title = DoneCommand,
                Value = DoneCommand
            };
        }
    }
}
```

In Listing 10-8, *MessageReceivedAsync* uses *GrooveService* to get a list of genre's, prompts the user to pick one, and processes the result in *ResumeAfterGenreAsync*. Next the code calls *GetHeroCardsForTracks* and *GetThumbnailCardForDone*, which we'll discuss soon.

Right now, look at how the return values from these methods get added to *Attachments*. *GetHeroCardsForTracks* returns a *List<HeroCard>*. Each card, whether it's a Hero card or any other type of card, has a *ToAttachment* method, to convert it to an attachment, and that's what's happening in the LINQ *ForEach* statement on the *List<CardAction>*. Each card gets converted to an *Attachment* and added to the *Attachments* collection.

The default layout for cards is vertical, but you can change that, as this demo does, by setting the *AttachmentLayout* property of the *reply Activity* to *AttachmentLayoutTypes.Carousel*. This makes the attachments scroll horizontally.

Now, each card has become an attachment and gets sent to the user in the same way you leared how to send an attachment earlier, via *SendToConversationAsync*. Next, we'll look at a couple of examples of how to create cards.

Creating Hero Cards

The *GetHeroCardsForTracks*, repeated below for convenience, builds a *List<HeroCard>*. These Hero cards display song tracks, using *GrooveService*, from the genre that the user selected. They also have a *Buy* button that takes the user to a Web page where they can read more info and/or buy the song.

```
List<HeroCard> GetHeroCardsForTracks(string genre)
{
    List<Item> tracks = new GrooveService().GetTracks(genre);

    var cards =
        (from track in tracks
         let artists =
             string.Join(", ",
                 from artist in track.Artists
                 select artist.Artist.Name)
         select new HeroCard
         {
             Title = track.Name,
             Subtitle = artists,
             Images = new List<CardImage>
             {
                 new CardImage
                 {
                     Alt = track.Name,
                     Tap = BuildBuyCardAction(track),
                     Url = track.ImageUrl
                 }
             },
             Buttons = new List<CardAction>
             {
                 BuildBuyCardAction(track)
             }
         })
        .ToList();
    return cards;
}
```

The *GetHeroCardsForTracks* is mostly a LINQ statement that builds a Hero card for each track returned by *GetTracks*. There could be multiple artists, so the *let* clause combines the artist names into a comma separated list. The *select* clause projects a new *HeroCard* instance.

Each *HeroCard* has a *Title*, *Subtitle*, *Images*, and *Buttons*. The *Images* property only takes a single image, of type *CardImage* where *Alt* is a description of the image and *Url* is the location of the image, both of which come from the *Item* type returned by the *GrooveService*. The *Tap* property executes a *CardAction*, returned by the *BuildBuyCardAction* method. The code assigns the return value of *BuildBuyCardAction*, repeated below from Listing 10-8, to the *List<CardAction>* assigned to *Buttons* too:

```
CardAction BuildBuyCardAction(Item track)
{
    return new CardAction
    {
        Type = ActionTypes.OpenUrl,
        Title = "Buy",
        Value = track.Link
    };
}
```

This is the same *CardAction* type explained in the previous section on Suggested Actions. This time, its *Type* is *ActionTypes.OpenUrl*, which opens a Web browser and navigates to the page specified by *track.Link*, which is assigned to *Value*.

A similar card is the Thumbnail card, discussed next.

Creating Thumbnail Cards

BrowseDialog adds a Thumbnail card to the end of the list so the user can click *Done* and go back to the previous dialog in the stack. Here's the implementation of *GetThumbnailCardForDone*, repeated from Listing 10-8:

```
ThumbnailCard GetThumbnailCardForDone()
{
    return new ThumbnailCard
    {
        Title = DoneCommand,
        Subtitle = "Click/Tap to exit",
        Images = new List<CardImage>
        {
            new CardImage
            {
                Alt = "Smile",
                Tap = BuildDoneCardAction(),
                Url = new FileService().GetBinaryUrl("Smile.png")
            }
        },
        Buttons = new List<CardAction>
        {
            BuildDoneCardAction()
        }
    };
}
```

This code is nearly identical to the *HeroCard*, except the type is *ThumbnailCard*. It has a single image—this time using *FileService* to get a *Smile.png* file from the local file system and returning a URL. Both *CardImage.Tap* and the *CardAction* for *Buttons* call *BuildCardDoneAction*, shown here:

```
CardAction BuildDoneCardAction()
{
    return new CardAction
    {
        Type = ActionTypes.PostBack,
        Title = DoneCommand,
```

```
                Value = DoneCommand
            };
        }
```

This is the same *CardAction* type used for the *HeroCard* demo, except that its *Type* is *ActionTypes.PostBack* and the *Title* and *Value* are set to the *BrowseDialog DoneCommand* constant, creating a button that says *Done*. Since *ResumeAfterGenreAsync* is the current method at the top of the dialog stack, Bot Builder calls that if a user selects the *Done* buton, the *ResumeAfterGenreAsync* method checks to see if that's the text of the current *IMessageActivity* and, if so, will pop the dialog stack to return control to the previous dialog.

Now you've seen how cards work, with a couple of examples. The next section discusses implementation of *PlaylistDialog* with another example of how to create cards.

Implementing *PlaylistDialog*

The *PlaylistDialog* lets a user select a genre and then receive a list of cards that let them play songs in that genre. Listing 10-9 shows *PlaylistDialog*, which implements the diagram in Figure 10-4.

LISTING 10-9 The Music Chatbot *PlaylistDialog*

```
using Microsoft.Bot.Builder.Dialogs;
using Microsoft.Bot.Connector;
using MusicChatbot.Models;
using MusicChatbot.Services;
using System;
using System.Collections.Generic;
using System.Linq;
using System.Threading.Tasks;

namespace MusicChatbot.Dialogs
{
    [Serializable]
    public class PlaylistDialog : IDialog<object>
    {
        const string DoneCommand = "Done";

        public Task StartAsync(IDialogContext context)
        {
            context.Wait(MessageReceivedAsync);
            return Task.CompletedTask;
        }

        Task MessageReceivedAsync(IDialogContext context, IAwaitable<object> result)
        {
            List<string> genres = new GrooveService().GetGenres();
            genres.Add("Done");
            PromptDialog.Choice(
                context, ResumeAfterGenreAsync, genres, "Which music category?");
            return Task.CompletedTask;
        }
```

```csharp
        async Task ResumeAfterGenreAsync(IDialogContext context, IAwaitable<string> result)
        {
            string genre = await result;

            if (genre == DoneCommand)
            {
                context.Done(this);
                return;
            }

            var reply = (context.Activity as Activity)
                .CreateReply($"## Viewing Top 5 Tracks in {genre} genre");

            List<AudioCard> cards = GetAudioCardsForPreviews(genre);
            cards.ForEach(card =>
                reply.Attachments.Add(card.ToAttachment()));

            ThumbnailCard doneCard = GetThumbnailCardForDone();
            reply.Attachments.Add(doneCard.ToAttachment());

            reply.AttachmentLayout = AttachmentLayoutTypes.Carousel;

            await
                new ConnectorClient(new Uri(reply.ServiceUrl))
                    .Conversations
                    .SendToConversationAsync(reply);

            context.Wait(MessageReceivedAsync);
        }

        ThumbnailCard GetThumbnailCardForDone()
        {
            return new ThumbnailCard
            {
                Title = DoneCommand,
                Subtitle = "Click/Tap to exit",
                Images = new List<CardImage>
                {
                    new CardImage
                    {
                        Alt = "Smile",
                        Tap = BuildDoneCardAction(),
                        Url = new FileService().GetBinaryUrl("Smile.png")
                    }
                },
                Buttons = new List<CardAction>
                {
                    BuildDoneCardAction()
                }
            };
        }

        List<AudioCard> GetAudioCardsForPreviews(string genre)
        {
```

```
            var grooveSvc = new GrooveService();
            List<Item> tracks = grooveSvc.GetTracks(genre);

            var cards =
                (from track in tracks
                 let artists =
                     string.Join(", ",
                         from artist in track.Artists
                         select artist.Artist.Name)
                 let preview = grooveSvc.GetPreview(track.Id)
                 select new AudioCard
                 {
                     Title = track.Name,
                     Subtitle = artists,
                     Media = new List<MediaUrl>
                     {
                         new MediaUrl(preview.Url)
                     }
                 })
                .ToList();

            return cards;
        }

        CardAction BuildDoneCardAction()
        {
            return new CardAction
            {
                Type = ActionTypes.PostBack,
                Title = DoneCommand,
                Value = DoneCommand
            };
        }
    }
}
```

Just like in the previous *BrowseDialog* code, the *PlaylistDialog* in Listing 10-9 asks for a genre and processes the user's choice in *ResumeAfterGenreAsync*. The code then calls *GetAudioCardsForPreviews*, adds the result to the *Attachments* collection of a *reply*, and calls *SendToConversationAsync* to show those Audio cards to the user. The next section discusses how to create the Audio cards.

Creating Audio Cards

An Audio card allows the user to listen to Audio recordings. In Music Chatbot, the user goes to the *PlaylistDialog*, can see a list of Audio cards with songs, and can listen. The following excerpt, from Listing 10-9, shows how to create Audio cards:

```
List<AudioCard> GetAudioCardsForPreviews(string genre)
{
    var grooveSvc = new GrooveService();
    List<Item> tracks = grooveSvc.GetTracks(genre);

    var cards =
```

```
            (from track in tracks
             let artists =
                 string.Join(", ",
                     from artist in track.Artists
                     select artist.Artist.Name)
             let preview = grooveSvc.GetPreview(track.Id)
             select new AudioCard
             {
                 Title = track.Name,
                 Subtitle = artists,
                 Media = new List<MediaUrl>
                 {
                     new MediaUrl(preview.Url)
                 }
             })
            .ToList();

        return cards;
    }
```

The *GetAudioCardsForPreview* method uses a LINQ statement to build a list of *AudioCard*. Instead of an *Image*, an *AudioCard* has a *Media* collection, which is a *List<MediaUrl>*. This method retrieved a list of preview items from *GrooveService*, where a preview is about 30 seconds of a song. This *preview* has a *Url*, that is passed to the *MediaUrl* constructor. As with other aspects of attachments, files are passed as URLs. Figure 10-8 shows what the Audio cards look like.

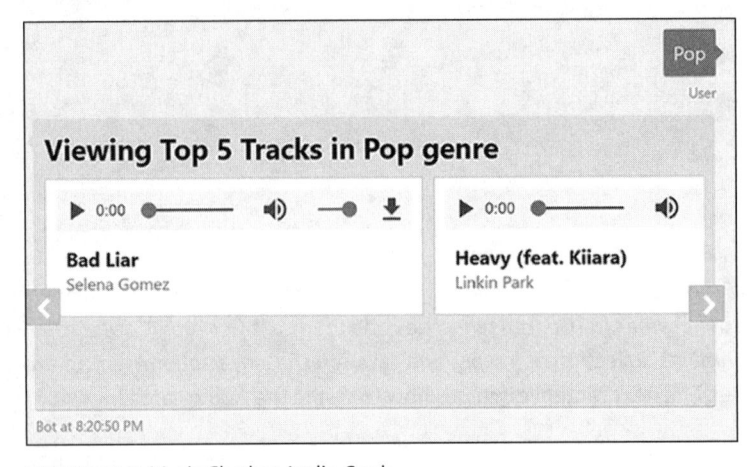

FIGURE 10-8 Music Chatbot Audio Cards.

One of the things you might notice in Figure 10-8 is that the Audio card is full-featured with go/pause button, timer progress, volume and so on. It uses the URL passed to *MediaUrl* to stream the music. Functionality like this really highlights a benefit of cards for accomplishing sophisticated tasks like this that might not have been as well suited (or possible) with text alone.

Creating Other Cards

One of the patterns you might have observed in previous card demos is that all of the cards, while having a different purpose are very similar in how you approach using them. There will typically be some back-end functionality, like *GrooveService* helping with data acquisition and other services, but the task of adding a card to a chatbot will be similar. Table 10-1 lists some of the other cards that you can use with a chatbot.

TABLE 10-1 Other Card Types

Card Type	Description
ReceiptCard	Has properties for holding item entries, taxes, and totals that is designed to provide a user with a receipt after purchase.
SignInCard	Indicates to a user that they need to sign in and the code will open a URL to a separate Web page or other authentication service to let a user sign in.
VideoCard	Plays videos. Like an Audio card, but accepts a URL to a video a user can watch.

> **Note** Whether a card is built-in or adaptive, different channels can render differently. You'll want to verify the rendering on each channel the chatbot surfaces on to verify your intended experience.

While the built-in cards are generally simple and quick, they might not always meet your needs. A more flexible card type is called Adaptive cards, which are discussed next.

Adaptive Cards

Adaptive cards let you include rich UI in a chatbot. They're designed to be cross-platform, with a goal of adding consistency between technology UI, foster greater sharing, and enabling a potential 3rd party market. This section shows how to create an Adaptive card for a Bot Framework chatbot.

> **Tip** Visit *http://adaptivecards.io/* for more information on Adaptive Cards.

The Adaptive cards feature list is huge and likely to increase over time. What this section does is highlight three useful areas that help you understand how to build these cards: layout, controls, and actions. *SearchDialog*, shown in Listing 10-10, performs a search function with the Groove API and uses Adaptive cards to do it. *SearchDialog* also implements the design in Figure 10-5.

LISTING 10-10 The Music Chatbot *SearchDialog*

```csharp
using AdaptiveCards;
using Microsoft.Bot.Builder.Dialogs;
using Microsoft.Bot.Connector;
using MusicChatbot.Models;
using MusicChatbot.Services;
using Newtonsoft.Json;
using Newtonsoft.Json.Linq;
using System;
using System.Collections.Generic;
using System.Threading.Tasks;

namespace MusicChatbot.Dialogs
{
    [Serializable]
    public class SearchDialog : IDialog<object>
    {
        public Task StartAsync(IDialogContext context)
        {
            context.Wait(MessageReceivedAsync);
            return Task.CompletedTask;
        }

        async Task MessageReceivedAsync(IDialogContext context,
IAwaitable<IMessageActivity> result)
        {
            var card = new AdaptiveCard();

            card.Body.AddRange(
                new List<CardElement>
                {
                    new Container
                    {
                        Items = BuildHeader()
                    },
                    new TextBlock { Text = "Query (max 200 chars):"},
                    new TextInput
                    {
                        Id = "query",
                        MaxLength = 200,
                        IsRequired = true,
                        Placeholder = "Query"
                    },
                    new TextBlock { Text = "Max Items (1 to 25):"},
                    new NumberInput
                    {
                        Id = "maxItems",
                        Min = 1,
                        Max = 25,
                        IsRequired = true
                    },
                    new TextBlock { Text = "Filters:"},
                    new ChoiceSet
                    {
```

```csharp
                    Id = "filters",
                    Choices = BuildFilterChoices(),
                    IsRequired = false,
                    Style = ChoiceInputStyle.Compact
                },
                new TextBlock { Text = "Source:"},
                new ChoiceSet
                {
                    Id = "source",
                    Choices = BuildSourceChoices(),
                    IsMultiSelect = false,
                    IsRequired = false,
                    Style = ChoiceInputStyle.Expanded
                }
        });

        card.Actions.Add(new SubmitAction
        {
            Title = "Search"
        });

        Activity reply = (context.Activity as Activity).CreateReply();
        reply.Attachments.Add(
            new Attachment()
            {
                ContentType = AdaptiveCard.ContentType,
                Content = card
            });

        await
            new ConnectorClient(new Uri(reply.ServiceUrl))
                .Conversations
                .SendToConversationAsync(reply);

        context.Wait(PerformSearchAsync);
    }

    async Task PerformSearchAsync(IDialogContext context, IAwaitable<IMessageActivity> result)
    {
        IMessageActivity activity = await result;

        string values = activity.Value?.ToString();
        var searchArgs = JsonConvert.DeserializeObject<SearchArguments>(values);

        var results = new GrooveService().Search(searchArgs);

        context.Done(this);
    }

    List<Choice> BuildFilterChoices()
    {
        return new List<Choice>
        {
            new Choice
            {
```

```csharp
                    Title = "artists",
                    Value = "artists"
                },
                new Choice
                {
                    Title = "albums",
                    Value = "albums"
                },
                new Choice
                {
                    Title = "tracks",
                    Value = "tracks"
                },
                new Choice
                {
                    Title = "playlists",
                    Value = "playlists"
                }
    };
}

List<Choice> BuildSourceChoices()
{
    return new List<Choice>
    {
        new Choice
        {
            Title = "catalog",
            Value = "catalog"
        },
        new Choice
        {
            Title = "collection",
            Value = "collection"
        }
    };
}

List<CardElement> BuildHeader()
{
    string contentUrl = new FileService().GetBinaryUrl("Smile.png");

    return new List<CardElement>
    {
        new ColumnSet
        {
            Columns = new List<Column>
            {
                new Column
                {
                    Items = new List<CardElement>
                    {
                        new TextBlock()
                        {
                            Text = "Music Search",
                            Size = TextSize.Large,
                            Weight = TextWeight.Bolder
```

```
                    },
                    new TextBlock()
                    {
                        Text = "Fill in form and click Search button.",
                        Color = TextColor.Accent
                    }
                }
            },
            new Column
            {
                Items = new List<CardElement>
                {
                    new Image()
                    {
                        Url = contentUrl
                    }
                }
            }
        }
    };
}
```

To get started, you'll need to add a reference to the *Microsoft.AdaptiveCards* NuGet package. *SearchDialog* is a typical *IDialog<T>* where *MessageReceivedAsync* handles the initial message after the user has chosen the *Search* option from *RootDialog*.

MessageReceivedAsync starts by instantiating an *AdaptiveCard* type. Quickly skimming *MessageReceivedAsync*, you can see that the *AdaptiveCard* type has a *Body* property, holding the content of the card. *AdaptiveCard* also has an *Actions* property that gives the user abilities to execute various commands.

Some of the *Adaptive* card features run together in the code, which will be normal when building any chatbot, so the following sections navigate the code, explaining core concepts, starting with layout.

Layout with Containers

The built-in cards specify layout for you, but this might not meet your needs, and would signify a reason to approach Adaptive cards. With Adaptive cards, you can use a container to wrap other controls. Additionally, you can wrap containers in other containers and specify rows and columns. As shown in the following excerpt from Listing 10-10, the *Body* property lets you build the card:

```
card.Body.AddRange(
    new List<CardElement>
    {
        new Container
        {
            Items = BuildHeader()
        },
        // …
    });
```

Body takes a *List<CardElement>* and *CardElement* is the base class of controls that can reside in an Adaptive card. Here, I've cut out the other controls assigned to the body, but the default layout of these controls is vertical, where the controls appear from top to bottom in the order they appear in code. The first *CardElement* in the list is a *Container*, which has an *Items* property of type *List<CardElement>*. The idea here is that the first part of the card is a header and the *BuildHeader* method, next, creates that *List<CardElement>* representing how the header should look:

```
List<CardElement> BuildHeader()
{
    string contentUrl = new FileService().GetBinaryUrl("Smile.png");

    return new List<CardElement>
    {
        new ColumnSet
        {
            Columns = new List<Column>
            {
                new Column
                {
                    Items = new List<CardElement>
                    {
                        new TextBlock()
                        {
                            Text = "Music Search",
                            Size = TextSize.Large,
                            Weight = TextWeight.Bolder
                        },
                        new TextBlock()
                        {
                            Text = "Fill in form and click Search button.",
                            Color = TextColor.Accent
                        }
                    }
                },
                new Column
                {
                    Items = new List<CardElement>
                    {
                        new Image()
                        {
                            Url = contentUrl
                        }
                    }
                }
            }
        }
    };
}
```

BuildHeader instantiates a *List<CardElement>* that is highly nested to show the container structure. The *ColumnSet* is a container for columns and the header is divided into two columns. *Columns* line up from left to right with the first column in code being the furthest most left.

Each *Column* has an *Items* collection of type *List<CardElement>*. The first column has two *TextBlock* controls to display the specified text. The second column has an image. That gives us a header with text on the left and an image on the right.

This is a first glimpse at controls, but you can also see how option-rich Adaptive cards can be by looking at the font styling properties of the *TextBlock*. The next section discusses controls in more detail.

Using Controls

In the previous section, you saw how *Image* and *TextBlock* controls fit into a container. These controls were primarily for display. Additionally, you can add input controls for text, combo boxes, radio buttons, and more. This excerpt from Listing 10-10 shows how to populate the *AdaptiveCard Body* property with controls that follow the header:

```csharp
card.Body.AddRange(
    new List<CardElement>
    {
        new Container
        {
            Items = BuildHeader()
        },
        new TextBlock { Text = "Query (max 200 chars):"},
        new TextInput
        {
            Id = "query",
            MaxLength = 200,
            IsRequired = true,
            Placeholder = "Query"
        },
        new TextBlock { Text = "Max Items (1 to 25):"},
        new NumberInput
        {
            Id = "maxItems",
            Min = 1,
            Max = 25,
            IsRequired = true
        },
        new TextBlock { Text = "Filters:"},
        new ChoiceSet
        {
            Id = "filters",
            Choices = BuildFilterChoices(),
            IsRequired = false,
            Style = ChoiceInputStyle.Compact
        },
        new TextBlock { Text = "Source:"},
        new ChoiceSet
        {
            Id = "source",
            Choices = BuildSourceChoices(),
            IsMultiSelect = false,
            IsRequired = false,
```

```
            Style = ChoiceInputStyle.Expanded
        }
    });
```

After adding the header, the remaining controls fit into the card vertically from the top down. The choice made here was to place the controls in pairs with a leading *TextBlock* to explain the purpose of the control. Each of these input controls correspond to a parameter for the Groove API search service.

The first and common part of all input controls is the *Id* property. It's important that you populate the *Id* property because when the user submits the card, you'll use that *Id* to know what each received value means. As a generally good coding practice, the *Id* you come up with for each control should be descriptive for the nature of the data the control captures. e.g. the *NumberInput* for the maximum number of items has an *Id* named *maxItems*. Also, when the user submits the form, the resulting value will be a JSON document, so you might find that it makes sense to use JSON propery naming conventions.

Another common property of controls is the *IsRequired* property, that if true makes it mandatory for the user to set a value in that control.

The *ChoiceSets* are interesting because they represent either combo boxes, radio button lists, or check box lists. Setting *IsMultiSelect* to *true* creates a check box list. Setting *Style* to *ChoiceInputStyle. Compact* creates a combo box and *ChoiceInputStyle.Expanded* creates an open list where the user can see all items at one time.

Controls have a plethora of options and the best way to learn about how they work is to use this chapter's code or create your own example and start experimenting with the options. That's how you add controls to an *AdaptiveCard*. Now, let's see what it takes to submit those values back to the chatbot.

Handling Actions

Much like *CardActions*, Adaptive cards allow a card to take various actions. This example uses the *SubmitAction*, shown in the exerpt below from Listing 10-10, to send the values from the card back to the chatbot:

```
card.Actions.Add(new SubmitAction
{
    Title = "Search"
});

Activity reply = (context.Activity as Activity).CreateReply();
reply.Attachments.Add(
    new Attachment()
    {
```

```
            ContentType = AdaptiveCard.ContentType,
            Content = card
    });

    await
        new ConnectorClient(new Uri(reply.ServiceUrl))
            .Conversations
            .SendToConversationAsync(reply);

    context.Wait(PerformSearchAsync);
```

All you need to do is add a *SubmitAction* instance to the *AdaptiveCard Actions* property. Adaptive cards are shared with the user as *Attachments* and this example instantiates a new *Attachment*, setting its *ContentType* to *AdaptiveCard.ContentType* and setting *Content* to the *AdaptiveCard* instance, *card*.

Next, look at the *context.Wait* and how it sets *PerformSearchAsync* as the next method on the stack to call. When the user submits the card, by clicking the button associated with the *SubmitAction*, the message goes back to the current method of the chatbot, which is *PerformSearchAsync*, shown here:

```
async Task PerformSearchAsync(IDialogContext context, IAwaitable<IMessageActivity> result)
{
    IMessageActivity activity = await result;

    string values = activity.Value?.ToString();
    var searchArgs = JsonConvert.DeserializeObject<SearchArguments>(values);

    var results = new GrooveService().Search(searchArgs);

    context.Done(this);
}
```

In *PerformSearchAsync*, the *Value* property on the *Activity* is a string that holds the JSON object containing the user's choices. The *SearchArguments* class holds properties corresponding the the *Ids* of the input controls in the *AdaptiveCard*. This is why it was important to specify the *Ids* on the input controls. With the proper parameters, the *GrooveService Search* method performs the search and returns results, which is a string containing the JSON object with results. This would be a great opportunity for you to practice using those results to build creative Adaptive cards for the chatbot to display to the user.

Figure 10-9 shows what the Adaptive card for the Groove API search looks like.

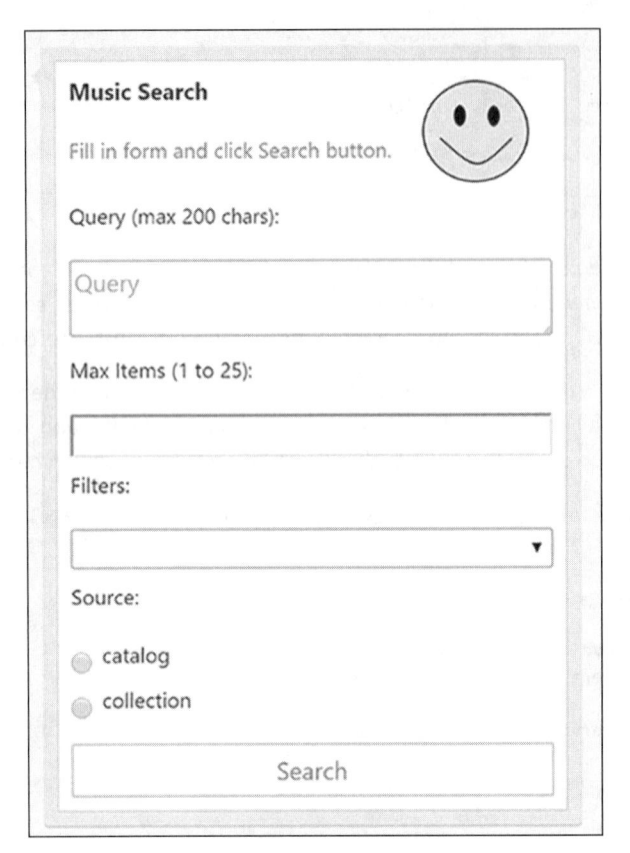

FIGURE 10-9 The Music Chatbot Search Adaptive Card

Summary

This chapter showed how to use cards to enhance the graphical appearace of a chatbot. It introduced Music Chatbot–a chatbot that uses the Microsoft Groove API to browse, play, and search for music.

You learned about some basic card features: *CardAction* and *Attachment*. We call them building blocks because they're essential parts of how cards work. You assign *CardAction* instances to Suggested Actions to give a user quick commands to select. *CardActions* are also used in other cards. *Attachments* let you exchange files with users and are the vehicle for sending cards to a user.

The demos showed you how to add Hero cards, Thumbnail cards, and Audio cards to display information and let a user listen to music. Remember that while the functionality of cards differ, the way you approach creating them is very similar.

Adaptive cards are cross-platform, reusable code that allows you to present rich content to a user. They're more flexible than the built-in cards because you can manage their layout, configure controls, and have various actions. The examples in this chapter showed you how to build an Adaptive card for searching the Groove API. Adaptive cards are nice because you can be creative and provide users with a nice interface.

Now that you know how to how to create cards and attach them to activities, you'll want to see how they appear in different channels. The next chapter shows how to set up and deploy chatbots to channels.

CHAPTER 11

Configuring Channels

Having read the previous chapters and having a working chatbot, the next step is to deploy this chatbot to one or more channels. A channel is where a user interacts with a chatbot. Often, chatbots reside on messaging channels and users invite the chatbot into their list on the platform the channel is designed for. Here are some of the major channels that are available now:

- Bing
- Email
- Facebook Messenger
- Skype
- Teams
- Twilio (SMS)
- Web Chat

In Chapter 12, Creating Email, SMS, and Web Chatbots, we'll cover email, Twilio, and the Web Chat control. In this chapter, you'll learn how to configure channels. Because there are a growing list of channels, it isn't practical to list them all here, so you'll see a couple that are available to get an idea of how the channel configuration process works. You'll get a general overview of channels and then look at Teams and Bing. The chapter wraps up with a discussion of the Channel Inspector and how it helps visualize specific features.

Channel Overview

Each chatbot has its own page where you can configure channels, view channel analytics, and modify chatbot settings. To get started, visit *https://dev.botframework.com/*, visit *My Bots*, and select the chatbot you want to deploy. This shows a list of channels, as shown in Figure 11-1.

> **Note** You can learn how to register a chatbot in Chapter 2; the *My Bots* list shows any registered chatbots.

The first page you'll see is the *Channel* page, discussed next.

279

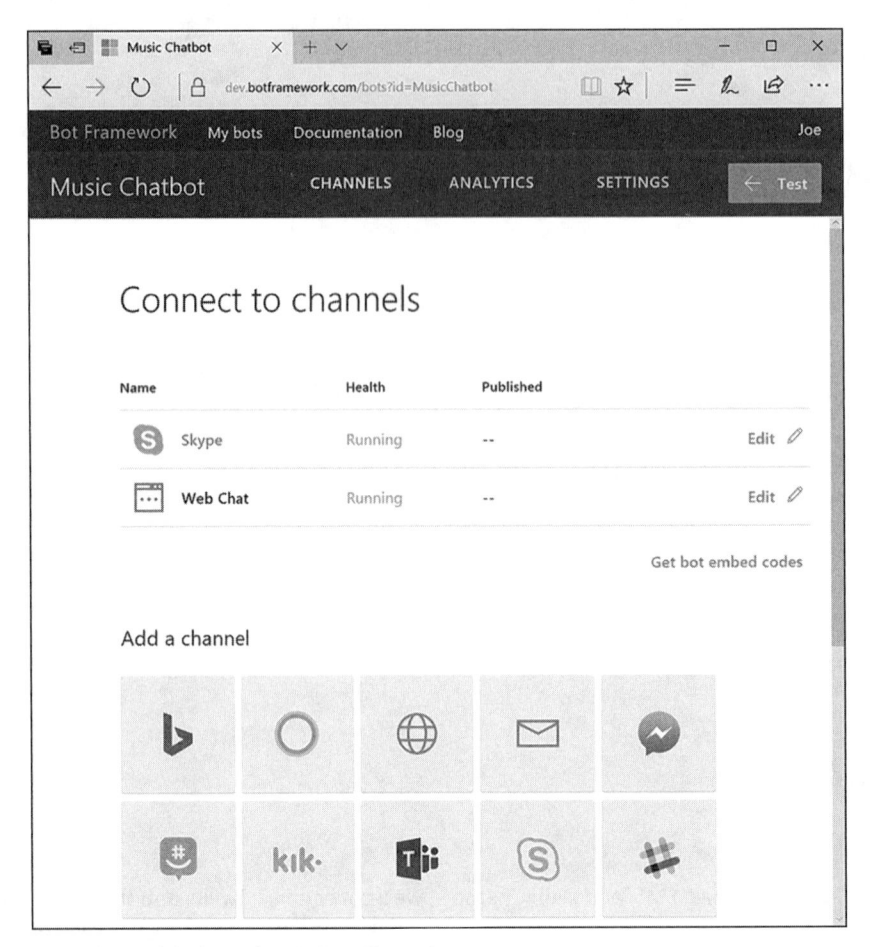

FIGURE 11-1 This shows the *My Bots* Channel page.

The Channel Configuration Page

Figure 11-1 shows that the Skype and Web Chat channels are running. The *Edit* link on the right lets you change the configuration settings. Clicking *Get Bot Embed Codes* pops up a window with an HTML snippet for each configured channel. Below *Add A Channel*, there's a list of available channels that aren't yet configured where you click the button for a channel to begin the configuration process.

Chatbot Analytics

The Analytics tab, as its name suggests, displays usage metrics for the chatbot. It doesn't work automatically and you need to set up Azure Application Insights. Application Insights is an Azure service that collects telemetry information and you can query that information to view the health of a chatbot or collect metrics as this section explains. To set up analytics, you need to configure Visual Studio, obtain keys and IDs, and update settings.

Configuring Visual Studio

Application Insights needs to be a part of a Visual Studio project so it can add the appropriate instrumentation to the project. To add Application Insights, right-click the chatbot project, select **Add | Application Insights Telemetry**, as shown in Figure 11-2.

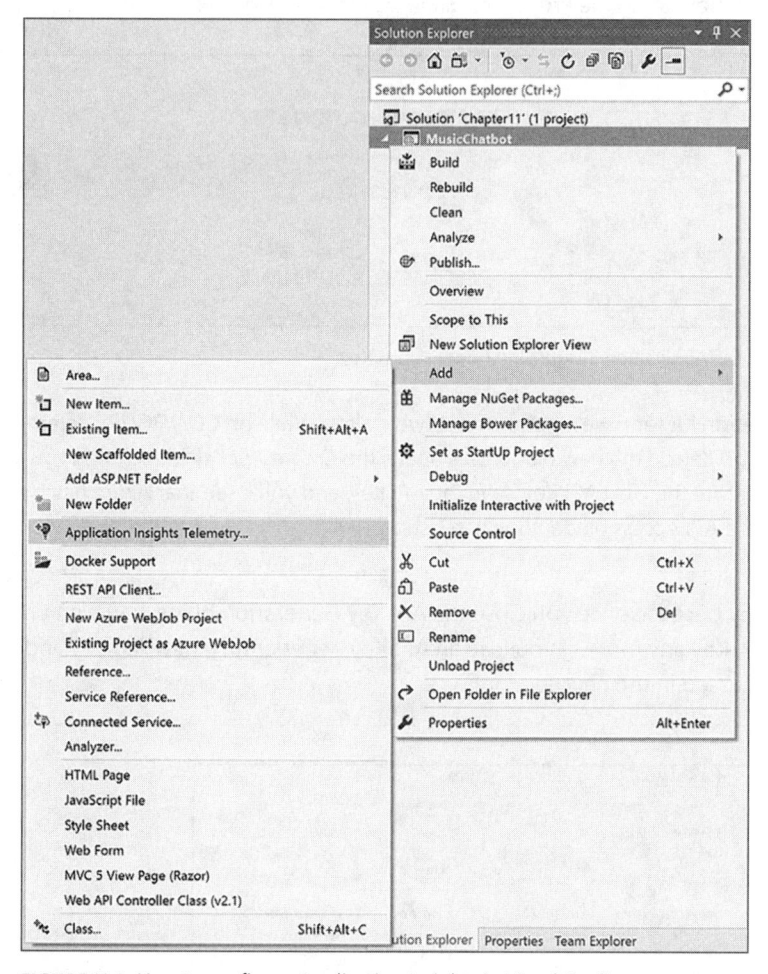

FIGURE 11-2 How to configure Application Insights in Visual Studio.

After selecting Application Insights Telemetry, you'll see windows that ask to log in to Azure and select various Azure service parameters, similar to the type of questions encountered during deployment in Chapter 2. Keep track of the name you used for Application Insights for this app because that will help you find it in the next section. After that you'll have new DLLs and other configuration items in the project to support Application Insights.

After adding Application Insights, deploy the chatbot to Azure, as explained in Chapter 2. If you've previously deployed the chatbot, you should re-deploy. Analytics only works if the chatbot is deployed where the telemetry collection will work.

Finding Application Insights Claims

Analytics requires three claims for registering Application Insights: Instrumentation Key, API Key, and Application ID. To get the Instrumentation Key, log in to your Azure portal, click the Applications Insights service (that you created in the previous section) for this chatbot, select **CONFIGURE | Properties**, and copy Instrumentation Key in the Properties blade, as shown in Figure 11-3.

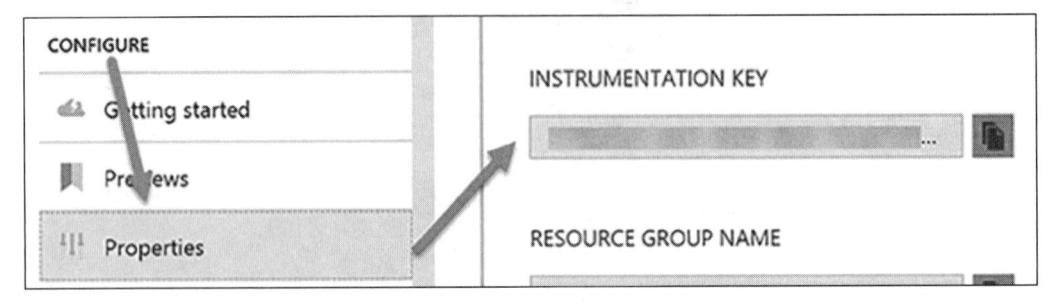

FIGURE 11-3 How to obtain the Application Insights Instrumentation Key.

Save that Instrumentation Key and we'll get the next two claims. While on *CONFIGURE*, select *API Access*, and click *Create API Key* on the new blade. This opens the *Create API Key* blade to let you generate a new API Key. Generate the New API Key, save the API Key, and you'll see that Azure has created a new API Key entry on the API Access blade, shown in Listing 11-4.

> **Tip** Save your API Key because once you close the API Key Generation blade, you won't be able to access the API Key any more. If you lose an API Key, regenerate a new API Key and update Settings (discussed in the next section).

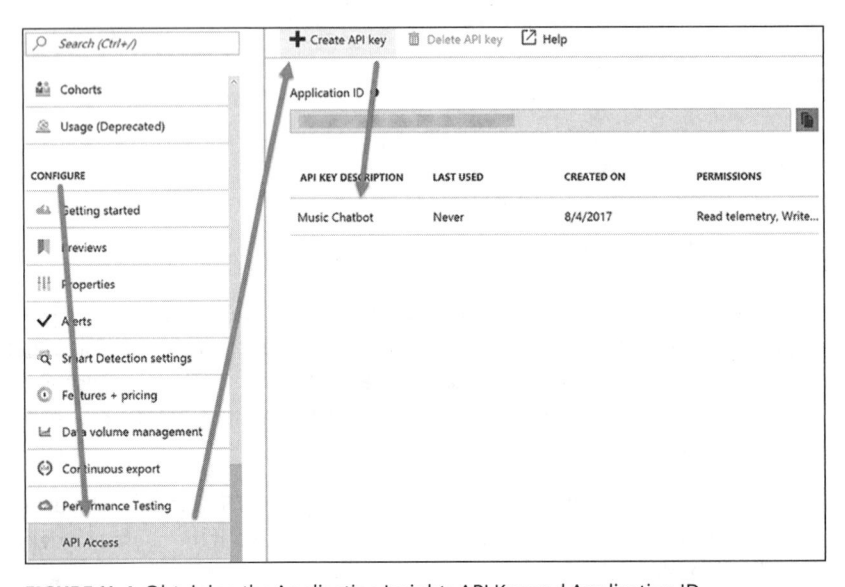

FIGURE 11-4 Obtaining the Application Insights API Key and Application ID.

Notice from Figure 11-4 that the API Acess blade also has an entry for *Application ID*. Copy the *Application ID*. You now have Instrumentation Key, API Key, and Application ID. Now you can update the chatbot settings.

Updating Settings

Once you have all the Application Insights claims, go back to the chatbot, click Settings and add those claims in the Analytics section, as shown in Figure 11-5.

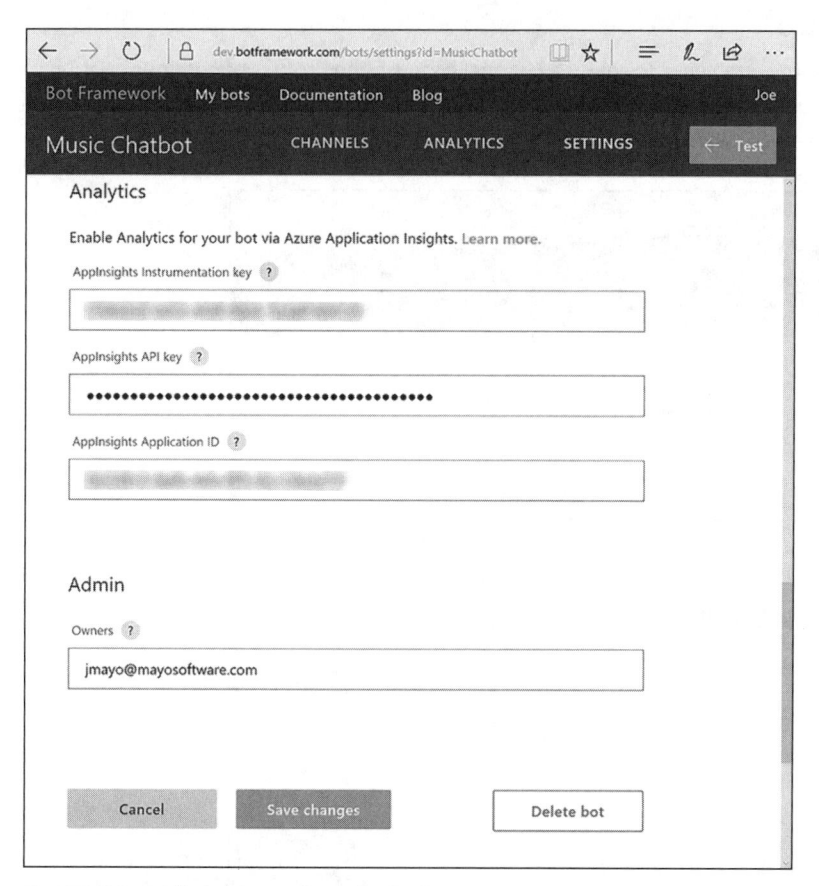

FIGURE 11-5 Adding claims to Analytics Settings.

Make sure you click Save Changes before leaving Settings. Now, any communication with a chatbot via a channel records metrics and stats you can view via the Analytics page.

In the following sections, you'll learn about setting up specific channels that will result in analytics data whenever people communicate with the chatbot.

Configuring Teams

Teams is a platform that allows people to chat and collaborate. It's currently part of the Office 365 suite of tools and integrates with office apps. It also supports developers through chatbots, tabs, and extensions. The chatbot part is what this book discusses because you can build a Bot Framework chatbot and let it participate in conversations through Teams. Let's look at getting set up first.

Channel Setup

At first glance, it might sound easy to set up a channel because the process of configuring a Teams channel consists of clicking the *Teams* button on the Channels page. Figure 11-6 shows a configured Teams channel for Music Chatbot.

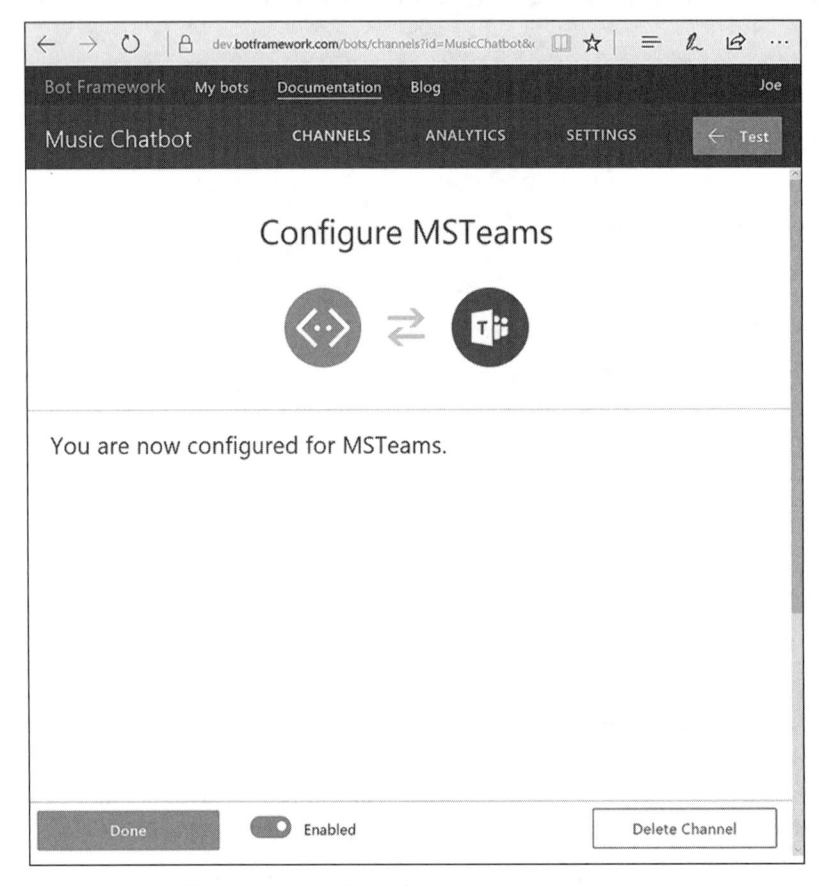

FIGURE 11-6 Configuring a Teams channel.

While Music Chatbot is configured for the Teams channel via the Bot Framework, you still need to perform Office 365 configuration. To do so, visit Office 365, log in as Administrator, select the **Settings | Services & Add-Ins**, and click Microsoft Teams, as shown in Figure 11-7.

> **Note** You need an Office 365 account and the ability to log into that account as Administrator to configure Teams.

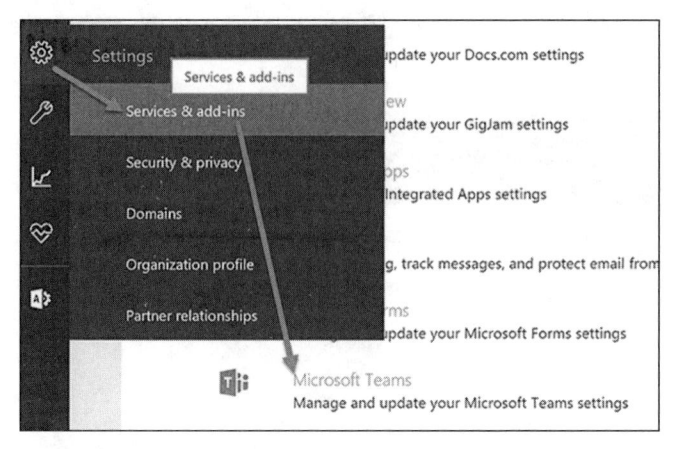

FIGURE 11-7 Finding the Teams configuration page in Office 365.

This opens the Microsoft Teams settings page. Set the *Turn Microsoft Teams On Or Off For Your Entire Organization* to *On*. Expand the *Apps* panel, as shown in Figure 11-8, and ensure those options are toggled On: *Allow External Apps In Microsoft Teams*, *Enable New External Apps By Default*, and *Allow Sideloading Of External Apps*.

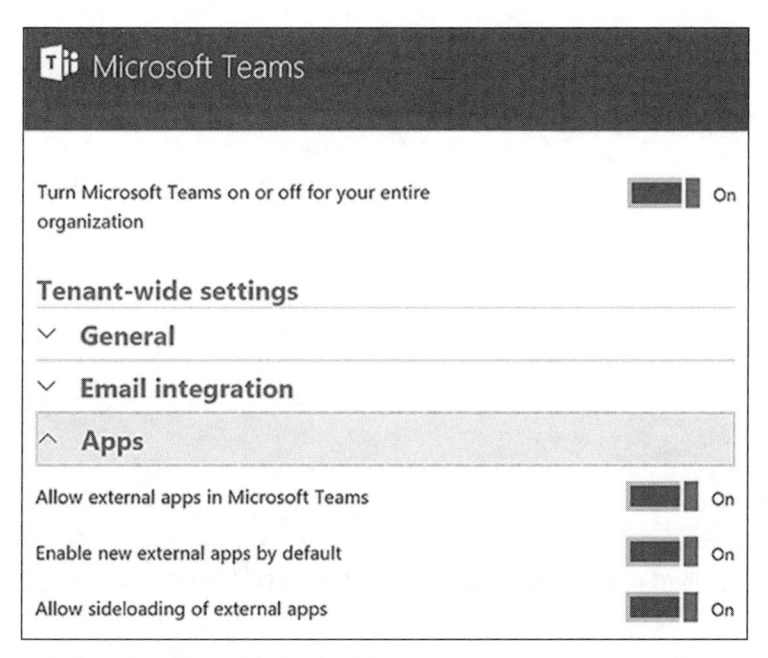

FIGURE 11-8 Configuring Teams in Office 365.

With the Office 365 configuration completed, you can start using the chatbot in Teams.

CHAPTER 11 **Configuring Channels** 285

Using the Chatbot

You can get started with Teams via their website: *https://teams.microsoft.com/*. The easiest way to start testing with Microsoft Teams is to visit the Bot Framework Channels page and click on the *Microsoft Teams* link (after configured as described above). This opens Microsoft Teams with the *Chat* opened to the chatbot. Figure 11-9 shows a conversation with Music Chatbot.

FIGURE 11-9 Testing in Microsoft Teams.

Figure 11-9 shows the *Done* thumbnail card in the *Browse* carousel for the *Pop genre*. As you can see, the channels render differently than the Bot Emulator.

Configuring Bing

Deploying to Bing makes a chatbot discoverable. When people do Bing searches, your chatbot can appear in relevant listings. This allows users to interact with the chatbot directly in Bing.

> **Tip** For some businesses, a Bing chatbot might introduce a competitive advantage, especially if competitors don't have a chatbot.

Channel Setup

The Bing Channel Setup page has several sections for chatbot information, categories, publisher information, and privacy/terms of use pages. Figure 11-10 shows a few of those fields, and most are self explanatory.

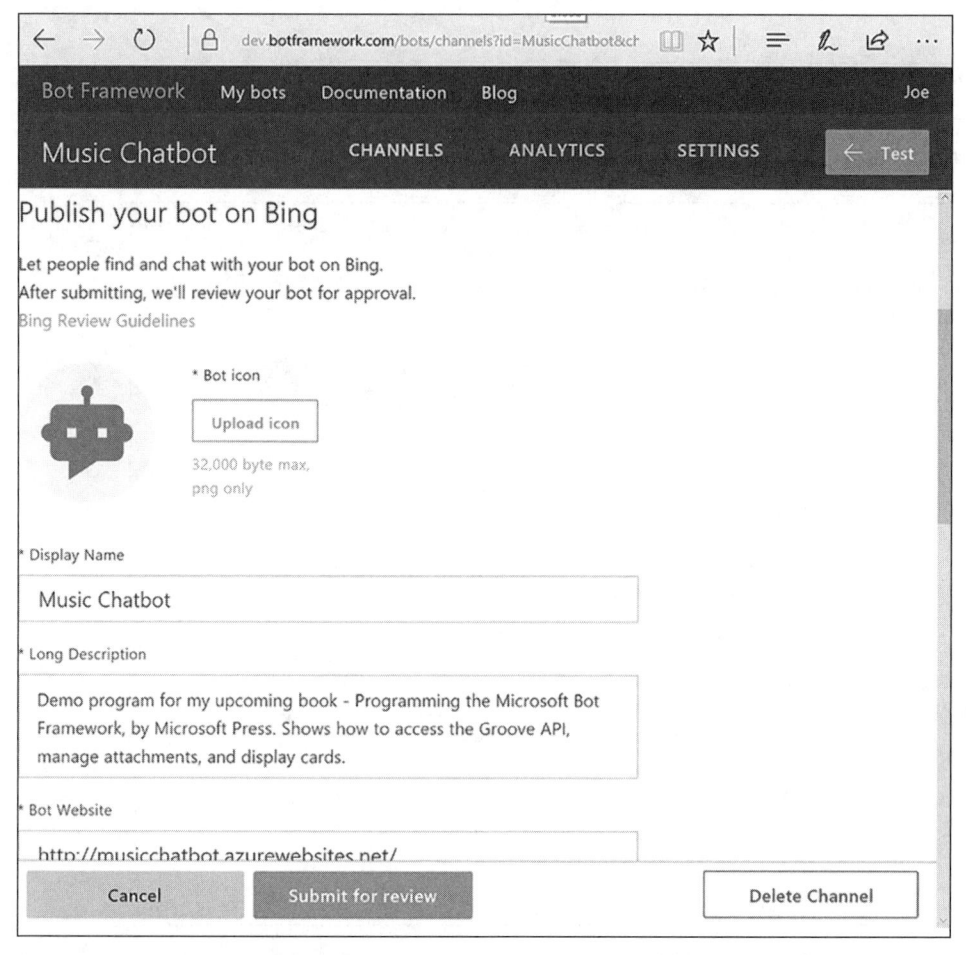

FIGURE 11-10 Configuring the Bing Channel.

You should click the *Bing Review Guidelines* for more information to ensure a successful submission.

On the *Terms Of Use* And *Privacy Statement*, you can have your own attorney do yours, or do a Bing search for *Free Terms of Use* and *Free Privacy Statement*, which returns many entries. The *Bing Review Guidelines* might have additional resources. We've included modified versions of these documents in the accompanying source code for Music Chatbot that we've used to successfully submit a chatbot and will give you an idea of what these documents can contain.

> **Tip** Microsoft has streamlined the submission guidelines for each channel they support to minimize the amount of work you need to do. However, the requirements for each channel vary and you should visit the channel sites and read their requirements thoroughly to make the submission process easier.

Using in Search

After submitting a chatbot for review you'll see a link to click to test the chatbot on Bing, as shown in Figure 11-11.

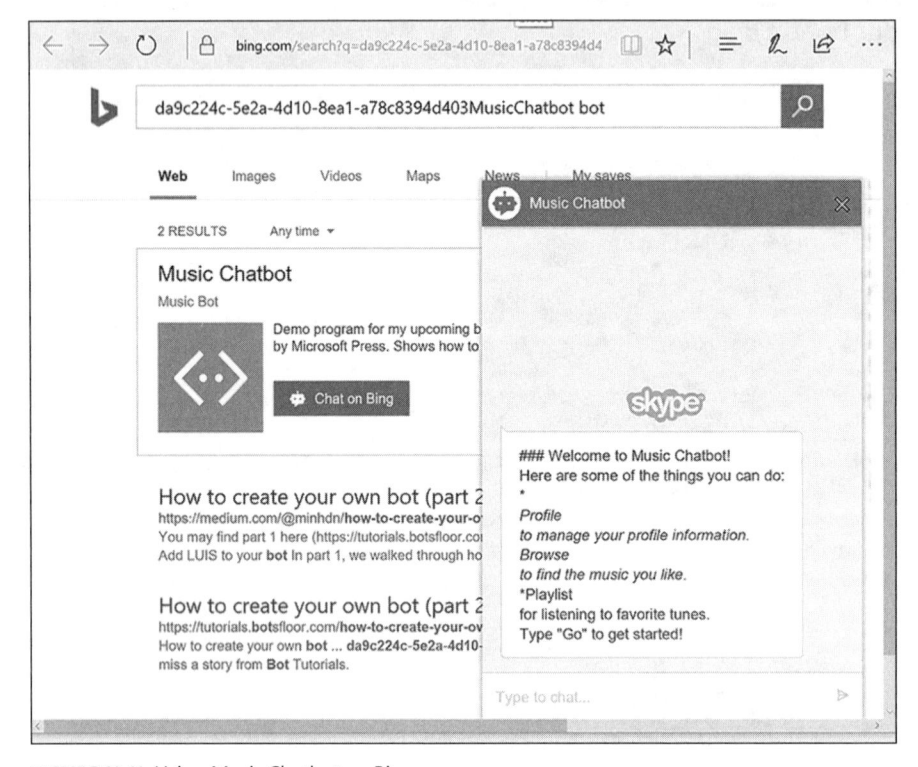

FIGURE 11-11 Using Music Chatbot on Bing.

As shown in Figure 11-11, when Bing shows a chatbot in search results, there's a *Chat On Bing* button. Clicking *Chat On Bing* opens a chat window where users can chat directly on the browser search page. There is no need to navigate, move out of context, or wonder – a user can ask a quick question and be closer to determining whether your chatbot or business meets their needs.

The Channel Inspector

The Bot Framework has a tool called a Channel inspector. What this does is show how each feature of a chatbot looks for a given channel. This can help when performing user experience design. Figure 11-12 shows what Channel Inspector looks like.

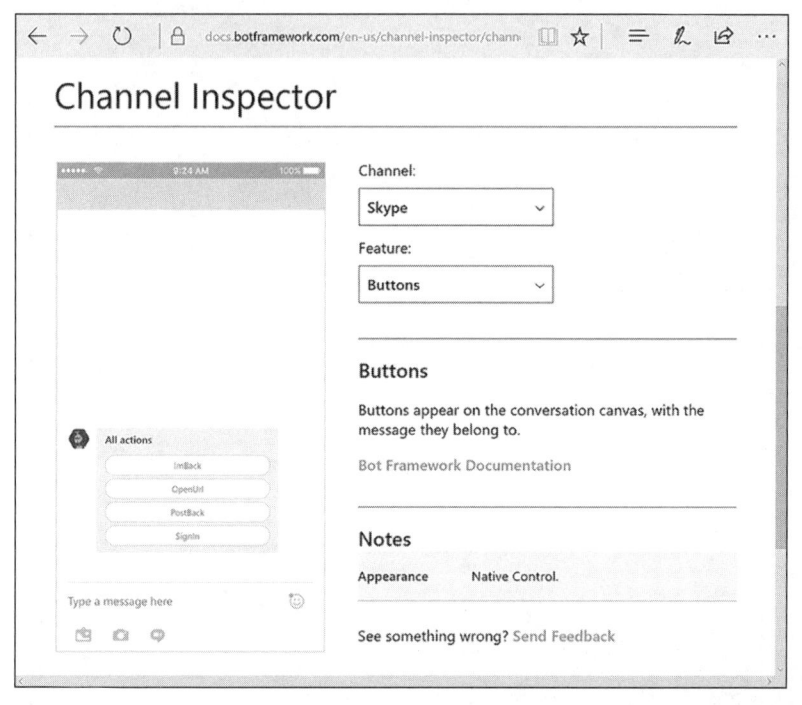

FIGURE 11-12 The Channel Inspector shows how various features of a chatbot look like on a channel.

The *Channel* drop-down in Figure 11-12 contains options for each channel that the Bot Framework supports. The *Feature* drop-down has options for various ways a chatbot can display content, such as Carousel, Hero Card, markdown, and more.

Summary

Building on the registration process from Chapter 2, this chapter discusses how to configure channels for a chatbot. The specific examples were for Microsoft Teams and Bing, but this was to give you an idea about how channels work and a few tips as you approach configuring other channels.

You learned how to configure analytics, instrumenting a project, and gathering keys from Application Insights. The Bot Framework page for a chatbot has an Analytics tab you can visit to learn a few things about the amount of traffic a chatbot receives. You also learned about Channel Inspector, which is a tool to help visualize various chatbot features on specific channels. The next chapter continues the journey into Email, SMS, and Web Chat channels.

CHAPTER 12

Creating Email, SMS, and Web Chatbots

Chapter 11 showed how channel configuration works, concentrating mostly on technologies that reach out to a single application. This chapter builds on that, but the theme is on the channels that go beyond a messaging app. There are some people and parts of the world that don't have smart phones, but they do have email, SMS, or some way to read a web page. This chapter shows how to set up those channels too.

Emailing Chatbot Conversations

The Bot Framework offers an email channel, which allows users to communicate with a chatbot via email. To use the Email channel, create an Office 365 email account and configure the email on the Bot Framework web page for the chatbot.

> **Tip** While you can communicate with a chatbot via any type of email account, the Email channel requires an Office 356 account.

Creating the Email Account

To create an email account, visit the Office 365 website as Administrator. Go to *Admin tools*, click **Users | Active Users**, and click the *Add User* button. Fill in the required information on the *Add User* blade, shown in Figure 12-1, and click *Add*.

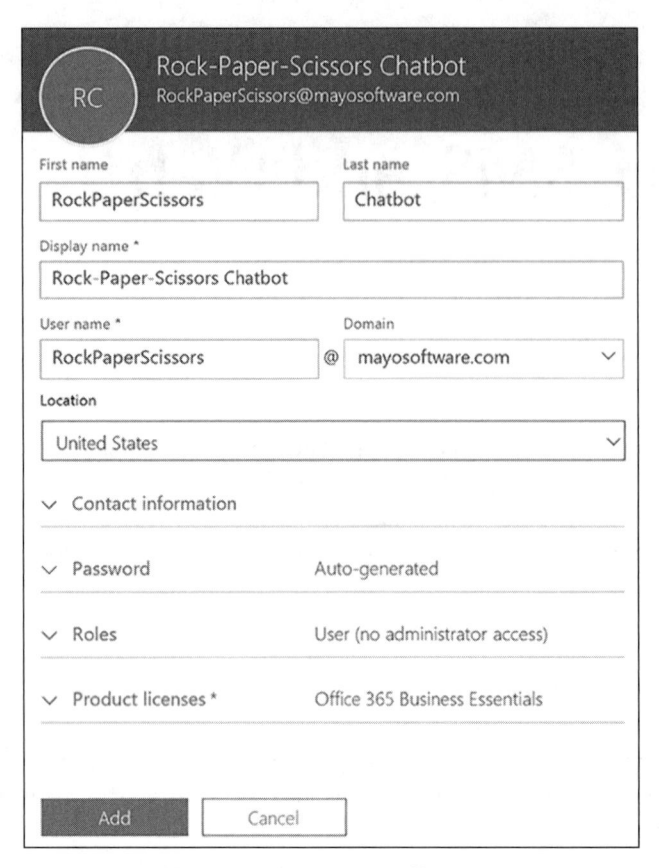

FIGURE 12-1 Adding an Email account to Office 365.

After clicking *Add*, keep track of the password for the new email account because you need it for configuring the Email channel, discussed next.

> **Tip** The initial Office 365 email password is temporary and valid for only 90 days. This is fine for a test where you intend to delete the email address, but you'll want to log onto that account and change the password to avoid hard to figure out problems when the chatbot stops working over email in 90 days.

Configuring the Email Channel

To configure the email channel, visit the chatbot on the Bot Framework site and click the email channel. As shown in Figure 12-2, add an *Email Address* and *Email Password*, and click *Save*.

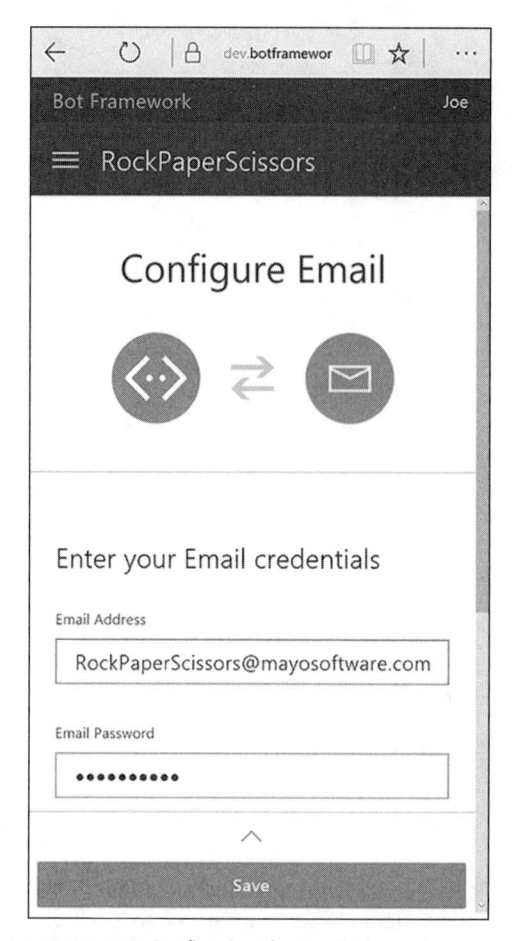

FIGURE 12-2 Configuring the Email Channel.

Now, a user can communicate with the *RockPaperScissors* chatbot by sending an email to the configured address, *RockPaperScissors@mayosoftware.com*, in this example.

Texting a Chatbot

You can let users communicate via SMS text messages with the Twilio channel. To start, visit the chatbot on the Bot Framework site and click the Twilio channel, shown in Figure 12-3.

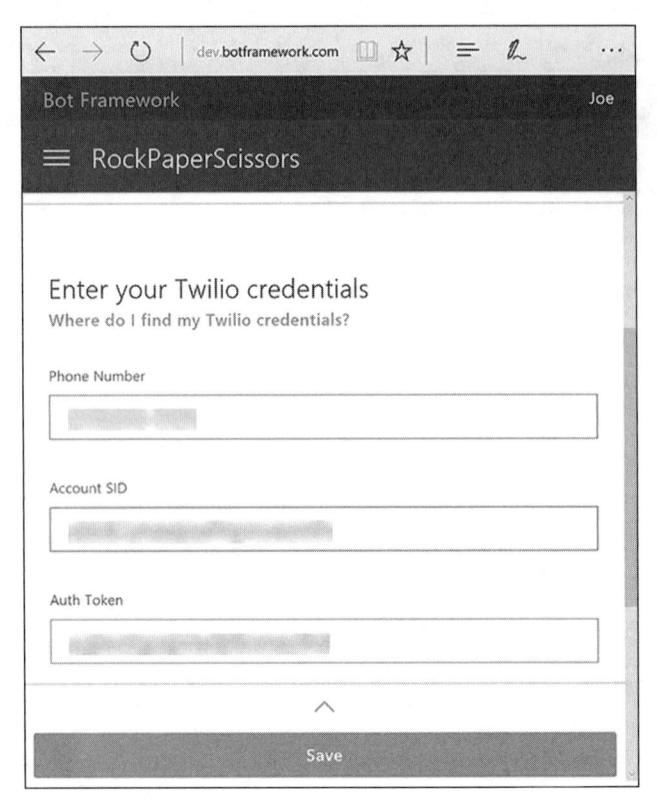

FIGURE 12-3 Configuring the Twilio channel.

Click the *Where Do I Find My Twilio Credentials* link. It has detailed instructions on how to register a new TwilioML app, select a phone number, and obtain credentials. Twilio offers a free level, which is ideal for testing.

> **Note** Some channels have extensive setup requirements for the third-party platform supported. The Bot Framework team keeps these documents up-to-date and they're typically very good at minimizing the amount of work you need to do. e.g. the Twilio setup procedures have links you can click that will take you to the exact page you need to be at, saving you from needing to search yourself.

Click the *Save* button when you've added the Twilio information. After channel configuration, users can communicate with the chatbot by sending a text message to the phone number, Figure 12-3, that the chatbot is configured to communicate with.

Embedding the Webchat Control

The Webchat control lets you add a chatbot to any web page. Since it fits in an HTML *iframe*, you can put the Webchat control on any surface that supports HTML, maybe even in another application's web control. This example adds a Webchat control to the *default.htm* file that comes with the default Bot Application template. This section shows one way to write the HTML for a web page to hold the Webchat control. The technique used here is to use JavaScript on the client to communicate with a server Web API. I'll explain why I recommend this approach, or something similar, in the next section.

Adding the Webchat IFrame Placeholder

To use the Webchat control, visit the Channels page for the chatbot, and add the Webchat channel, like other channels. If the Webchat channel is already added, click the edit button and you'll see a list of sites. If there aren't any sites, click the *Add New Site* button, give the site a name, and click *Done*. Selecting that site in the list, you'll see a screen similar to Figure 12-4.

> **Tip** Adding multiple sites allows you to have different credentials for each location where the Webchat control deploys.

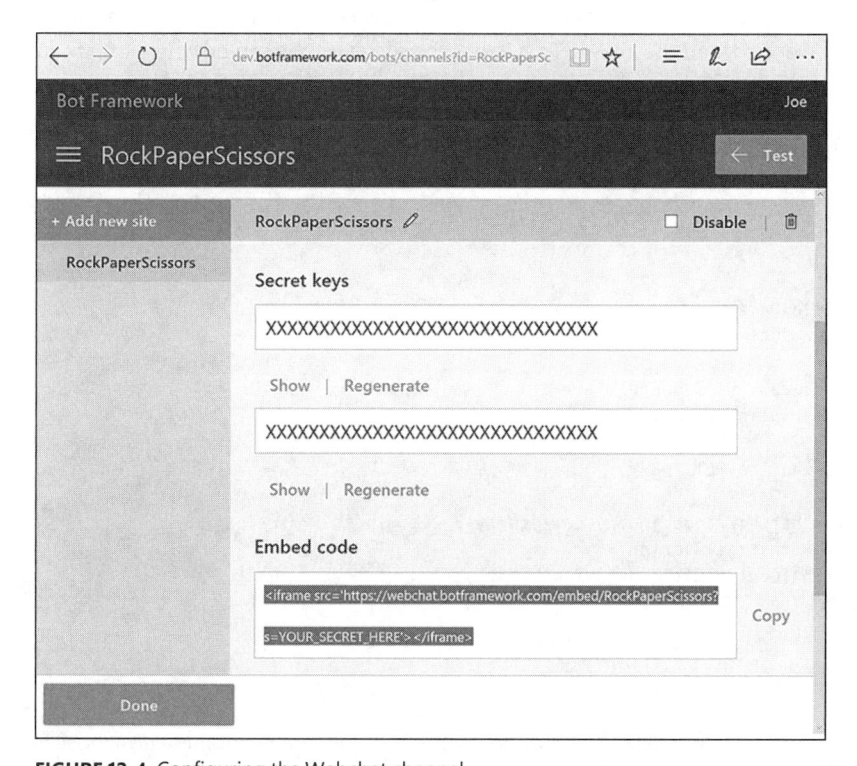

FIGURE 12-4 Configuring the Webchat channel.

CHAPTER 12 Creating Email, SMS, and Web Chatbots **295**

As shown in Figure 12-4, you get two secret keys, which is useful for scenarios where you might want one for production and the other for testing, prototyping, or anywhere someone needs to use the chatbot without needing to share the production key. You can always click the *Regenerate* button to create a new key if one is compromised or doesn't need to be shared anymore.

The *Embed code* is the *iframe* with the *src* for the Webchat control. Notice that the parameter to the URL, *s*, is for the secret key, repeated below:

```
<iframe src='https://webchat.botframework.com/embed/RockPaperScissors?s=YOUR_SECRET_HERE'>
</iframe>
```

If you add this *iframe* to a web page and use one of the secret keys for the *s* parameter, this will work and users can interact with the Webchat control on a web page just fine. However, this could be bad because it compromises your secret key. Anyone can look at the source code in the browser and read the secret key. You should do your own threat assessment to determine the impact of this, but I'm not going to show how to do that because of the potential problems it can cause you, starting in the next section.

Client-Side Coding

There are various ways to securely deploy the Webchat control without exposing secrets and I'll show you one of those. The approach in this section is to place a *div* placeholder in the HTML page and use jQuery to call the server and load the *iframe*. That keeps the secret on the server, where users can't see it. Listing 12-1 is the HTML page that does this.

LISTING 12-1 The HTML page for the Webchat control: *default.htm*.

```
<!DOCTYPE html>
<html>
<head>
    <title></title>
    <meta charset="utf-8" />
</head>
<body style="font-family:'Segoe UI'">
    <h1>Rock Paper Scissors</h1>

    <div id="webChatControl"/>

    <script src="https://code.jquery.com/jquery-3.2.1.min.js"></script>
    <script type="text/javascript">
        $("#webChatControl").load("api/WebChat");
    </script>
</body>
</html>
```

Listing 12-1 shows the modified default.htm page. Notice that the body contains a *div* whose *id* is *webChatControl*, which is a placeholder. Below that are a couple of script tags – the first loading jQuery. The second script tag contains a single jQuery statement that selects the *#webChatControl div* and loads the contents that comes back from an HTTP GET request to *api/WebChat*. The jQuery load function specifies a relative address, using the Web API convention of adding a first segment of *api*. The second segment specifies the controller and under Web API conventions, this would map to the controller class named *WebChatController*.

That's how the client obtains and loads the Webchat control onto the page. Next, let's examine *WebChatController* and how it responds to the request from the page.

> **Note** The Web Chat control is open-source and you can visit it at *https://github.com/Microsoft/BotFramework-WebChat*. In addition to the techniques I cover here, the GitHub site has additional ways to integrate the Web Chat control into yor site, such as with a React.js wrapper.

Handling the Server Request

As stated earlier, the benefit of avoiding putting the *iframe* directly in the HTML is to avoid putting secret keys on the page where any user can access it. Instead, we want to return an *iframe* that doesn't contain the secret key. The Bot Framework supports this by having a REST endpoint that exchanges the secret key for a token. This means we can pass the *iframe* back to the caller with the token, rather than the secret key. Each instance of the Webchat control gets a separate token and the token expires after a period of non-use, meaning that a user can't reuse that token like they could with the secret key. In this section, you'll see how to create the *WebChatController*, addressed in the client in the previous section, to handle swapping the secret key for a token and returning a more secure version of the *iframe*.

To start, copy one of the secret keys from the Webchat channel configuration to *web.config*. Here's what the *appSettings* element in *web.config* might look like:

```
<appSettings>
  <!-- update these with your BotId, Microsoft App Id and your Microsoft App Password-->
  <add key="BotId" value="RockPaperScissors" />
  <add key="MicrosoftAppId" value="Your App ID" />
  <add key="MicrosoftAppPassword" value="Your App Password" />
  <add key="WebChatSecret" value="Your Secret Key" />
</appSettings>
```

WebChatSecret holds the key and is read by the *WebChatController*.

To add a new controller, right-click the *Controllers* folder, select **Add | Controller**, click *Web API 2 Controller – Empty*, click the *Add* button, and the controller *WebChatController*, and click the *Add* button. This produces a new *WebChatController.cs* file. Listing 12-2 shows the *WebChatController* class with code that sends a new *iframe* back to the calling client (web page) and Listing 12-3 shows the *WebChatTokenResponse* that *WebChatController* uses to parse the token response from the Bot Framework.

LISTING 12-2 The controller that swaps secret keys for tokens: *WebChatController.cs*.

```csharp
using Newtonsoft.Json;
using RockPaperScissors3.Models;
using System.Configuration;
using System.Net;
using System.Net.Http;
using System.Text;
using System.Threading.Tasks;
using System.Web.Http;

namespace RockPaperScissors3.Controllers
{
    public class WebChatController : ApiController
    {
        public async Task<HttpResponseMessage> Get()
        {
            string webChatSecret = ConfigurationManager.AppSettings["WebChatSecret"];

            string result = await GetIFrameViaPostWithToken(webChatSecret);

            HttpResponseMessage response = Request.CreateResponse(HttpStatusCode.OK);
            response.Content = new StringContent(result, Encoding.UTF8, "text/html");
            return response;
        }

        async Task<string> GetIFrameViaPostWithToken(string webChatSecret)
        {
            var request = new HttpRequestMessage(
                HttpMethod.Post, "https://webchat.botframework.com/api/conversations");
            request.Headers.Add("Authorization", "BOTCONNECTOR " + webChatSecret);

            HttpResponseMessage response = await new HttpClient().SendAsync(request);
            string responseJson = await response.Content.ReadAsStringAsync();
            WebChatTokenResponse webChatResponse =
                JsonConvert.DeserializeObject<WebChatTokenResponse>(responseJson);

            return
                $"<iframe width='600px' height='500px' " +
                $"src='https://webchat.botframework.com/embed/RockPaperScissors" +
                $"?t={webChatResponse.Token}'>" +
                $"</iframe>";
        }
    }
}
```

LISTING 12-3 The class for helping parse the token: WebChatTokenResponse.cs

```
namespace RockPaperScissors3.Models
{
    public class WebChatTokenResponse
    {
        public string ConversationID { get; set; }
        public string Token { get; set; }
    }
}
```

The *Get* method in Listing 12-2 reads the secret key from configuration, calls the *GetIFrameViaPostWithToken* method, and sends the response back to the caller, which is the *default.htm* page in this example.

There are three main sections of *GetIFrameViaPostWithTokenMethod*: setting request parameters, requesting the response, and formatting the return string. The first section configures the request parameters:

```
var request = new HttpRequestMessage(
    HttpMethod.Post, "https://webchat.botframework.com/api/conversations");
request.Headers.Add("Authorization", "BOTCONNECTOR " + webChatSecret);
```

This request will send an HTTP POST to the *https://webchat.botframework.com/api/conversations* endpoint. The request must be authorized by appending the secret key to *BOTCONNECTOR* and assigning that to the *Authorization* header. Next, we need to make the HTTP request and parse the results:

```
HttpResponseMessage response = await new HttpClient().SendAsync(request);
string responseJson = await response.Content.ReadAsStringAsync();
WebChatTokenResponse webChatResponse =
    JsonConvert.DeserializeObject<WebChatTokenResponse>(responseJson);
```

This code calls *SendAsync* to post the request and reads the results into the *responseJson* string. Since this is a JSON document, we call *JsonConvert.DeserializeObject* to convert the JSON string into the *WebChatTokenResponse* object, whose type is defined in Listing 12-3. Next, we format the response:

```
return
    $"<iframe width='600px' height='500px' " +
    $"src='https://webchat.botframework.com/embed/RockPaperScissors" +
    $"?t={webChatResponse.Token}'>" +
    $"</iframe>";
```

This is an *iframe* similar to the one given by the Webchat control channel. The first difference is that its *height/width* is 600x500. You can change the dimensions how you like for your page. The second difference is that the s parameter is now *t*. Also, the *t* parameter value is the token returned from the REST call just made, *webChatResponse.Token*. Figure 12-5 shows what this looks like on the web page.

FIGURE 12-5 The Webchat control on a web page.

Summary

Now you know how to set up a few more channels. You learned how to configure the email channel. When working with email, remember that you need admin access to Office 365 to create an email account that the Bot Framework uses.

The Twilio channel supports configuring a chatbot to communicate over SMS text. The example we used, RockPaperScissors chatbot, was ideal for this type of communication because it's text only. This also provides access to users in the world who might not have a smart phone.

Finally, you learned how to add the Webchat control to a page. This is useful to put a chatbot anywhere an HTML page can go. The example focused on a secure way to save secret keys on the server, making the chatbot more secure.

You've seen examples of how to use pre-configured channels, but sometimes you might want to surface a chatbot in your own applications. The next chapter shows how to use Direct Line to create custom channels.

PART IV

APIs, Integrations, and Voice

CHAPTER 13	Coding Custom Channels with the Direct Line API 303
CHAPTER 14	Integrating Cognitive Services 321
CHAPTER 15	Adding Voice Services 345

This final part of the book is all about reaching outside of the Bot Framework and using related technologies to enhance a chatbot. You'll learn how to create custom channels.   Chatbots are also excellent applications for artificial intelligence (AI) services and you'll learn how to use Microsoft's Cognitive Services in this space. Finally, you've seen the benefits of conversation and natural language processing (NLP) for understanding human language. Now, you'll be able to add voice to that equation.

When existing channels don't meet your needs, you can read Chapter 13, *Coding Custom Channels with the Direct Line API*. Chapter 14, *Integrating Cognitive Services* shows how to add AI capabilities to a chatbot. The examples include a few ways to use Microsoft Cognitive Services . You'll even learn how to use the Microsoft QnA Maker for question and answer chatbots, which also uses NLP to understand user questions. Chapter 15, *Adding Voice Services* tops off the Bot Framework story by showing how to enable voice interaction with a chatbot. The underlying technology for this is a Bot Framework chatbot using the Corana channel, integrating with Cortana Skills Kit.

301

Combined with everything else you've learned in this book, this last part illustrates the true potential and vision of chatbots. You'll be able to write an intelligent chatbot, surfaced anywhere, that users can talk to.

CHAPTER 13

Coding Custom Channels with the Direct Line API

As you've seen in previous chapters, there are several channels to surface a chatbot upon. If the messaging channels don't fit your needs, there are email, SMS, and Webchat control channels. However, there are requirements where even those channels won't fit, and that's where the Direct Line API comes in.

The Direct Line API lets developers create their own custom channels. For example, you can use direct line to add a chatbot to a mobile app, so in addition to its basic functionality, the app can have a page where users can communicate with a chatbot. There are different applications for the enterprise too, with many programs written for Windows Forms and Windows Presentation Foundation (WPF), and there might be a requirement to add a tab or window to host a chatbot.

The Direct Line API is based on a REST interface, but this chapter uses the Microsoft Direct Line SDK, which is a NuGet package that any project can reference. The fact that this is a REST interface means that any platform and any language can use the Direct Line API. Essentially, you can surface a Bot Framework chatbot literally anywhere there's a capability to communicate via HTTP over the Internet. This epitomizes the multi-platform nature of the Microsoft Bot Framework. This chapter demonstrates the run anywhere concept via a custom Console Channel and the next section kicks off the chapter by talking about how that works.

Overview of the Console Channel

The example program for this chapter is the Console Channel. As its name suggests, this is a channel that exposes a chatbot on the command line. Before you dismiss the utility of a Console channel as a theoretical exercise, think about the patterns and trends developers increasingly engage in. There's a renewed focus on the command line as we work with project automation and various scripting shells. Think about how much time Windows administrators and developers spend in PowerShell. Recently, Microsoft has added the Windows Subsystem for Linux (WSL) on Windows 10, opening the path to various flavors of Linux operating systems such as Ubuntu and SUSE. Also, consider that .NET Core not only runs on Windows, but supports applications in both Linux and MacOS – places where command line work is common. Imagine a main frame developer or operator telnetting into a computer that hosts the Console channel and it becomes even more believable that this could have a purpose and make chatbots available to anyone on any platform.

Console Channel Components

In the downloadable source code the *ConsoleChannel* project has the logic that uses the Direct Line API to communicate with Wine Bot. The same solution also has an updated version of *WineBotLuis*, from Chapter 8, as the chatbot to communicate with. While this chapter uses Wine Bot as the example chatbot, a few quick changes allows the same code to work with any chatbot.

The design of Console Channel is based on the ability to divide the program into parts that make it easy to explain. Though it's likely that you and others would organize the code differently, the organization here is on understanding how to use the Direct Line API and making the explanation as simple as possible. Figure 13-1 shows the organization of the code into six major modules: *Program, Authenticate, Listen, Configure,* and *Prompt*. Each of these modules is a C# class in a .NET console application.

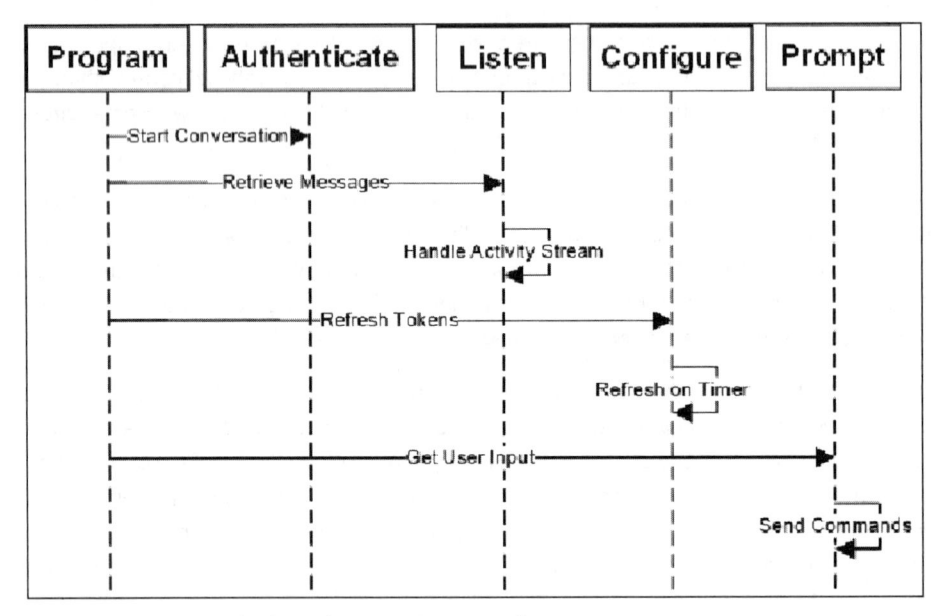

FIGURE 13-1 The *ConsoleChannel* Program Sequence diagram.

As shown in Figure 13-1, The *Program* class drives the whole application and calls into the other classes. *Authenticate* starts the conversation and returns values that the other classes use, including an expiring token. *Listen* starts a stream that processes chatbot messages as they arrive. Because the token received in *Authenticate* expires, *Configure* takes the responsibility for periodically refreshing that token to keep it alive while the program runs. Once all the other infrastructure is running, *Prompt* starts taking user input and sending it to the chatbot. Figure 13-2 shows the Console Channel in action.

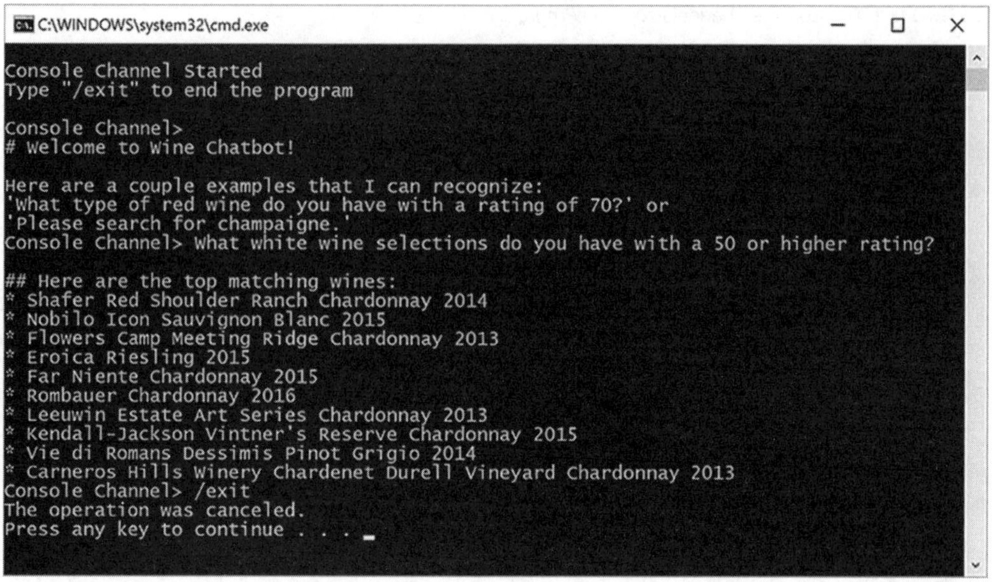

FIGURE 13-2 A user session with the console channel.

You should run this program to really see the sequence of events, but we explain what you'll be seeing. The program has prompts, *Console Channel>*, that are where the user can communicate with the chatbot. Before the first prompt, the Console Channel program shows a message that also includes the **/exit** command. Before the user has a chance to type anything (the first prompt is empty) Wine Bot responds with the welcome message. After that, the user types **What white wine selections do you have with a 50 or higher rating?** As an aside, this is the beauty of NLP with LUIS because it recognized a phrase it wasn't explicitly trained for. The response shows a list of available wines. Also notice the format of the Wine Bot responses. This is a benefit of markdown because it looks great on a surface that translates it graphically, but is also readable in plain text. The user ends the program with the **/exit** command and the program response indicates that the operation was cancelled—a subtle indicator of how the program ends the conversation.

Examining Console Channel Code

The Console Channel program, Listing 13-1, follows the same logic described in the Figure 13-1 sequence diagram. It asynchronously calls methods of each of the classes, representing the modules of the application. This program references the *Microsoft.Bot.Connector.DirectLine* NuGet package – a Microsoft library for accessing the Direct Line API. Many of the types used in this program are from this package.

CHAPTER 13   Coding Custom Channels with the Direct Line API   **305**

LISTING 13-1 The Console Channel Program: *Program.cs*

```csharp
using Microsoft.Bot.Connector.DirectLine;
using System.Threading;
using System.Threading.Tasks;

namespace ConsoleChannel
{
    class Program
    {
        static void Main() => new Program().MainAsync().GetAwaiter().GetResult();

        async Task MainAsync()
        {
            var cancelSource = new CancellationTokenSource();

            AuthenticationResults results =
                await Authenticate.StartConversationAsync(cancelSource.Token);

            await Listen.RetrieveMessagesAsync(results.Conversation, cancelSource);

            await Configure.RefreshTokensAsync(results.Conversation, results.Client, cancelSource.Token);

            await Prompt.GetUserInputAsync(results.Conversation, results.Client, cancelSource);
        }
    }
}
```

In Listing 13-1, *Main* invokes the *MainAsync* method. The first thing *MainAsync* does is instantiate a *CancellationTokenSource*, *cancelSource*, which is an argument to each method, providing a means to end the program. *StartConversationAsync* starts a new conversation with Direct Line and returns an instance of *AuthenticationResults*, *results*, containing the Direct Line library type, *results.Conversation*, holding several values the rest of the program needs to operate. The results variable also contains the *DirectLineClient* instance, *results.Client*, so other modules can share the same instance. The remaining methods operate via the general description given for Figure 13-1.

An important aspect of this program's design is multi-threading. This program runs on three threads:

1. The *Program*, *Authenticate*, and *Prompt* class methods run on the main thread.
2. The *Listen* class methods run on a second thread.
3. The *Configure* class methods run on a third thread.

This simplifies the program because each thread has a unique responsibility. The first thread starts the conversation, calls each method, and makes its way to *GetUserInputAsync*, which you'll learn later runs a loop to continuously accept user input until they choose to end the program. The *RetrieveMessagesAsync* method starts a new thread that opens the input stream from the chatbot and waits to process each message as it arrives. *RefreshTokensAsync* starts a new thread that delays and wakes up in time to refresh the current conversation token so the user can seamlessly continue communicating with the chatbot.

> **Note** The alternative to the multi-threaded design is to weave the functionality of these three threads together into one and perform polling, which would be inherently inefficient. When you consider the type of error handling and instrumentation required to make a real-world implementation work, a single threaded approach has the potential for more complexity than what you might think at first glance. Also, remember that this is a console application and the implementation will be much different in another technology like WPF, UWP, or a multi-platform mobile toolkit like Xamarin.

Some of the modules use a shared class, *Message*, that holds some common code, shown in Listing 13-2. Again, your choice for shared data might differ, but this is a simplicity-first approach, rather than attempting to adhere to one of the many opinions on the subject.

LISTING 13-2 The Console Channel Program: *Message.cs*

```csharp
using System;

namespace ConsoleChannel
{
    static class Message
    {
        public const string ClientID = "ConsoleChannel";
        public const string ChatbotID = "WineChatbot";

        volatile static string watermark;

        public static string Watermark
        {
            get { return watermark; }
            set { watermark = value; }
        }

        public static void WritePrompt() => Console.Write("Console Channel> ");
    }
}
```

Listing 13-2 has two constants: *ClientID* and *ChatbotID*. *ConsoleChannel* uses *ClientID* as the *From* ID when communicating with the chatbot. This is hard-coded, rather than writing code for users to log in. Your program might have users that log in and you'll have a user name and ID to use instead. The *ChatbotID*, *WineChatbot*, is the registered handle with the Bot Framework for Wine Bot. If creating a generic channel that can communicate with any chatbot, this would be configurable.

An important part of communicating with a chatbot via Direct Line is the *Watermark*, which is a property, wrapping the *watermark* field. The *watermark* indicates which message we received from the chatbot and helps avoid duplicate messages. Both of the threads that listen for new messages and accept user input use *Watermark*. Because it's a shared field, used by multiple fields, we used the *volatile* modifier to minimize incomplete reads and writes. Depending on your implementation, this might or might not work for you. In this example, we only care that we read the value consistently. Knowing that there's still potential for duplicates, Channel Console has mitigating code, which you'll see later in this chapter, to prevent showing any duplicates to the user.

> **Tip** For more information on multi-threaded programming, Microsoft Press has an excellent book: CLR via C# (*https://aka.ms/clrcsbook*). The book's author, Jeffrey Richter, is one of the foremost experts on multi-threading in the Windows and .NET domains and this book has a thorough discussion on the topic for .NET developers.

Various parts of the program call *WritePrompt*, which is a single location for a consistent user prompt. The remaining sections of this chapter explain how each of these modules work, with the next section explaining how to start a conversation.

Starting a Conversation

As mentioned earlier, this application uses the Microsoft Direct Line NuGet package. The code in Listing 13-3 shows how to use that package to instantiate a *DirectLineClient* and start a conversation. We've called the containing class *Authenticate* because starting a conversation also authenticates at the same time.

LISTING 13-3 The Console Channel Program: *Authenticate.cs*

```
using Microsoft.Bot.Connector.DirectLine;
using System.Configuration;
using System.Threading;
using System.Threading.Tasks;

namespace ConsoleChannel
{
    class AuthenticationResults
    {
        public Conversation Conversation { get; set; }
        public DirectLineClient Client { get; set; }
    }

    class Authenticate
    {
        public static async Task<AuthenticationResults>
            StartConversationAsync(CancellationToken cancelToken)
        {
            System.Console.WriteLine(
                "\nConsole Channel Started\n" +
                "Type \"/exit\" to end the program\n");
```

```
            Message.WritePrompt();

            string secret = ConfigurationManager.AppSettings["DirectLineSecretKey"];
            var client = new DirectLineClient(secret);
            Conversation conversation =
                await client.Conversations.StartConversationAsync(cancelToken);

            return
                new AuthenticationResults
                {
                    Conversation = conversation,
                    Client = client
                };
        }
    }
}
```

StartConversationAsync accepts a *CancellationToken* and returns an instance of *AuthenticationResults*, which contains *Conversation* and *DirectLineClient*. We discuss the *cancelToken* in a later part of the chapter on ending a conversation, but you'll notice that it's passed to all async method calls throughout the program.

The first part of the *StartConversationAsync* method lets a user know that the program started and sends the prompt to the screen, letting the user know that the program is ready for input.

The *app.config* configuration file contains an *appSettings* key for *DirectLineSecretKey*. You can get this key by visiting the Bot Framework site for the chatbot you want to communicate with, Wine Bot, and creating a Direct Line channel. Chapters 11 and 12 showed how to configure several channels and this process is similar. Just create the channel and copy a secret key into an *appSettings* entry, like below:

```
<appSettings>
  <add key="DirectLineSecretKey" value="Your secrect key goes here"/>
</appSettings>
```

The *StartConversationAsync* method passes that secret key as an argument to instantiate a new *DirectLineClient*. This is the *DirectLineClient* instance returned to the caller.

The *DirectLineClient* type has a *Conversations* property, representing the */conversations* segment of the REST endpoint URL. Calling *StartConversationAsync*, the Direct Like API returns a *Conversations* object, holding several values required in the rest of the program. You'll see what these values are and how they're used in context in the following sections of this chapter.

With a started conversation, the program starts a new thread for listening for new activities from the chatbot.

Listening for New Activities

Direct Line offers a couple of ways to receive chatbot messages: polling and stream. The polling approach uses a *GetActivitiesAsync* method that returns a set of activities from the chatbot. The thing is that you need to continue polling and the activities returned are as fresh as the time between polls. Polling too frequently results in wasted bandwidth and might slow down a program and polling with too much of a delay leaves a less than ideal experience for the user.

The most efficient technique for receiving new activities from a chatbot is via a stream, which is an implementation of Web Sockets. This means that the program receives a response as soon as the chatbot sends it out, ensuring efficient operation and a responsive user experience. Fortunately, the .NET Framework has support for Web Sockets that the Console Channel program uses, as shown in Listing 13-4.

LISTING 13-4 The Console Channel Program: *Listen.cs*

```
using Microsoft.Bot.Connector.DirectLine;
using Newtonsoft.Json;
using System;
using System.Collections.Generic;
using System.Linq;
using System.Net.WebSockets;
using System.Text;
using System.Threading;
using System.Threading.Tasks;

namespace ConsoleChannel
{
    class Listen
    {
        public static async Task RetrieveMessagesAsync(
            Conversation conversation, CancellationTokenSource cancelSource)
        {
            const int ReceiveChunkSize = 1024;

            var webSocket = new ClientWebSocket();
            await webSocket.ConnectAsync(
                new Uri(conversation.StreamUrl), cancelSource.Token);

            var runTask = Task.Run(async () =>
            {
                try
                {
                    while (webSocket.State == WebSocketState.Open)
                    {
                        var allBytes = new List<byte>();
                        var result = new WebSocketReceiveResult(0, WebSocketMessageType.Text, false);

                        byte[] buffer = new byte[ReceiveChunkSize];

                        while (!result.EndOfMessage)
                        {
```

```csharp
                        result = await webSocket.ReceiveAsync(
                            new ArraySegment<byte>(buffer), cancelSource.Token);

                        allBytes.AddRange(buffer);
                        buffer = new byte[ReceiveChunkSize];
                    }

                    string message = Encoding.UTF8.GetString(allBytes.ToArray()).Trim();
                    ActivitySet activitySet = JsonConvert.DeserializeObject<ActivitySet>(message);

                    if (activitySet != null)
                        Message.Watermark = activitySet.Watermark;

                    if (CanDisplayMessage(message, activitySet, out List<Activity> activities))
                    {
                        Console.WriteLine();
                        activities.ForEach(activity => Console.WriteLine(activity.Text));

                        Message.WritePrompt();
                    }
                }
            }
            catch (OperationCanceledException oce)
            {
                Console.WriteLine(oce.Message);
            }
        });
    }

    static bool CanDisplayMessage(string message, ActivitySet activitySet, out List<Activity> activities)
    {
        if (activitySet == null)
            activities = new List<Activity>();
        else
            activities =
                (from activity in activitySet.Activities
                 where activity.From.Id == Message.ChatbotID &&
                     !string.IsNullOrWhiteSpace(activity.Text)
                 select activity)
                .ToList();

        SuppressRepeatedActivities(activities);

        return !string.IsNullOrWhiteSpace(message) && activities.Any();
    }

    static Queue<string> processedActivities = new Queue<string>();
    const int MaxQueueSize = 10;

    static void SuppressRepeatedActivities(List<Activity> activities)
    {
        foreach (var activity in activities)
```

```
            {
                if (processedActivities.Contains(activity.Id))
                {
                    activities.Remove(activity);
                }
                else
                {
                    if (processedActivities.Count >= 10)
                        processedActivities.Dequeue();

                    processedActivities.Enqueue(activity.Id);
                }
            };
        }
    }
}
```

Listing 13-4 is quite extensive, as you might expect because it launches a new thread that implements a loop, handling the Direct Line stream through Web Sockets. The *RetrieveMessagesAsync* method starts by instantiating a *WebSocket* and connecting, as shown below:

```
var webSocket = new ClientWebSocket();
await webSocket.ConnectAsync(
    new Uri(conversation.StreamUrl), cancelSource.Token);
```

The *Uri* argument to *ConnectAsync* uses the *StreamUrl* from the conversation parameter. This is the same *Conversation* instance that *StartConversation* returned. The code uses the Task Parallel Library (TPL) *Task.Run* to start the rest of the code on a new thread. There's also a *while* loop that iterates as long as the *WebSocket* is in the open state.

The Direct Line API returns chunks of text in blocks of 1024 bytes, so we need a loop that collects each block until the entire set of activities delivers, shown below:

```
var allBytes = new List<byte>();
var result = new WebSocketReceiveResult(0, WebSocketMessageType.Text, false);

byte[] buffer = new byte[ReceiveChunkSize];

while (!result.EndOfMessage)
{
    result = await webSocket.ReceiveAsync(
        new ArraySegment<byte>(buffer), cancelSource.Token);

    allBytes.AddRange(buffer);
    buffer = new byte[ReceiveChunkSize];
}
```

The *result* variable is an instance of *WebSocketReceiveResult*. The *ReceiveAsync* method waits for the next available set of activities from the chatbot and returns a *WebSocketReceiveResult* to indicate current status, causing the loop to continue until *EndOfMessage* is *true*. *ReceiveAsync* also populates *buffer*, through the *ArraySegment* instance. The *allBytes* collects all of the bytes returned for later deserialization.

Notice that the last line of the loop re-instantiates *buffer*, which is important because buffer retains the contents of the previous call to *ReceiveAsync*. You would receive garbage on any final loop, from the previous buffer contents, for a set of activities where the number of bytes are less than 1024 bytes, which is frequent.

When the code receives a full activity, *result.EndOfMessage* is *true*, it stops executing the while loop and processes the activity, repeated below:

```
string message = Encoding.UTF8.GetString(allBytes.ToArray()).Trim();
ActivitySet activitySet = JsonConvert.DeserializeObject<ActivitySet>(message);

if (activitySet != null)
    Message.Watermark = activitySet.Watermark;

if (CanDisplayMessage(message, activitySet, out List<Activity> activities))
{
    Console.WriteLine();
    activities.ForEach(activity => Console.WriteLine(activity.Text));
    Message.WritePrompt();
}
```

The Direct Line API returns data in a UTF-8 format and the code uses that to convert bytes to a string, which is a JSON object. This JSON object represents a set of activities, called an *ActivitySet*, which is a type in the Direct Line library.

ActivitySet also has a *Watermark* property. If you recall from the *Message* class, Listing 13-2, discussion, *Watermark* keeps track of messages to help avoid duplicates. In our case, we're minimizing duplicates and you'll see how that happens soon. The *CanDisplayMessage*, shown below, accepts the *message* and *activitySet* arguments and returns a *List<Activity>*, *activities*, that can be displayed to the user:

```
static bool CanDisplayMessage(string message, ActivitySet activitySet, out List<Activity> activities)
{
    if (activitySet == null)
        activities = new List<Activity>();
    else
        activities =
            (from activity in activitySet.Activities
             where activity.From.Id == Message.ChatbotID &&
                 !string.IsNullOrWhiteSpace(activity.Text)
             select activity)
            .ToList();

    SuppressRepeatedActivities(activities);

    return !string.IsNullOrWhiteSpace(message) && activities.Any();
}
```

The Direct Line API sends empty *messages/activities* as a form of keep-alive message to keep the stream open during periods of inactivity. So, the code has to detect these keep-alive messages. When *activitySet* has values in its *Activities* property, the LINQ statement filters to make sure that only activities with text can appear. It also ensures that the message received is from the chatbot, rather than the Bot Framework, or replays of the user's message. The method returns *false* if the message was a keep-alive message or none of the activities passed the LINQ filter. Prior to returning, the code also makes sure we don't show the user any duplicates that might have slipped through, via the *SuppressRepeatedActivities* method as follows:

```csharp
static Queue<string> processedActivities = new Queue<string>();
const int MaxQueueSize = 10;

static void SuppressRepeatedActivities(List<Activity> activities)
{
    foreach (var activity in activities)
    {
        if (processedActivities.Contains(activity.Id))
        {
            activities.Remove(activity);
        }
        else
        {
            if (processedActivities.Count >= 10)
                processedActivities.Dequeue();

            processedActivities.Enqueue(activity.Id);
        }
    };
}
```

> **Note** In addition to message activities, Direct Line supports a large subset of other activity types. It doesn't support *contactRelationUpdate*. Also, since the Bot Connector takes care of *conversationUpdate*, you don't need to (so, *conversationUpdate* isn't available either). The *typing* activity is only available via web sockets, but not polling (HTTP GET). All other activities are supported via both polling and web sockets.

The Direct Line API tends to be overprotective to ensure it doesn't lose any messages, so it's probable that you'll receive duplicates. This program keeps track of the most recent messages and doesn't show a message that has already been shown to a user. The implementation simulates a circular buffer with a *Queue<string>*, *processedActivities*. We chose the size to be 10 because more than 10 duplicates for Wine Bot is unlikely. Generally, you want to minimize the size to prevent resource waste from too many queue searches, yet large enough to cover message bursts from the chatbot to avoid duplicates. The code checks *processedActivities* for each of the activities and removes duplicates. If the queue is larger than max, it dequeues the oldest activity ID before enqueueing the current activity ID.

That was how you can receive messages from a chatbot. Next, let's look at how to keep the conversation going.

Keeping the Conversation Open

The *Authenticate* module, discussed previously, started the conversation. One of the outputs of that process was the Direct Line type, *Conversation*, which holds *ConversationId* and *ExpiresIn* properties. Because the conversation expires, in the number of seconds indicated by *ExpiresIn*, this program takes a pro-active approach and keeps the conversation alive. An alternative is to wait until the conversation closes and re-open. The pro-active approach minimizes interruption to the user because disconnections typically involve timeouts and network latency that might result in a degraded user experience. It's your choice and this discussion is to help you think about the trade-offs when designing a custom channel. The *Configure* class in Listing 13-5 implements the *Configure* module from Figure 13-1, starting a new thread to handle token refreshes.

LISTING 13-5 The Console Channel Program: *Configure.cs*

```csharp
using Microsoft.Bot.Connector.DirectLine;
using System;
using System.Threading;
using System.Threading.Tasks;

namespace ConsoleChannel
{
    class Configure
    {
        public static async Task RefreshTokensAsync(
            Conversation conversation, DirectLineClient client, CancellationToken cancelToken)
        {
            const int ToMilliseconds = 1000;
            const int BeforeExpiration = 60000;

            var runTask = Task.Run(async () =>
            {
                try
                {
                    int millisecondsToRefresh =
                        ((int)conversation.ExpiresIn * ToMilliseconds) - BeforeExpiration;

                    while (true)
                    {
                        await Task.Delay(millisecondsToRefresh);

                        await client.Conversations.ReconnectToConversationAsync(
                            conversation.ConversationId,
                            Message.Watermark,
                            cancelToken);
                    }
                }
                catch (OperationCanceledException oce)
                {
                    Console.WriteLine(oce.Message);
                }
            });
```

```
            await Task.FromResult(0);
        }
    }
}
```

The *RefreshTokensAsync* method in Listing 13-5 starts a new thread with *Task.Run* that loops and periodically updates the tokens associated with the current conversation. The *DirectLineClient* instance has a *Tokens* property that contains the tokens obtained during the call to *StartConversationAsync*. The program shares the same *DirectLineClient* instance with multiple methods, preventing the need to re-instantiate a new object every time we want to call a method.

The *ToMilliseconds* constant helps convert the *ExpiresIn* seconds to milliseconds. Subtracting the *BeforeExpiration* constant sets the number of milliseconds to wait before refreshing the token. The goal here is to do the refresh for a certain amount of time before expiration to avoid ending the conversation and you might experiment with this value to account for any delays or network latency between your code and the Direct Line API. You can see the *Task.Delay*, taking *millisecondsToRefresh* as a makeshift timer between refreshes in the *while* loop.

ReconnectToConversationAsync performs the refresh and keeps the conversation open. Notice that it's also passing the *Watermark* as an argument. This is why you want to think about how to synchronize access to *Watermark* because there are two threads reading or writing to it. See the discussion in the previous section for more information on how *Watermark* is used there.

Now that we're receiving real-time messages from the chatbot and are keeping the conversation open, let's look at how to send user input to the chatbot.

Sending Activities

While separate threads are running for receiving tasks and keeping the program alive, the main thread makes its way down to the *Prompt* class, which implements the *Prompt* module from Figure 13-1. *Prompt* starts a loop that takes user input and sends that input to the chatbot. Listing 13-6 shows the *Prompt* implementation.

LISTING 13-6 The Console Channel Program: *Prompt.cs*

```
using Microsoft.Bot.Connector.DirectLine;
using System;
using System.Net.Http;
using System.Threading;
using System.Threading.Tasks;

namespace ConsoleChannel
{
    class Prompt
    {
        public static async Task GetUserInputAsync(
```

```csharp
            Conversation conversation, DirectLineClient client, CancellationTokenSource cancelSource)
        {
            string input = null;

            try
            {
                while (true)
                {
                    input = Console.ReadLine().Trim().ToLower();

                    if (input == "/exit")
                    {
                        await EndConversationAsync(conversation);
                        cancelSource.Cancel();
                        await Task.Delay(500);
                        break;
                    }

                    if (string.IsNullOrWhiteSpace(input))
                    {
                        Message.WritePrompt();
                    }
                    else
                    {
                        IMessageActivity activity = Activity.CreateMessageActivity();
                        activity.From = new ChannelAccount(Message.ClientID);
                        activity.Text = input;

                        await client.Conversations.PostActivityAsync(
                            conversation.ConversationId,
                            activity as Activity,
                            cancelSource.Token);
                    }
                }
            }
            catch (OperationCanceledException oce)
            {
                Console.WriteLine(oce.Message);
            }
        }

        static async Task EndConversationAsync(Conversation conversation, DirectLineClient client)
        {
            IEndOfConversationActivity activity = Activity.CreateEndOfConversationActivity();
            activity.From = new ChannelAccount(Message.ClientID);

            await client.Conversations.PostActivityAsync(
                conversation.ConversationId, activity as Activity);
        }
    }
}
```

The *GetUserInputAsync* method in Listing 13-6 handles the task of obtaining user input and sending it to the chatbot. The *while* loop continuously prompts the user for input until the user types **/exit** and we'll discuss how that code works in the next section.

The *if* statement prevents sending empty text to the chatbot and writes a new prompt. If user input does contain text, the code builds a new *Activity* instance. The *From* property receives a *ChannelAccount* instance, with an *ID* set to *Message.ClientID*. If you recall from previous discussions, your implementation might have a user log in, thus passing your user ID and user name in the *ChannelAccount* instance. Finally, *PostAsync* sends the activity, containing user input, to the chatbot.

So far, you've seen how this program receives chatbot messages, keeps the conversation alive, and sends user input to the chatbot. Next, let's discuss what happens when the program ends.

Ending Conversations

.NET has formalized support for cancelling async and multi-threaded applications through the *CancellationTokenSource* and *CancellationToken* types. This program uses that cancellation support for handling when a user wants to exit the conversation. In this section, we'll discuss the .NET cancellation support and then show how ending a conversation works.

Examining *CancellationTokenSource* and *CancellationToken*

Listing 13-1 shows how *MainAsync* creates a *CancellationTokenSource* instance, *cancelSource*. This *CancellationTokenSource* instance is responsible for notifying all threads when they should stop running. In this example, the purpose of stopping threads is to close the program.

Notice that *RetrieveMessagesAsync* and *GetUserInputAsync* accept the *cancelSource* instance, giving them the ability to cancel all threads in addition to being notified when their threads should be canceled. *StartConversationAsync* and *RefreshTokensAsync* receive *cancelSource.Token*, where *Token* is type *CancellationToken*. Having a *CancellationToken* gives the methods the ability to know when their thread is being canceled. Let's examine how a couple of these methods use *CancellationToken*.

The *StartConversationAsync* method, Listing 13-3, passes the *cancelToken* parameter to the Direct Line *StartConversationAsync*.

```
Conversation conversation =
    await client.Conversations.StartConversationAsync(cancelToken);
```

The *RefreshTokenAsync* method, Listing 13-5, passes the *cancelToken* parameter to the Direct Line *ReconnectToConversationAsync*.

```
await client.Conversations.ReconnectToConversationAsync(
    conversation.ConversationId,
    Message.Watermark,
    cancelToken);
```

In each of these cases when the *CancellationTokenSource* is told to cancel, each of the async methods, if currently running, receives a notification through *cancelToken* that they should stop running. This lets the program stop running all threads simultaneously and shut down gracefully.

> **Note** How graceful a thread shuts down upon receiving a cancellation notice, is up to you for the code you've written. 3rd party developer libraries might also have guidance on how to end their threads.

Next, let's discuss what happens when the user wants to end the conversation.

Handling User Exits

When a user wants to stop the conversation, they type **/exit**. This command goes to the *GetUserInputAsync* method, which handles all user input. The logic to handle the **/exit** command is below, repeated from Listing 13-6:

```
if (input == "/exit")
{
    await EndConversationAsync(conversation);
    cancelSource.Cancel();
    await Task.Delay(500);
    break;
}
```

After calling *EndConversationAsync*, the code uses its *CancellationTokenSource* parameter, *cancelSource*, to *Cancel* all threads. Since all methods receive the same *CancellationTokenSource* instance or the *CancellationToken* from that instance, they will all be notified to cancel. The previous section showed how the *RetrieveMessagesAsync* method handles cancellation when this happens. This code also adds a 500 millisecond delay before exiting, just in case the other threads need a little more time because the break causes the current loop to exit, return to *MainAsync*, return to *Main* and end the program. You might want to tweak this, depending on how you decide to manage the lifetime of your own threads.

The *EndConversationAsync* method, repeated from Listing 13-6 below, shows how to let Direct Line know that the user wants to end the conversation:

```
static async Task EndConversationAsync(Conversation conversation, DirectLineClient client)
{
    IEndOfConversationActivity activity = Activity.CreateEndOfConversationActivity();
    activity.From = new ChannelAccount(Message.ClientID);

    await client.Conversations.PostActivityAsync(
        conversation.ConversationId, activity as Activity);
}
```

The Direct Line library, from the NuGet package, has methods to handle most scenarios, but doesn't handle ending a conversation. So, *EndConversationAsync* uses *client.Conversations.PostActivityAsync* to send an *IEndOfConversationActivity* to the chatbot.

> **Tip** You can use the technique in *EndConversationAsync* to send other types of non-message activities to a chatbot.

> **Note** In addition to the services discussed in this chapter, you can attach files also by using the *client.Conversations.UploadAsync* method, where client is an instance of the *DirectLineClient* type.

Summary

Now you know how to build a client with the Bot Framework Direct Line API. The demo program in this chapter is a Console Channel. While minimal to make the demo simple, this program demonstrates that you can expose a chatbot on literally any platform where code can make HTTP calls.

This code was organized for learning and there were several tips along the way to help consider how to design your own custom channel implementation. In particular this code runs on three threads where the main thread does set up and then gets user input. The other two threads receive user input and keep the conversation open. Finally, you saw how to exit the program using the standard .NET *CancellationTokenSource* for graceful shutdown.

The last few chapters have discussed built-in, 3rd party, and custom channels. In the next chapter we'll move from client side, back to server side development and discuss how to add intelligence to a chatbot with Cognitive Services.

CHAPTER 14

Integrating Cognitive Services

Developers can write plenty of chatbots with logic that presents options, accepts a user's response, and implements logic to respond to a user's request. This code doesn't fall into the category of algorithms considered as Artificial Intelligence (AI). Most of the chatbots in previous chapters of this book don't use any AI. The exception was Chapter 8, which used Natural Language Processing (NLP) with Language Understanding Intelligence Service (LUIS). NLP is a form of AI and Wine Bot used LUIS to translate human sentences into computer readable intents.

LUIS is one of many services that Microsoft offers under a suite of AI REST Application Programming Interfaces (APIs) called the Cognitive Services. The Cognitive Services are organized into several categories including Vision, Speech, Language, Knowledge, and Search. As this chapter is written, there are already over 25 services and more on the way.

In practical terms, all of the Cognitive Services and their details can't be covered in a single chapter. Rather, this chapter takes a small sub-set of what's available to explain how Cognitive Services can enhance chatbots written with the Bot Framework. We'll start off in the Search category, requesting news about a particular music artist. Then you'll see how to use the description capabilities in the Vision category to obtain a narrative for a music album cover. If you're interested in supporting multiple locales, the Speech category has text translation APIs and you'll see examples of how that works with Wine queries. To round out the chapter, you'll learn how to add a FAQ to a chatbot by using the QnA Maker.

Searching with Bing

Search is a useful technique for various types of chatbots. The Cognitive Services use Bing for search, enhanced with capabilities for different types of searches. For example, you can search videos, images, news, and more.

These search APIs are ideal for enhancing an existing chatbot with content specific queries. Another technique you'll see with a lot of chatbots is handling the cases where users ask questions that the chatbot doesn't know how to answer, falling back to a search based on the user's query. You've probably seen Cortana fall back to a search when unable to answer a question directly.

This section uses search to enhance Music Bot, reused from earlier chapters. It adds a News button to the Browse response, returning News articles on the music artist when clicked. To get started, you'll need to create a service in Azure by clicking *Create New Service, AI + Cognitive Services, Bing Search*

APIs, and fill out the forms to create the service. Listing 14-1 is the *CognitiveService* class, showing how to call that service.

LISTING 14-1 Calling the Bing Search Service: *CognitiveService.cs*

```csharp
using AILib.Models;
using Newtonsoft.Json;
using System;
using System.Configuration;
using System.Linq;
using System.Net.Http;
using System.Text;
using System.Threading.Tasks;
using System.Xml.Linq;

namespace AILib
{
    [Serializable]
    public class CognitiveService
    {
        const string AccessKey = "Ocp-Apim-Subscription-Key";

        public async Task<NewsArticles> SearchForNewsAsync(string artistName)
        {
            const string BaseUrl = "https://api.cognitive.microsoft.com/bing/v5.0";
            string url = $"{BaseUrl}/news/search?" +
                $"q={Uri.EscapeUriString(artistName)}&" +
                $"category=Entertainment_Music&" +
                $"count=5";

            string accessKey = ConfigurationManager.AppSettings["SearchKey"];
            var client = new HttpClient();
            client.DefaultRequestHeaders.Add(AccessKey, accessKey);

            string response = await client.GetStringAsync(url);

            NewsArticles articles = JsonConvert.DeserializeObject<NewsArticles>(response);
            return articles;
        }
    }
}
```

In the downloadable source code, you'll find an *AILib* project containing the *CognitiveService* class, shown in Listing 14-1 along with supporting model classes. I removed methods from this class for clarity, but you'll see them in subsequent listings when we discuss other Cognitive Services.

The *BaseUrl* shows how the services are versioned, the current version being *5.0* and that will likely change with future enhancements. Remember to escape URI strings, as we did with *artistName*. Each of these parameters and more are documented clearly with the Cognitive Services API documentation. In addition to the query, *q*, the *category* says to look in the *Music* sub-category of the *Entertainment* category and the *count* is set to return only 5 results.

The Azure service blade for this service has a *Manage Keys* link, shown in Figure 14-1, that you should click and copy the key into a *web.config appSettings* element, named *SearchKey*. The *AccessKey*, *Ocp-Apim-Subscription-Key* is an HTTP header sent with the request for authorization. The code deserializes the response into a custom *NewsArticles* type, containing the list of articles.

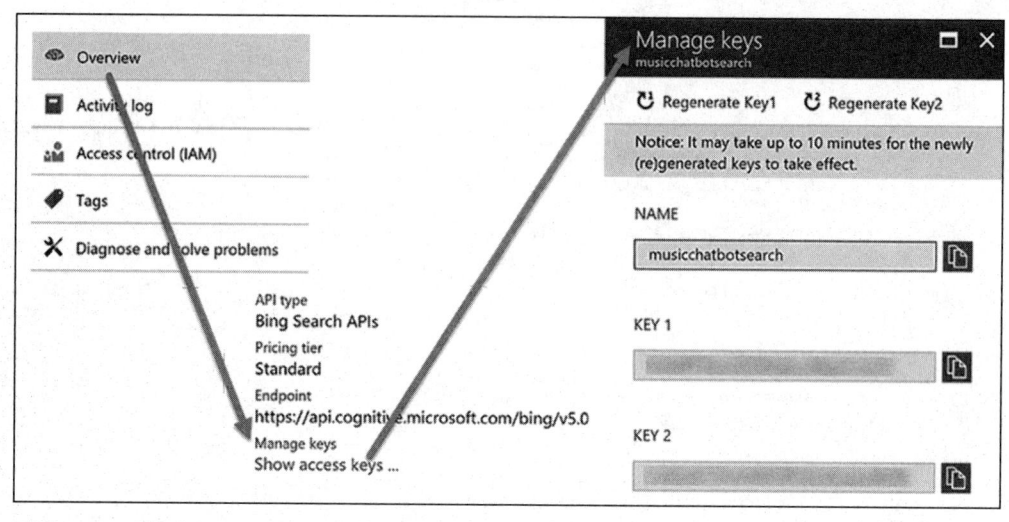

FIGURE 14-1 Obtaining Keys from Cognitive Services on Azure.

Listing 14-2 shows the *NewsDialog*, which calls *SearchForNewsAsync*. The *NewsDialog* displays a carousel of cards with news items. The user can click any of the articles to read the news article through their browser.

LISTING 14-2 Searching the Bing Service: *NewsDialog.cs*

```
using AILib;
using AILib.Models;
using Microsoft.Bot.Builder.Dialogs;
using Microsoft.Bot.Connector;
using MusicChatbot.Models;
using MusicChatbot.Services;
using Newtonsoft.Json;
using System;
using System.Collections.Generic;
using System.Linq;
using System.Threading.Tasks;

namespace MusicChatbot.Dialogs
{
    [Serializable]
    public class NewsDialog : IDialog<object>
    {
        const string DoneCommand = "Done";
        readonly string artistName;

        public NewsDialog(string artistName)
```

CHAPTER 14 Integrating Cognitive Services 323

```csharp
        {
            this.artistName = artistName;
        }

        public Task StartAsync(IDialogContext context)
        {
            context.Wait(MessageReceivedAsync);
            return Task.CompletedTask;
        }

        async Task MessageReceivedAsync(IDialogContext context, IAwaitable<object> result)
        {
            var activity = await result as IMessageActivity;

            if (activity.Text == DoneCommand)
            {
                context.Done(this);
                return;
            }

            NewsArticles articles = await new CognitiveService().SearchForNewsAsync(artistName);

            var reply = (context.Activity as Activity)
                .CreateReply($"## Reading news about {artistName}");

            List<ThumbnailCard> cards = GetHeroCardsForArticles(articles);
            cards.ForEach(card =>
                reply.Attachments.Add(card.ToAttachment()));

            ThumbnailCard doneCard = GetThumbnailCardForDone();
            reply.Attachments.Add(doneCard.ToAttachment());

            reply.AttachmentLayout = AttachmentLayoutTypes.Carousel;

            await
                new ConnectorClient(new Uri(reply.ServiceUrl))
                    .Conversations
                    .SendToConversationAsync(reply);

            context.Wait(MessageReceivedAsync);
        }

        List<ThumbnailCard> GetHeroCardsForArticles(NewsArticles articles)
        {
            var cards =
                (from article in articles.Value
                 select new ThumbnailCard
                 {
                     Title = article.Name,
                     Subtitle = "About: " + article.About.FirstOrDefault()?.Name,
                     Text = article.Description,
                     Images = new List<CardImage>
                     {
                         new CardImage
                         {
                             Alt = article.Description,
```

```csharp
                        Tap = BuildViewCardAction(article.Url),
                        Url = article.Image.Thumbnail.ContentUrl
                    }
                },
                Buttons = new List<CardAction>
                {
                    BuildViewCardAction(article.Url)
                }
            })
        .ToList();
    return cards;
}

CardAction BuildViewCardAction(string url)
{
    return new CardAction
    {
        Type = ActionTypes.OpenUrl,
        Title = "Read",
        Value = url
    };
}

ThumbnailCard GetThumbnailCardForDone()
{
    return new ThumbnailCard
    {
        Title = DoneCommand,
        Subtitle = "Click/Tap to exit",
        Images = new List<CardImage>
        {
            new CardImage
            {
                Alt = "Smile",
                Tap = BuildDoneCardAction(),
                Url = new FileService().GetBinaryUrl("Smile.png")
            }
        },
        Buttons = new List<CardAction>
        {
            BuildDoneCardAction()
        }
    };
}

CardAction BuildDoneCardAction()
{
    return new CardAction
    {
        Type = ActionTypes.PostBack,
        Title = DoneCommand,
        Value = DoneCommand
    };
}
```

Listing 14-2 shows a carousel of cards, which you've seen in earlier chapters for Music Bot. The difference is that the news cards are type *ThumbnailCard* with buttons for viewing each article. The *BuildViewCardAction* method creates a *CardAction* for *OpenUrl* on the URL associated with the news article.

Listing 14-3 shows how *NewsDialog* opens. It's the *BrowseDialog* from earlier chapter, with the addition of a *CardAction* for viewing news for an artist.

LISTING 14-3   The News Browsing Option: *BrowseDialog.cs*

```
using AILib;
using Microsoft.Bot.Builder.Dialogs;
using Microsoft.Bot.Connector;
using MusicChatbot.Models;
using MusicChatbot.Services;
using Newtonsoft.Json;
using System;
using System.Collections.Generic;
using System.Linq;
using System.Threading.Tasks;

namespace MusicChatbot.Dialogs
{
    [Serializable]
    public class BrowseDialog : IDialog<object>
    {
        const string DoneCommand = "Done";
        const string NewsCommand = "News";

        public Task StartAsync(IDialogContext context)
        {
            context.Wait(MessageReceivedAsync);
            return Task.CompletedTask;
        }

        async Task MessageReceivedAsync(IDialogContext context, IAwaitable<object> result)
        {
            var activity = await result as IMessageActivity;
            string heroCardValue = activity?.Text;
            if (!string.IsNullOrWhiteSpace(heroCardValue) && heroCardValue.StartsWith("{"))
            {
                var news = JsonConvert.DeserializeObject<Item>(heroCardValue);
                string artistName = news.Artists.First().Artist.Name;
                await context.Forward(
                    new NewsDialog(artistName),
                    MessageReceivedAsync,
                    activity);
            }
            else
            {
                List<string> genres = new GrooveService().GetGenres();
                genres.Add("Done");
                PromptDialog.Choice(context, ResumeAfterGenreAsync, genres, "Which music category?");
```

```csharp
        }
    }

    async Task ResumeAfterGenreAsync(IDialogContext context, IAwaitable<string> result)
    {
        string genre = await result;

        if (genre == DoneCommand)
        {
            context.Done(this);
            return;
        }

        var reply = (context.Activity as Activity)
            .CreateReply($"## Browsing Top 5 Tracks in {genre} genre");

        List<HeroCard> cards = await GetHeroCardsForTracksAsync(genre);
        cards.ForEach(card =>
            reply.Attachments.Add(card.ToAttachment()));

        ThumbnailCard doneCard = GetThumbnailCardForDone();
        reply.Attachments.Add(doneCard.ToAttachment());

        reply.AttachmentLayout = AttachmentLayoutTypes.Carousel;

        await
            new ConnectorClient(new Uri(reply.ServiceUrl))
                .Conversations
                .SendToConversationAsync(reply);

        context.Wait(MessageReceivedAsync);
    }

    async Task<List<HeroCard>> GetHeroCardsForTracksAsync(string genre)
    {
        List<Item> tracks = new GrooveService().GetTracks(genre);

        var cogSvc = new CognitiveService();

        foreach (var track in tracks)
            track.ImageAnalysis = await cogSvc.AnalyzeImageAsync(track.ImageUrl);

        var cards =
            (from track in tracks
             let artists =
                 string.Join(", ",
                     from artist in track.Artists
                     select artist.Artist.Name)
             select new HeroCard
             {
                 Title = track.Name,
                 Subtitle = artists,
                 Text = track.ImageAnalysis.Description.Captions.First().Text,
                 Images = new List<CardImage>
                 {
```

```csharp
                new CardImage
                {
                    Alt = track.Name,
                    Tap = BuildBuyCardAction(track),
                    Url = track.ImageUrl
                }
            },
            Buttons = new List<CardAction>
            {
                BuildBuyCardAction(track),
                BuildNewsCardAction(track)
            }
        })
        .ToList();
    return cards;
}

CardAction BuildBuyCardAction(Item track)
{
    return new CardAction
    {
        Type = ActionTypes.OpenUrl,
        Title = "Buy",
        Value = track.Link
    };
}

CardAction BuildNewsCardAction(Item track)
{
    return new CardAction
    {
        Type = ActionTypes.PostBack,
        Title = NewsCommand,
        Value = JsonConvert.SerializeObject(track)
    };
}

ThumbnailCard GetThumbnailCardForDone()
{
    return new ThumbnailCard
    {
        Title = DoneCommand,
        Subtitle = "Click/Tap to exit",
        Images = new List<CardImage>
        {
            new CardImage
            {
                Alt = "Smile",
                Tap = BuildDoneCardAction(),
                Url = new FileService().GetBinaryUrl("Smile.png")
            }
        },
        Buttons = new List<CardAction>
        {
            BuildDoneCardAction()
        }
    };
```

```
            }

            CardAction BuildDoneCardAction()
            {
                return new CardAction
                {
                    Type = ActionTypes.PostBack,
                    Title = DoneCommand,
                    Value = DoneCommand
                };
            }
        }
    }
```

Look at the *BuildNewsCardAction* method and how its type is *PostBack*, repeated below:

```
        CardAction BuildNewsCardAction(Item track)
        {
            return new CardAction
            {
                Type = ActionTypes.PostBack,
                Title = NewsCommand,
                Value = JsonConvert.SerializeObject(track)
            };
        }
```

In particular, notice that the *Value* property of *CardAction* receives the serialized *track* instance. This might seem peculiar because at first thought, *Value* takes a single value to post back with. However, since *Value* is type *string*, we can pass complex values in the form of JSON objects.

To determine which method the *PostBack* goes to, look at the last statement in *ResumeAfterGenreAsync* and how it does a *Wait*, making *MessageReceivedAsync* the next method in the dialog stack, repeated as follows:

```
            async Task MessageReceivedAsync(IDialogContext context, IAwaitable<object> result)
            {
                var activity = await result as IMessageActivity;
                string heroCardValue = activity?.Text;
                if (!string.IsNullOrWhiteSpace(heroCardValue) && heroCardValue.StartsWith("{"))
                {
                    var news = JsonConvert.DeserializeObject<Item>(heroCardValue);
                    string artistName = news.Artists.First().Artist.Name;
                    await context.Forward(
                        new NewsDialog(artistName),
                        MessageReceivedAsync,
                        activity);
                }
                else
                {
                    List<string> genres = new GrooveService().GetGenres();
                    genres.Add("Done");
                    PromptDialog.Choice(context, ResumeAfterGenreAsync, genres, "Which music category?");
                }
            }
```

MessageReceivedAsync determines whether the *activity.Text* has a left curly brace—an indicator that this is a JSON object. If so, it deserializes into an object it can work with, extracts the name of the artist, and calls *Forward* to pass control to *NewsDialog*. The *NewsDialog* parameter passes the artist name as an argument.

> **Tip** You can assign an entire object, serialized as a string, to the *Value* property of *CardAction* to send multiple values back to a dialog when a user clicks an *ImBack* or *PostBack* card button.

Back at *NewsDialog*, Listing 14-2, notice that it has a constructor accepting artist name and it saves the artist name in a field. Also, the call to *SearchForNewsAsync* on the *CognitiveService* instance in the *MessageReceivedAsync* method uses that artist name to perform the search.

That was the search, and the next section enhances Music Bot with a Vision service.

Interpreting an Image

Cognitive Services offers several APIs in the Vision category including description, video, face, and emotion detection. The example in this section uses the Vision API to analyze and describe an album cover image. Applications of this capability might be to display a set of images with captions, alt tags in HTML images, or indexable content for subsequent image search. To use the vision service, create a new service in the Azure portal by clicking *Create New Service*, *AI + Cognitive Services*, *Computer Vision API*, and fill out the form to complete creating the service. Listing 14-4 shows the code to call the Vision API to obtain a description of an album cover.

LISTING 14-4 Calling the Vision Service: *CognitiveService.cs*

```
using AILib.Models;
using Newtonsoft.Json;
using System;
using System.Configuration;
using System.Linq;
using System.Net.Http;
using System.Text;
using System.Threading.Tasks;
using System.Xml.Linq;

namespace AILib
{
    [Serializable]
    public class CognitiveService
    {
        const string AccessKey = "Ocp-Apim-Subscription-Key";

        public async Task<ImageAnalysis> AnalyzeImageAsync(string imageUrl)
        {
```

```
                const string BaseUrl = "https://westus.api.cognitive.microsoft.com/vision/
v1.0";
                string url = $"{BaseUrl}/analyze?visualFeatures=Description";

                string accessKey = ConfigurationManager.AppSettings["VisionKey"];
                var client = new HttpClient();
                client.DefaultRequestHeaders.Add(AccessKey, accessKey);

                var content = new StringContent(
                    $"{{ \"url\": \"{imageUrl}\" }}", Encoding.UTF8, "application/json");

                HttpResponseMessage response = await client.PostAsync(url, content);
                string jsonResult = await response.Content.ReadAsStringAsync();

                ImageAnalysis analysis = JsonConvert.DeserializeObject<ImageAnalysis>(jsonRes
ult);
                return analysis;
            }
        }
}
```

The *AnalyzeImageAsync* method in Listing 14-4 calls the Azure vision service. You can find the keys on the Azure service blade, titled Manage Keys, as illustrated in Figure 14-1, and then copy a key to *web. config appSettings* as *VisionKey*, matching the *ConfigurationManager.AppSettings["VisionKey"]* call in the code.

Looking at how *url* is constructed, this example is only requesting a *Description* of the image. Other options include *Categories*, *Tags*, *ImageType,* and more. The API documentation has details on all of the options and contains any new features added in the future. There are two ways to supply the image: via URL or binary attachment. This example builds a *StringContent* with a JSON object, specifying the URL, with a value being the *imageUrl* parameter.

The code posts the request and deserializes results into the custom *ImageAnalysis* instance, which contains the description of the image. The *BrowseDialog* uses the *ImageAnalysis* to display the caption in *HeroCards,* and the following excerpt from Listing 14-3 shows how:

```
async Task<List<HeroCard>> GetHeroCardsForTracksAsync(string genre)
{
    List<Item> tracks = new GrooveService().GetTracks(genre);

    var cogSvc = new CognitiveService();

    foreach (var track in tracks)
        track.ImageAnalysis = await cogSvc.AnalyzeImageAsync(track.ImageUrl);

    var cards =
        (from track in tracks
         let artists =
             string.Join(", ",
                 from artist in track.Artists
                 select artist.Artist.Name)
         select new HeroCard
         {
```

```
                    Title = track.Name,
                    Subtitle = artists,
                    Text = track.ImageAnalysis.Description.Captions.First().Text,
                    Images = new List<CardImage>
                    {
                        new CardImage
                        {
                            Alt = track.Name,
                            Tap = BuildBuyCardAction(track),
                            Url = track.ImageUrl
                        }
                    },
                    Buttons = new List<CardAction>
                    {
                        BuildBuyCardAction(track),
                        BuildNewsCardAction(track)
                    }
                })
                .ToList();
            return cards;
        }
```

We added a property named *ImageAnalysis* of type *ImageAnalysis* to the *Item* class, which holds each music track. The *foreach* statement calls *AnalyzeImageAsync* to get a description for each track, assigning the results to the *ImageAnalysis* property for that track. With descriptions assigned to each track, the LINQ statement can build a *HeroCard*, assigning the first caption (image description) to the *Text* property. The reason that *Captions* is a collection is because there could be multiple captions, each with a confidence score, with the first being the highest number.

As with all the Cognitive Service categories, there are many more Vision APIs. Next, we'll look at one of the Text category services.

> **Tip** Examples in this chapter use *HttpClient* to communicate with Cognitive Services. However, many of the Cognitive Services also have an SDK to make working with them easier. For example, the Vision API has an SDK, which you can find on NuGet (*https://www.nuget.org/packages/Microsoft.ProjectOxford.Vision/*). You can find other SDKs on NuGet or on the reference pages for the Cognitive Service you want to work with.

Translating Text

In many applications, the need for multi-cultural communication is essential. While there's support for internationalization/localization in ASP.NET Web API, chatbots pose a unique situation. The built-in localization support pre-supposes what an application's responses are. However, with chatbots and NLP, you can't predict what the user will say. Considering that chatbots also rely on external services for responses, we can't always predict what the chatbot will say. That's where the Cognitive Services Text Translation API can help, which is also the example in this section.

This example uses Wine Bot, from Chapter 8, using LUIS to handle natural language queries from the user. The strategy used here is to intercept a message as soon as it arrives at the chatbot, detect the language and translate to English if necessary. The code keeps track of the user's language for the current request and performs a translation of the response from English to the user's language if necessary. That way, the chatbot can communicate with a user via any language the user chooses, even switching languages in the middle of a conversation. To use the Translator Text API, create a new service in the Azure portal by clicking *Create New Service*, *AI + Cognitive Services*, *Translator Text API*, and fill out the form to complete creating the service. Listing 14-5 has the code that uses the Cognitive Services language detect and translate APIs.

LISTING 14-5 Calling the Text Translation Service: *CognitiveService.cs*

```csharp
using AILib.Models;
using Newtonsoft.Json;
using System;
using System.Configuration;
using System.Linq;
using System.Net.Http;
using System.Text;
using System.Threading.Tasks;
using System.Xml.Linq;

namespace AILib
{
    [Serializable]
    public class CognitiveService
    {
        const string AccessKey = "Ocp-Apim-Subscription-Key";

        public async Task<string> DetectLanguageAsync(string text)
        {
            const string BaseUrl = "https://api.microsofttranslator.com/V2/Http.svc/Detect";
            string encodedText = Uri.EscapeUriString(text);
            string url = $"{BaseUrl}?text={encodedText}";

            string accessKey = ConfigurationManager.AppSettings["TranslateKey"];
            var client = new HttpClient();
            client.DefaultRequestHeaders.Add(AccessKey, accessKey);

            string response = await client.GetStringAsync(url);
            response = XElement.Parse(response).Value;
            return response;
        }

        public async Task<string> TranslateTextAsync(string text, string language)
        {
            string encodedText = Uri.EscapeUriString(text);
            string encodedLang = Uri.EscapeUriString(language);
            const string BaseUrl = "https://api.microsofttranslator.com/V2/Http.svc/Translate";
            string url = $"{BaseUrl}?text={encodedText}&to={encodedLang}";
```

```
            string accessKey = ConfigurationManager.AppSettings["TranslateKey"];
            var client = new HttpClient();
            client.DefaultRequestHeaders.Add(AccessKey, accessKey);

            string response = await client.GetStringAsync(url);
            response = XElement.Parse(response).Value;
            return response;
        }
    }
}
```

On the Azure service blade, there's a Manage Keys entry, as illustrated in Figure 14-1, where you should copy the key into the web.config file *appSettings* element named *TranslateKey*. This matches the *ConfigurationManager.AppSettings* call in each method in Listing 14-5.

The *DetectLanguageAsync* method builds *url* with the *text* parameter, containing the user's message. Notice that unlike most of the Cognitive Service APIs, the Text Translation APIs return XML, hence the LINQ to XML to extract the value. The return value is a two-digit standards-compliant country code.

The *TranslateTextAsync* builds *url* with the *text* and *language* to convert that text to. Notice that both methods call *Uri.EscapeUriString* to encode input strings, which is essential to making sure the request goes through properly incase of special characters. The code then uses LINQ to XML to return the translated text to the caller.

That was the code to call Cognitive Services, and next you'll see the two use cases for translation in Wine Bot. Listing 14-6 shows how the chatbot in the *WineBotLuis* project for the code accompanying this chapter handles incoming message translations.

LISTING 14-6 Translating User Input: *MessagesController.cs*

```
using System.Net;
using System.Net.Http;
using System.Threading.Tasks;
using System.Web.Http;
using Microsoft.Bot.Builder.Dialogs;
using Microsoft.Bot.Connector;
using System;
using WineBotLuis.Dialogs;
using System.Linq;
using AILib;

namespace WineBotLuis
{
    [BotAuthentication]
    public class MessagesController : ApiController
    {
        public async Task<HttpResponseMessage> Post([FromBody]Activity activity)
        {
            if (activity.Type == ActivityTypes.Message)
            {
```

```csharp
            await DetectAndTranslateAsync(activity);
            await Conversation.SendAsync(activity, () => new Dialogs.WineBotDialog());
        }
        else
        {
            await HandleSystemMessageAsync(activity);
        }
        var response = Request.CreateResponse(HttpStatusCode.OK);
        return response;
    }

    async Task DetectAndTranslateAsync(Activity activity)
    {
        var cogSvc = new CognitiveService();

        string language = await cogSvc.DetectLanguageAsync(activity.Text);

        if (!language.StartsWith("en"))
        {
            activity.Text = await cogSvc.TranslateTextAsync(activity.Text, "en");
            activity.Locale = language;
        }
    }

    async Task HandleSystemMessageAsync(Activity message)
    {
        if (message.Type == ActivityTypes.ConversationUpdate)
        {
            if (message.Type == ActivityTypes.ConversationUpdate)
            {
                Func<ChannelAccount, bool> isChatbot =
                        channelAcct => channelAcct.Id == message.Recipient.Id;

                if (message.MembersAdded?.Any(isChatbot) ?? false)
                {
                    Activity reply = (message as Activity).CreateReply(
                        "# Welcome to Wine Chatbot!\n" + WineBotDialog.ExampleText);
                    var connector = new ConnectorClient(new Uri(message.ServiceUrl));
                    await connector.Conversations.ReplyToActivityAsync(reply);
                }
            }
        }
    }
}
```

As soon as the *Post* method in Listing 14-6 knows that this is a message from the user, it calls *DetectAndTranslateAsync*. The first statement of *DetectAndTranslateAsync* calls *DetectLanguageAsync* returning the language of the user's message. This chatbot knows how to work with English text, so it checks to see if the langage is English. If not, it calls *TranslateTextAsync* to covert from the detected language to English. The code then uses the current activity to set *Text* and *Locale* for subsequent processing.

When *DetectAndTranslateAsync* returns to *Post*, it calls *SendAsync* on a new instance of *WineBotDialog*. If you recall, *WineBotDialog* derives from *LuisDialog*, which makes a request to LUIS behind the scenes and returns to a matching intent. Therefore, it was important to do the translation prior to transferring control to *WineBotDialog*.

> **Tip** As you use Cognitive Services APIs, check for what their language capabilities are to avoid unnecessary API calls. One example is that LUIS supports a growing list of languages and a potential optimization in the *DetectAndTranslateAsync* method would be to check for not only English, but the other languages that LUIS supports, saving an extra call to the Translate API.

Now that we've examined handling user messages, next let's look at sending responses back to the user. Listing 14-7 has the *WineBotDialog*. In the intent handler methods, the code translates the response to the user's language.

LISTING 14-7 Translating User Responses: *WineBotDialog.cs*

```csharp
using System;
using System.Threading.Tasks;
using Microsoft.Bot.Builder.Dialogs;
using Microsoft.Bot.Builder.Luis.Models;
using Microsoft.Bot.Builder.Luis;
using WineBotLib;
using System.Collections.Generic;
using System.Text.RegularExpressions;
using System.Linq;
using Microsoft.Bot.Connector;
using AILib;
using AILib.Models;
using System.Web;

namespace WineBotLuis.Dialogs
{
    [LuisModel(
        modelID: "<your LUIS model ID>",
        subscriptionKey: "<your LUIS subscription key>")]
    [Serializable]
    public class WineBotDialog : LuisDialog<object>
    {
        public const string ExampleText = @"
Here are a couple examples that I can recognize:
'What type of red wine do you have with a rating of 70?' or
'Please search for champaigne.'";

        readonly CognitiveService cogSvc = new CognitiveService();

        [LuisIntent("")]
        public async Task NoneIntent(IDialogContext context, LuisResult result)
        {
            QnAAnswer qnaAnswer = await cogSvc.AskWineChatbotFaqAsync(result.Query);
```

```csharp
    string message =
        qnaAnswer.Score == 0 ?
            @"Sorry, I didn't get that. " + ExampleText :
            HttpUtility.HtmlDecode(qnaAnswer.Answer);

    message = await TranslateResponseAsync(context, message);

    await context.PostAsync(message);
    context.Wait(MessageReceived);
}

[LuisIntent("Searching")]
public async Task SearchingIntent(IDialogContext context, LuisResult result)
{
    if (!result.Entities.Any())
    {
        await NoneIntent(context, result);
        return;
    }

    int wineCategory;
    int rating;
    ExtractEntities(result, out wineCategory, out rating);

    var wines = await new WineApi().SearchAsync(
        wineCategory, rating, inStock: true, searchTerms: string.Empty);

    string message = wines.Any() ?
        "Here are the top matching wines" :
        "Sorry, No wines found matching your criteria.";

    message = await TranslateResponseAsync(context, message);

    if (wines.Any())
        message =
            $"## {message}:\n" +
            $"{ string.Join("\n", wines.Select(w => $"* {w.Name}"))}";

    await context.PostAsync(message);

    context.Wait(MessageReceived);
}

void ExtractEntities(LuisResult result, out int wineCategory, out int rating)
{
    const string RatingEntity = "builtin.number";
    const string WineTypeEntity = "WineType";

    rating = 1;
    EntityRecommendation ratingEntityRec;
    result.TryFindEntity(RatingEntity, out ratingEntityRec);
    if (ratingEntityRec?.Resolution != null)
        int.TryParse(ratingEntityRec.Resolution["value"] as string, out rating);

    wineCategory = 0;
```

```csharp
            EntityRecommendation wineTypeEntityRec;
            result.TryFindEntity(WineTypeEntity, out wineTypeEntityRec);

            if (wineTypeEntityRec != null)
            {
                string wineType = wineTypeEntityRec.Entity;

                wineCategory =
                    (from wine in WineTypeTable.Keys
                     let matches = new Regex(WineTypeTable[wine]).Match(wineType)
                     where matches.Success
                     select (int)wine)
                    .FirstOrDefault();
            }
        }

        async Task<string> TranslateResponseAsync(IDialogContext context, string message)
        {
            var activity = context.Activity as IMessageActivity;
            if (!activity.Locale.StartsWith("en"))
                message = await cogSvc.TranslateTextAsync(message, activity.Locale);
            return message;
        }

        Dictionary<WineType, string> WineTypeTable =
            new Dictionary<WineType, string>
            {
                [WineType.ChampagneAndSparkling] = "champaign and sparkling|champaign|sparkling",
                [WineType.DessertSherryAndPort] = "dessert sherry and port|desert|sherry|port",
                [WineType.RedWine] = "red wine|red|reds|cabernet|merlot",
                [WineType.RoseWine] = "rose wine|rose",
                [WineType.Sake] = "sake",
                [WineType.WhiteWine] = "white wine|white|whites|chardonnay"
            };
    }
}
```

You've seen the majority of this code in Chapter 8, when discussing how LUIS works. What's different is that the intent handler methods, *NoneIntent* and *SearchingIntent,* call *TranslateResponseAsync* to convert the response to the user's language and assign the translation to the message response sent back to the user.

TranslateResponseAsync checks to see if the user's language is English and performs the translation if not. This prevents wasting a call to the Translation API when the text is already English.

Notice how *SearchingIntent* handles the response from *TranslateResponseAsync* if wines.Any() is *true*. The result is formatted with markdown with a ## prefix for the title. The # character doesn't encode and causes an HTTP 400 error when passed to the Translation API. Therefore, you want to avoid situations where markdown can be passed to the Translation API. If you do receive errors, like the HTTP 400, from the Translation API, immediately examine the URL for any special characters that aren't encoded for the text parameter.

Now that we've examined how to use the Translation API, let's look at additional ways to interpret and respond to user messages.

Building FAQ Chatbots with QnA Maker

Another set of use cases for chatbots is in the customer support arena. You've seen chatbots in earlier chapters that were games, entertainment, or e-commerce. In any of those scenarios and more, customers might want to ask additional questions. A mature chatbot would anticipate most questions and provide reasonable responses. Related to general questions is the concept of Frequently Asked Questions (FAQs). Instead of users visiting a web page or App menu item for a FAQ, it might be more convenient to just ask a chatbot.

This section draws upon that FAQ scenario by showing how to use another tool in the Cognitive Services suite called QnA Maker. The Q&A Maker is both an API and a website that consumes that API. This section explains how to use the web page to create a FAQ and then consume it with code. The specific example is an enhancement to Wine Bot, where if Wine Bot doesn't recognize a question, it passes the query to QnA Maker to see if there's a FAQ item related to the query.

To get started with QnA Maker, visit *https://qnamaker.ai/* and click the *Create New Service* tab (you'll need to log in). Figure 14-2 shows the form for creating the FAQ.

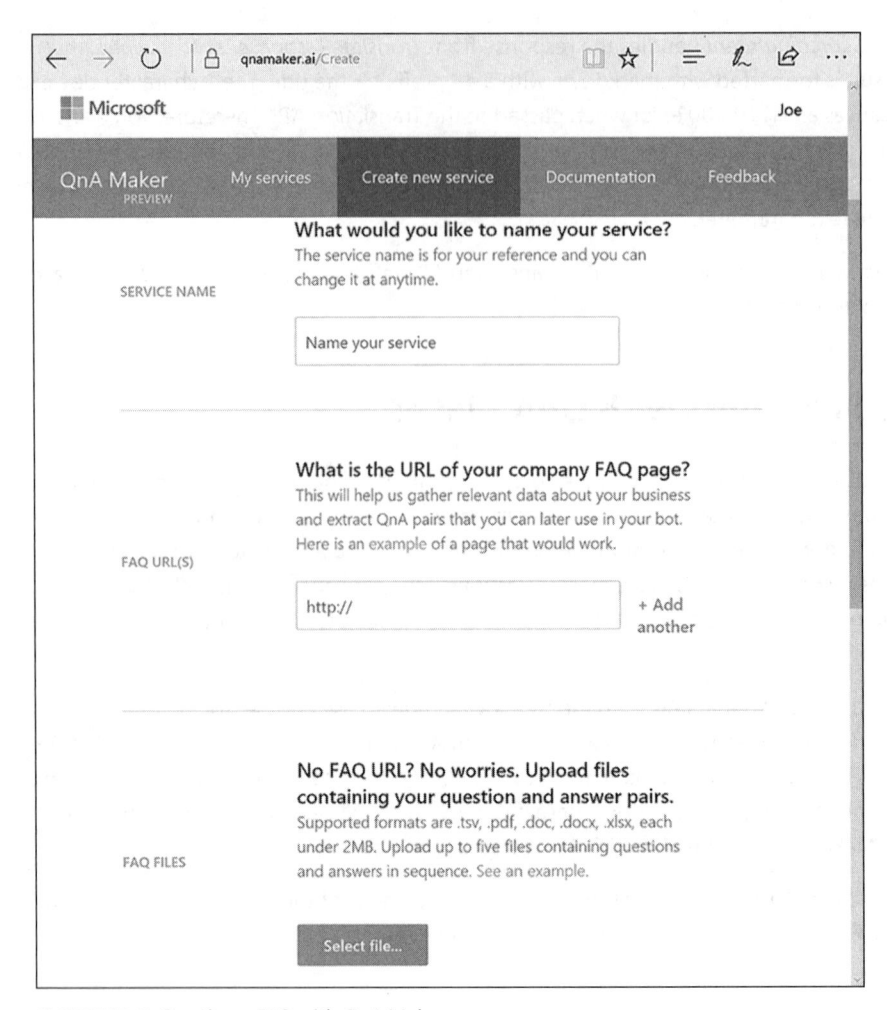

FIGURE 14-2 Creating a FAQ with QnA Maker.

As shown in Figure 14-2, you can name the service. You also have the option to point at an existing FAQ on the web with a URL, or upload a file in a supported format, like *.pdf*, *.doc*, and more. Alternatively, you can ignore either of these settings and edit the FAQ in the web UI.

After creating the FAQ, the *My Services* tab lists the new FAQ along with any others you've created. Clicking the *Edit* button on the FAQ opens an editor screen, similar to the Wine Bot screen in Figure 14-3.

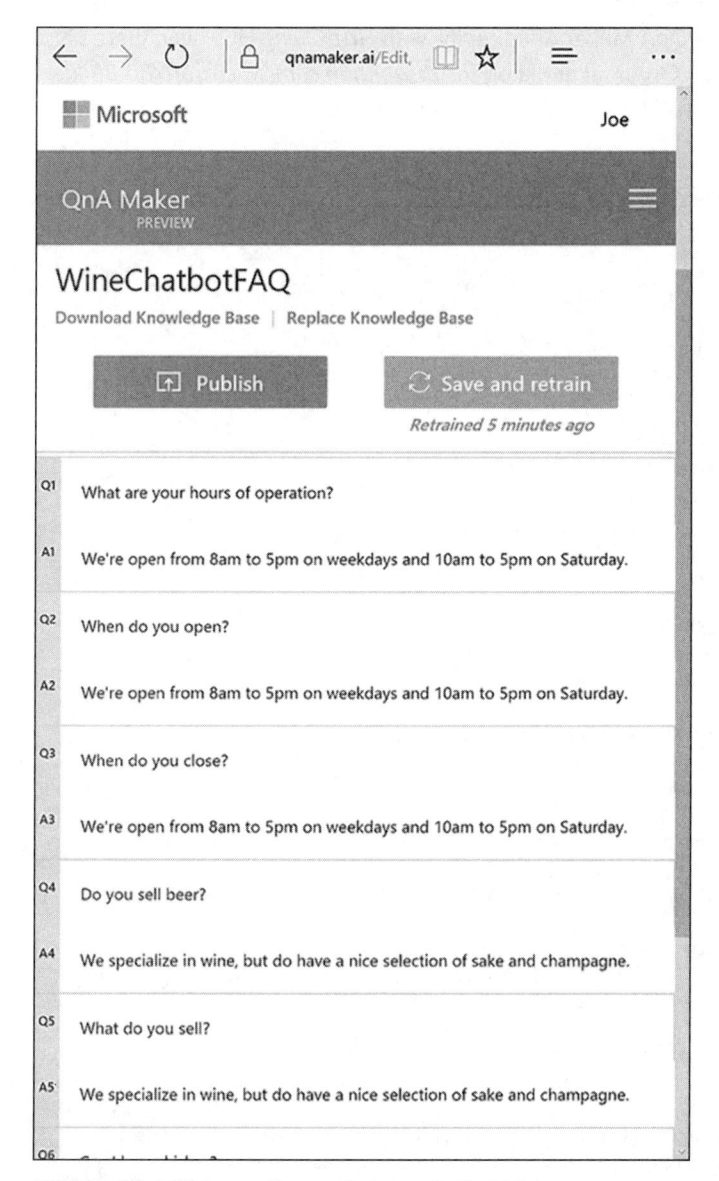

FIGURE 14-3 Editing questions and answers in QnA Maker.

Figure 14-3 shows several questions and answers. Notice that *Q1*, *Q2*, and *Q3* have the same answer, but different utterances. The UI might change between now and the time you read this, but you should look at this like LUIS, where you can have multiple utterances per Intent, except this has answers instead of Intents.

With a FAQ in place, click the *Save And Retrain* button to build a model that recognizes questions and responds with appropriate answers. When ready to use the service, click *Publish*. Once the FAQ is published, you'll be able to call the FAQ with an HTTP POST call, as shown in the *CognitiveService* class in Listing 14-8.

CHAPTER 14   Integrating Cognitive Services   **341**

> **Tip** This example called the QnA Maker API directly, with *HttpClient*. However, there's also an open-source project for a QnA Maker dialog (*https://github.com/Microsoft/BotBuilder-CognitiveServices/tree/master/CSharp/Samples/QnAMaker*) that might be more convenient.

LISTING 14-8 Calling the QnA Maker Service: *CognitiveService.cs*

```
using AILib.Models;
using Newtonsoft.Json;
using System;
using System.Configuration;
using System.Linq;
using System.Net.Http;
using System.Text;
using System.Threading.Tasks;
using System.Xml.Linq;

namespace AILib
{
    [Serializable]
    public class CognitiveService
    {
        const string AccessKey = "Ocp-Apim-Subscription-Key";

        public async Task<QnAAnswer> AskWineChatbotFaqAsync(string question)
        {
            const string BaseUrl = "https://westus.api.cognitive.microsoft.com/qnamaker/v2.0";
            string knowledgeBaseID = ConfigurationManager.AppSettings["QnAKnowledgeBaseID"];

            string url = $"{BaseUrl}//knowledgebases/{knowledgeBaseID}/generateAnswer";

            var client = new HttpClient();
            string accessKey = ConfigurationManager.AppSettings["QnASubscriptionKey"];
            client.DefaultRequestHeaders.Add(AccessKey, accessKey);

            var content = new StringContent(
                $"{{ \"question\": \"{question}\" }}", Encoding.UTF8, "application/json");

            HttpResponseMessage response = await client.PostAsync(url, content);
            string jsonResult = await response.Content.ReadAsStringAsync();

            QnAResponse qnaResponse = JsonConvert.DeserializeObject<QnAResponse>(jsonResult);

            return qnaResponse.Answers.FirstOrDefault();
        }
    }
}
```

AskWineChatbotFaqAsync in Listing 14-8 uses two claims from the QnA Maker FAQ: *Knowledgebase ID* and *Ocp-Apim-Subscription-Key*. You can find these values by visiting the *My Services* tab and clicking *View Code*, which opens the HTTP request specification shown in Figure 14-4.

```
                    Sample HTTP Request    Knowledgebase ID (query string)

                    POST /knowledgebases/▓▓▓▓▓▓▓▓▓▓▓▓▓▓▓▓/generateAnswer
                    Host: https://westus.api.cognitive.microsoft.com/qnamaker/v2.0
                    Ocp-Apim-Subscription-Key: ▓▓▓▓▓▓▓▓▓▓
                    Content-Type: application/json
                    {"question":"hi"}

                                                    HTTP Header

                              Copy              Close
```

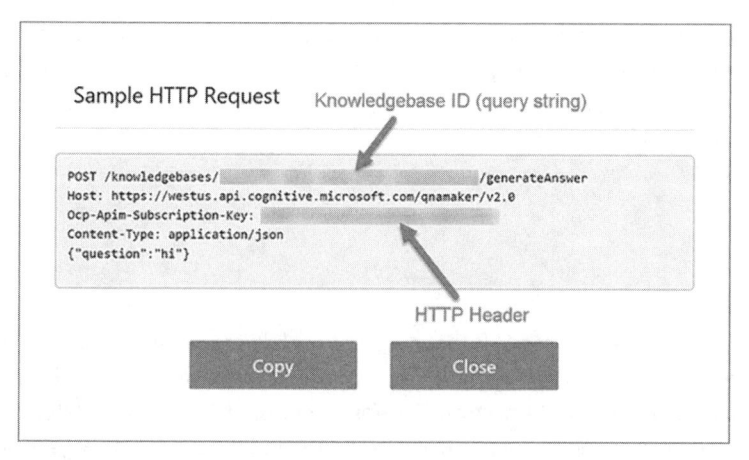

FIGURE 14-4 Obtaining Knowledgebase ID and Ocp-Apim-Subscription-Key from QnA Maker.

Figure 14-4 shows that you can find the *Knowledgebase ID* in the *POST* query string and the *Ocp-Apim-Subscription-Key* in the request header. Copy these values into web.config with the *QnAKnowledgeBaseID* and *QnASubscriptionKey* entries in the *appSettings* section, matching the corresponding keys in the code read with *ConfigurationSettings.AppSettings* calls.

The *AskWineChatbotFaqAsync* method performs a normal HTTP POST to QnA Maker, deserializing the results into the custom *QnAResponse* object, and returns the first item in the *Answers* collection. The *NoneIntent* method from the *WineBotDialog* class in Listing 14-7, repeated below calls *AskWineChatbotFaqAsync*, attempting to find an answer for an unrecognized intent.

```
[LuisIntent("")]
public async Task NoneIntent(IDialogContext context, LuisResult result)
{
    QnAAnswer qnaAnswer = await cogSvc.AskWineChatbotFaqAsync(result.Query);

    string message =
        qnaAnswer.Score == 0 ?
            @"Sorry, I didn't get that. " + ExampleText :
            HttpUtility.HtmlDecode(qnaAnswer.Answer);

    message = await TranslateResponseAsync(context, message);

    await context.PostAsync(message);
    context.Wait(MessageReceived);
}
```

The response from QnA Maker includes a set of answers, ranked by score, where the highest score represents the probability of matching the answer. When the score is 0, QnA Maker isn't able to match any questions, resulting in the *NoneIntent* code returning an explanation that it doesn't understand the question. When there is a good answer, with a score greater than 0 (it's a probability, which is always greater than 0), the code responds with the answer to the user. Notice how *HttpUtility* decodes the answer because QnA Maker can return HTML encoded answers that isn't a proper way to present answers to a user.

> **Note** You've seen various encodings throughout this chapter to make sure request data is in the proper format. It's easy for data with unexpected encodings to slip through processing, causing HTTP errors. It might be worth the time to do extensive testing with different URL, HTML, and Markdown related encodings.

Summary

This chapter covered a set of Cognitive Services APIs that might be useful in some of the chatbots you build. You saw how to use the Search API to find news articles associated with a music artist. The Vision API example created captions for a musical album cover. The Text processing API example detected the user's language and converted queries and responses from/to English and the user's language. Finally, you saw how the QnA Maker helps build FAQs to enhance a chatbot's ability to respond to user questions.

While this was only a small group of services, the entire Cognitive Services suite of tools offers a growing list of many more APIs. You can use these to enhance chatbots, adding additional features and leveraging AI technologies for smarter chatbots. We encourage you to visit the Azure website for Cognitive Services and learn about all the other services to see how they can help you.

Not only can you make your chatbots more intelligent with Cognitive Services, but the next chapter helps chatbots sound smarter with voice.

CHAPTER 15

Adding Voice Services

The Bot Framework runs on many platforms and all of the channels you've configured so far involve text interaction. Users type into an interface for the channel, often a messaging app, and the chatbot responds in text. However, text isn't the only medium available with chatbots – another is voice.

Using voice, users can talk to a chatbot directly without typing. The chatbot can respond with voice and/or text. Cortana is one such interface where users can summon Cortana and ask her to interact with a chatbot. This chapter starts off by showing how to add speech capabilities with different types of responses to a user. You'll see how to implement input hints to indicate the chatbot's listening state, and then you'll learn how to deploy the chatbot to Cortana.

Adding Speech to Activities

The first chatbot example in this book used the *connector.Conversations.ReplyToActivityAsync(reply)* to send a message back to a user. The *reply* argument was an *Activity* instance, whose *Text* property contained the message. You might have noticed this technique throughout the book, especially in building the welcome message during the *ConversationUpdate* event handling.

In addition to *Text*, the *Activity* type has a *Speak* property. Channels, like Cortana, use the text in the *Speak* property to talk to the user. Listing 15-1 is a modified version of the RockPaperScissors program, from Chapter 4, that adds speech to *Activities*.

LISTING 15-1 Adding Speech to Activities: *MessagesController.cs*

```
using System;
using System.Net;
using System.Net.Http;
using System.Threading.Tasks;
using System.Web.Http;
using Microsoft.Bot.Connector;
using RockPaperScissors4.Models;
using System.Web;

namespace RockPaperScissors4
{
    [BotAuthentication]
    public class MessagesController : ApiController
    {
```

```csharp
public async Task<HttpResponseMessage> Post([FromBody]Activity activity)
{
    HttpStatusCode statusCode = HttpStatusCode.OK;

    var connector = new ConnectorClient(new Uri(activity.ServiceUrl));

    if (activity.Type == ActivityTypes.Message)
    {
        string message = await GetMessage(connector, activity);

        Activity reply = activity.BuildMessageActivity(message);
        reply.Speak = message;
        reply.InputHint = InputHints.AcceptingInput;

        await connector.Conversations.ReplyToActivityAsync(reply);
    }
    else
    {
        try
        {
            await new SystemMessages().Handle(connector, activity);
        }
        catch (HttpException ex)
        {
            statusCode = (HttpStatusCode) ex.GetHttpCode();
        }
    }

    HttpResponseMessage response = Request.CreateResponse(statusCode);
    return response;
}

async Task<string> GetMessage(ConnectorClient connector, Activity activity)
{
    var state = new GameState();

    string userText = activity.Text.ToLower();
    string message = "";

    if (userText.Contains(value: "score"))
    {
        message = await state.GetScoresAsync(connector, activity);
    }
    else if (userText.Contains(value: "delete"))
    {
        message = await state.DeleteScoresAsync(activity);
    }
    else
    {
        var game = new Game();
        message = game.Play(userText);

        if (message.Contains(value: "Tie"))
        {
            await state.AddTieAsync(activity);
        }
```

```
            else
            {
                bool userWin = message.Contains(value: "win");
                await state.UpdateScoresAsync(activity, userWin);
            }
        }

        return message;
    }
}
```

The *Post* method, in Listing 15-1, adds speech to an *Activity*, repeated here.

```
Activity reply = activity.BuildMessageActivity(message);
reply.Speak = message;
```

This assigns the message, which is the user's reply text, to the *Speak* property of the *Activity*.

> **Tip** The examples in this chapter implement speech with plain text. You can also use Speech Synthesis Markup Language (SSML). Voice synthesizers can often mispronounce and misunderstand text to the point that some parts of a sentence might be unintelligible. SSML is an XML markup language that helps with special attributes that specify emphasis, prosidy, aliasing, and more to make the text more understandable when spoken. You can learn more at the Microsoft page at *https://aka.ms/ssmlref* or the W3C Standard at *http://www.w3.org/TR/speech-synthesis/*.

That was *Activity* and you can also add speech to prompts – discussed next.

Adding Speech with *SayAsync*

Another way to respond to the user is with *PromptAsync*. However, there's a speech equivalent, *SayAsync*, that allows you to specify both text and speech. Wine Bot, from Chapter 8, used *PromptAsync* and listing 15-2 shows how it was modified to use speech.

LISTING 15-2 Adding Speech with SayAsync: *WineBotDialog.cs*

```
using System;
using System.Threading.Tasks;
using Microsoft.Bot.Builder.Dialogs;
using Microsoft.Bot.Builder.Luis.Models;
using Microsoft.Bot.Builder.Luis;
using WineBotLib;
using System.Collections.Generic;
using System.Text.RegularExpressions;
using System.Linq;
```

CHAPTER 15 Adding Voice Services   **347**

```csharp
using Microsoft.Bot.Connector;

namespace WineBotLuis.Dialogs
{
    [LuisModel(
        modelID: "<your model ID goes here>",
        subscriptionKey: "<your subscription key goes here>")]
    [Serializable]
    public class WineBotDialog : LuisDialog<object>
    {
        [LuisIntent("")]
        public async Task NoneIntent(IDialogContext context, LuisResult result)
        {
            string message = @"
Sorry, I didn't get that.
Here are a couple examples that I can recognize:
'What type of red wine do you have with a rating of 70?' or
'Please search for champaigne.'";

            await context.SayAsync(text: message, speak: message,
                options: new MessageOptions
                {
                    InputHint = InputHints.AcceptingInput
                });
            context.Wait(MessageReceived);
        }

        [LuisIntent("Searching")]
        public async Task SearchingIntent(IDialogContext context, LuisResult result)
        {
            if (!result.Entities.Any())
            {
                await NoneIntent(context, result);
                return;
            }

            int wineCategory;
            int rating;
            ExtractEntities(result, out wineCategory, out rating);

            var wines = await new WineApi().SearchAsync(
                wineCategory, rating, inStock: true, searchTerms: string.Empty);
            string message;

            if (wines.Any())
                message = "Here are the top matching wines: " +
                            string.Join(", ", wines.Select(w => w.Name));
            else
                message = "Sorry, No wines found matching your criteria.";

            await context.SayAsync(text: message, speak: message,
                options: new MessageOptions
                {
                    InputHint = InputHints.AcceptingInput
                });
```

```csharp
            context.Wait(MessageReceived);
        }

        void ExtractEntities(LuisResult result, out int wineCategory, out int rating)
        {
            const string RatingEntity = "builtin.number";
            const string WineTypeEntity = "WineType";

            rating = 1;
            EntityRecommendation ratingEntityRec;
            result.TryFindEntity(RatingEntity, out ratingEntityRec);
            if (ratingEntityRec?.Resolution != null)
                int.TryParse(ratingEntityRec.Resolution["value"] as string, out rating);

            wineCategory = 0;
            EntityRecommendation wineTypeEntityRec;
            result.TryFindEntity(WineTypeEntity, out wineTypeEntityRec);

            if (wineTypeEntityRec != null)
            {
                string wineType = wineTypeEntityRec.Entity;

                wineCategory =
                    (from wine in WineTypeTable.Keys
                     let matches = new Regex(WineTypeTable[wine]).Match(wineType)
                     where matches.Success
                     select (int)wine)
                    .FirstOrDefault();
            }
        }

        Dictionary<WineType, string> WineTypeTable =
            new Dictionary<WineType, string>
            {
                [WineType.ChampagneAndSparkling] = "champaign and sparkling|champaign|sparkling",
                [WineType.DessertSherryAndPort] = "dessert sherry and port|desert|sherry|port",
                [WineType.RedWine] = "red wine|red|reds|cabernet|merlot",
                [WineType.RoseWine] = "rose wine|rose",
                [WineType.Sake] = "sake",
                [WineType.WhiteWine] = "white wine|white|whites|chardonnay"
            };
    }
}
```

Instead of *PromptAsync*, both *NoneIntent* and *SearchingIntent* use *SayAsync*, shown here:

```csharp
await context.SayAsync(text: message, speak: message);
```

SayAsync has parameters for both *text* and *speak*.

> **Tip** It's okay to use the speech related syntax, like *SayAsync*, even if you don't intend to use speech. It still works for text-only chatbots and can make the transition easier if you still want to deploy to a voice channel, like Cortana, later.

That's how to add speech to prompts. Next, we'll look at the *PromptDialog* types.

Adding Speech to *PromptDialog*

Another interaction with users comes via *PromptDialog*. The *PromptDialog.Choice* option offers voice options for both the basic message and the retry option. The Music Chatbot, from Chapter 10, uses *PromptDialog* in several places and Listing 15-3 shows how to implement speech.

LISTING 15-3 Adding Speech to PromptDialog: *ProfileDialog.cs*

```
using Microsoft.Bot.Builder.Dialogs;
using Microsoft.Bot.Connector;
using MusicChatbot.Models;
using MusicChatbot.Services;
using Newtonsoft.Json.Linq;
using System;
using System.IO;
using System.Linq;
using System.Net.Http;
using System.Threading.Tasks;

namespace MusicChatbot.Dialogs
{
    [Serializable]
    public class ProfileDialog : IDialog<object>
    {
        public string Name { get; set; }
        public byte[] Image { get; set; }

        public Task StartAsync(IDialogContext context)
        {
            context.Wait(MessageReceivedAsync);
            return Task.CompletedTask;
        }

        Task MessageReceivedAsync(IDialogContext context, IAwaitable<object> result)
        {
            ShowMainMenu(context);
            return Task.CompletedTask;
        }

        void ShowMainMenu(IDialogContext context)
        {
            var options = Enum.GetValues(typeof(ProfileMenuItem)).Cast<ProfileMenuItem>().ToArray();
```

```csharp
            string promptMessage = "What would you like to do?";
            string retryMessage = "I don't know about that option, please select an item in the list.";

            var promptOptions =
                new PromptOptions<ProfileMenuItem>(
                    prompt: promptMessage,
                    retry: retryMessage,
                    options: options,
                    speak: promptMessage,
                    retrySpeak: retryMessage);

            PromptDialog.Choice(
                context: context,
                resume: ResumeAfterChoiceAsync,
                promptOptions: promptOptions);
        }

        async Task ResumeAfterChoiceAsync(IDialogContext context, IAwaitable<ProfileMenuItem> result)
        {
            ProfileMenuItem choice = await result;

            switch (choice)
            {
                case ProfileMenuItem.Display:
                    await DisplayAsync(context);
                    break;
                case ProfileMenuItem.Update:
                    await UpdateAsync(context);
                    break;
                case ProfileMenuItem.Done:
                default:
                    context.Done(this);
                    break;
            }
        }

        Task UpdateAsync(IDialogContext context)
        {
            PromptDialog.Text(context, ResumeAfterNameAsync, "What is your name?");
            return Task.CompletedTask;
        }

        async Task ResumeAfterNameAsync(IDialogContext context, IAwaitable<string> result)
        {
            Name = await result;
            string message = "Please upload your profile image.";
            await context.SayAsync(text: message, speak: message,
                options: new MessageOptions
                {
                    InputHint = InputHints.AcceptingInput
                });
            context.Wait(UploadAsync);
        }
```

```csharp
async Task UploadAsync(IDialogContext context, IAwaitable<object> result)
{
    var activity = await result as Activity;

    if (activity.Attachments.Any())
    {
        Attachment userImage = activity.Attachments.First();
        Image = await new HttpClient().GetByteArrayAsync(userImage.ContentUrl);

        StateClient stateClient = activity.GetStateClient();
        BotData userData = await stateClient.BotState.GetUserDataAsync(activity.ChannelId, activity.From.Id);
        userData.SetProperty(nameof(Name), Name);
        userData.SetProperty(nameof(Image), Image);
        await stateClient.BotState.SetUserDataAsync(activity.ChannelId, activity.From.Id, userData);
    }
    else
    {
        string message = "Sorry, I didn't see an image in the attachment.";
        await context.SayAsync(text: message, speak: message,
            options: new MessageOptions
            {
                InputHint = InputHints.IgnoringInput
            });
    }

    ShowMainMenu(context);
}

async Task DisplayAsync(IDialogContext context)
{
    Activity activity = context.Activity as Activity;

    StateClient stateClient = activity.GetStateClient();
    BotData userData =
        await stateClient.BotState.GetUserDataAsync(
            activity.ChannelId, activity.From.Id);

    if ((userData.Data as JObject)?.HasValues ?? false)
    {
        string name = userData.GetProperty<string>(nameof(Name));

        await context.SayAsync(text: name, speak: name);

        byte[] image = userData.GetProperty<byte[]>(nameof(Image));

        var fileSvc = new FileService();
        string imageName = $"{context.Activity.From.Id}_Image.png";

        string imageFilePath = fileSvc.GetFilePath(imageName);
        File.WriteAllBytes(imageFilePath, image);

        string contentUrl = fileSvc.GetBinaryUrl(imageName);
        var agenda = new Attachment("image/png", contentUrl, imageName);
```

```
                Activity reply = activity.CreateReply();
                reply.Attachments.Add(agenda);

                await
                    new ConnectorClient(new Uri(reply.ServiceUrl))
                        .Conversations
                        .SendToConversationAsync(reply);
            }
            else
            {
                string message = "Profile not available. Please update first.";
                await context.SayAsync(text: message, speak: message);
            }

            ShowMainMenu(context);
        }
    }
}
```

The *ShowMainMenu*, from Listing 15-3, adds speech to *PromptDialog.Choice*, repeated below:

```
            string promptMessage = "What would you like to do?";
            string retryMessage = "I don't know about that option, please select an item in the list.";

            var promptOptions =
                new PromptOptions<ProfileMenuItem>(
                    prompt: promptMessage,
                    retry: retryMessage,
                    options: options,
                    speak: promptMessage,
                    retrySpeak: retryMessage);

            PromptDialog.Choice(
                context: context,
                resume: ResumeAfterChoiceAsync,
                promptOptions: promptOptions);
```

The way to implement speech with *PromptDialog* is via *PromptOptions*. Notice that the *PromptOptions* type has parameters for both *prompt* and *speak* and *retry* and *retrySpeak* pairs. Then use the overload of *PromptDialog.Choice* that accepts the *promptOptions* argument.

Now that you know various techniques for adding speech to responses, let's look at input hints.

Specifying Input Hints

Input hints communicate a chatbot's assumptions on when the user should communicate. The following *InputHints* class, from the Bot Framework, describes what input hints are available and what they mean:

```csharp
/// <summary>
/// Indicates whether the bot is accepting, expecting, or ignoring input
/// </summary>
public static class InputHints
{
    /// <summary>
    /// The sender is passively ready for input but is not waiting on a response.
    /// </summary>
    public const string AcceptingInput = "acceptingInput";

    /// <summary>
    /// The sender is ignoring input. Bots may send this hint if they are
    /// actively processing a request and will ignore input
    /// from users until the request is complete.
    /// </summary>
    public const string IgnoringInput = "ignoringInput";

    /// <summary>
    /// The sender is actively expecting a response from the user.
    /// </summary>
    public const string ExpectingInput = "expectingInput";
}
```

The default *InputHint* for an *Activity* or *SayAsync* call is *AcceptingInput*. Here's the *Activity* example from Listing 15-1, repeated below:

```csharp
Activity reply = activity.BuildMessageActivity(message);
reply.Speak = message;
reply.InputHint = InputHints.AcceptingInput;
```

Just set the *Activity.InputHint* to an *InputHints* member. *SayAsync* is a little different because there isn't an explicit parameter for input hints. Here's how to add an input hint to *SayAsync*, from Listing 15-2, repeated below:

```csharp
await context.SayAsync(text: message, speak: message,
    options: new MessageOptions
    {
        InputHint = InputHints.AcceptingInput
    });
```

In the case of *SayAsync*, we instantiate a *MessageOptions*, setting its *InputHint* property.

If a chatbot sends a message to a user and expects a response, they would use *InputHints.ExpectingInput*. An example of this would be the *PromptDialog* type, where it makes sense because the purpose of the prompt is to obtain a response from the user. The default for *PromptDialog* is always *InputHints.ExpectingInput*, making it unnecessary to explicity set an input hint.

Listing 15-3 has an example of how to use *InputHints.IgnoringInput*, repeated below:

```csharp
string message = "Sorry, I didn't see an image in the attachment.";
await context.SayAsync(text: message, speak: message,
    options: new MessageOptions
    {
        InputHint = InputHints.IgnoringInput
    });
```

In this example, the user didn't upload an image when building their profile. The message is information only and doesn't require a response, making *InputHints.IgnoringInput* the appropriate choice. Additionally, the next code to execute is *ShowMainMenu*, which does prompt the user and accepts input.

Setting up Cortana

Once a chatbot has speech capabilities, you can deploy it to a voice channel. In this example, we'll deploy the chatbot to Cortana. Like many of the other channels, setup isn't too difficult for Cortana, but there are a couple extra steps that are important to know. The first step is to make sure the chatbot is registered with the Bot Framework, as described in Chapter 2. After that, you can visit *My bots*, select the registered chatbot, and click the *Cortana* channel. You'll see a window similar to Figure 15-1.

> **Note** Remember to deploy the chatbot before testing with Cortana.

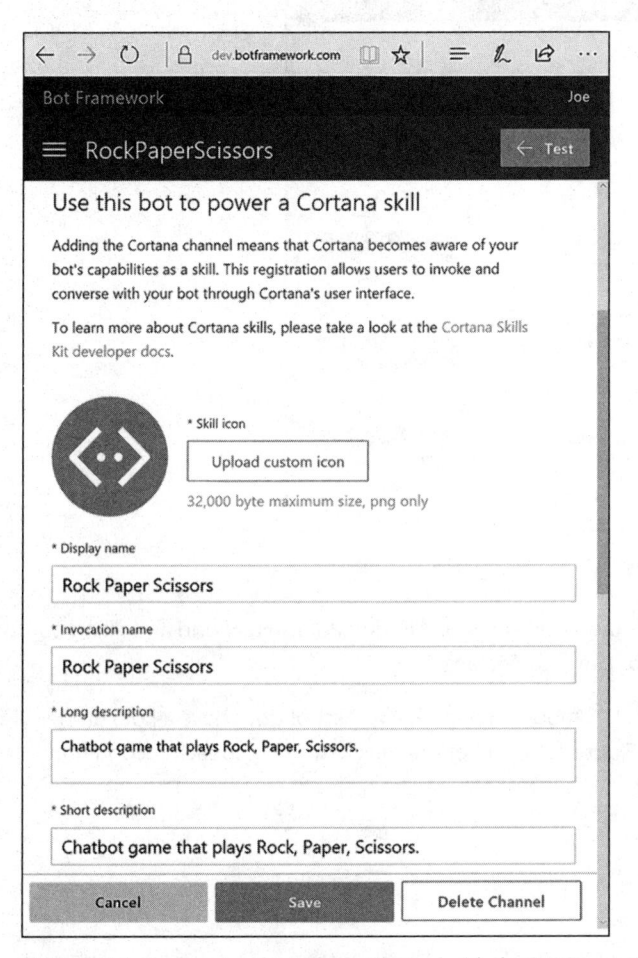

FIGURE 15-1 Configuring the Cortana Channel with the Bot Framework.

The Bot Framework pre-populates fields based on chatbot registration and you should review the values to ensure they're correct or accurate. The most interesting field here is *Invocation Name,* which is the name users say to Cortana to invoke the chatbot.

Clicking the *Save* button brings you back to the Channel page for the chatbot, where there's a link for the Cortana dashboard. Click the link that says Manage In Cortana Dashboard, and you'll see a window similar to Figure 15-2.

> **Note** You may be asked to create an account to manage Cortana Skills. If so, make sure you log in with the same Microsoft accout that you used to register the chatbot, allowing you to test prior to publishing.

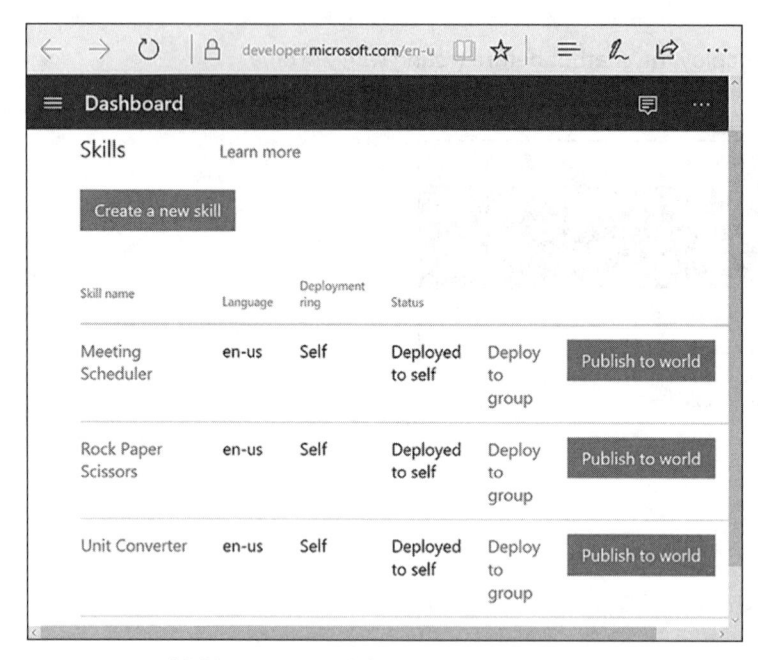

FIGURE 15-2 Publishing a Cortana Skill.

The RockPaperScissors chatbot is now a Cortana Skill. Clicking Publish to world and filling out the submission form makes it available for use with Cortana.

RockPaperScissors doesn't have a LUIS model. However, if a chatbot does have a LUIS model, visit the chatbot Settings page, shown in Figure 15-3, and prime the chatbot for speech recognition.

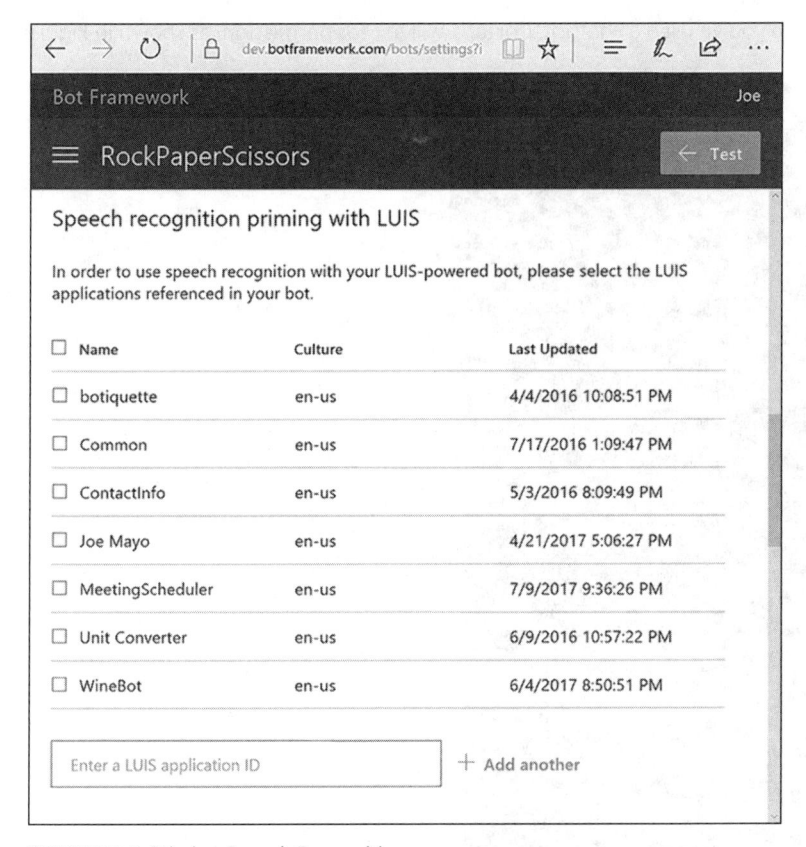

FIGURE 15-3 Priming Speech Recognition.

To prime speech recognition, visit the luis.ai web page to get the application ID for the LUIS model to prime, add that ID to the Enter A LUIS pplication ID box, in Figure 15-3, and check the box by the matching LUIS model. This helps Cortana, or any other voice channel recognized user utterances.

> **Note** To make Cortana available for testing, open the Cortana menu, on the task bar, click the Settings icon, and ensure that the Let Cortana Respond To "Hey Cortana" option is turned on.

To test the new chatbot, get Cortana's attention and use the chatbot invocation phrase to start the chatbot, by saying this:

Hey Cortana! Ask Rock Paper Scissors to play!

> **Tip** We've found that it works better if you use the invocation phrase immediately after saying Hey Cortana. If you hesitate too long between the two phrases, Cortana defaults to a Bing search.

CHAPTER 15 Adding Voice Services   357

If this is the first time you've used a chatbot, Cortana will ask for permission, as shown in Figure 15-4.

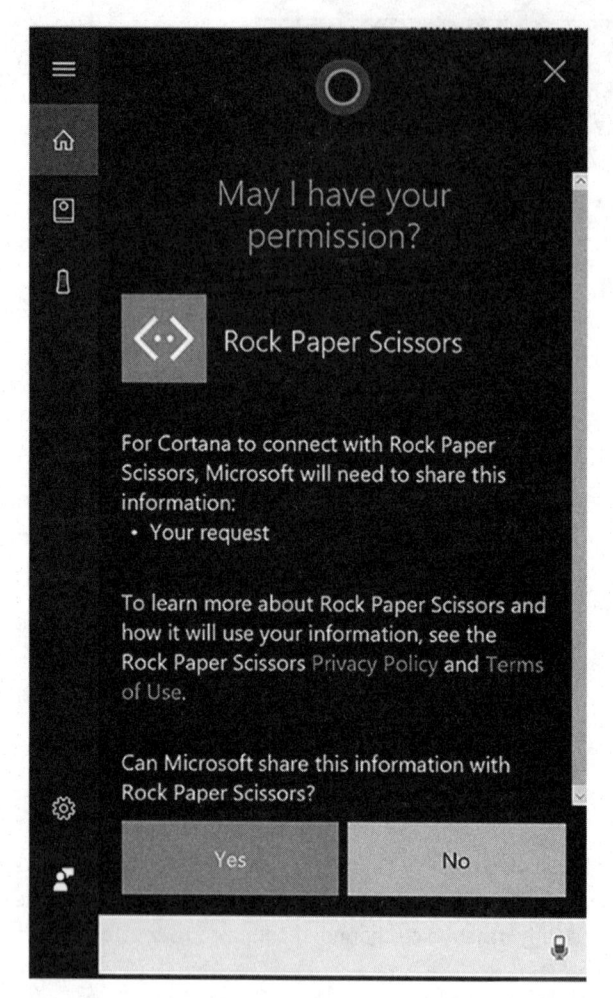

FIGURE 15-4 Giving Cortana permission to invoke a chatbot.

Clicking Yes in Figure 15-4 invokes the RockPaperScissors chatbot. Click the microphone to speak and Cortana translates the text to words and sends that through Bot Connector, which arrives at the RockPaperScissors chatbot as a new message activity. Figure 15-5 shows open possible output when the user says Rock.

FIGURE 15-5 Speaking with a chatbot via Cortana.

While Figure 15-5 shows the text on the screen, this isn't a normal messaging interface. In real life, you will hear Cortana read the text. Also, the comments don't scroll, but display the last thing Cortana says.

Summary

Now you know how to build a chatbot that users can interact with via voice. You saw how each of the main chatbots built throughout this book were refactored to implement speech. The RockPaperScissors chatbot adds speech to the *Speak* property of an Activity. Wine Bot replaced *PostAsync* with *SayAsync* and added a *speak* parameter. Music Chatbot added additional options to *PromptDialog* that included *speak* and *retrySpeak* parameters.

In addition to specifying speech, you can add input hints that tell when a chatbot is listening to user input.

Finally, you learned how to register with the Cortana channel and transform the chatbot into a Cortana Skill. You then learned how to invoke the chatbot via Cortana and then interact with it via voice.

In many ways, voice exemplifies the promise of chatbots. Rather than needing to learn a new user interface for apps of the past, users can now talk to their computers via chatbots. Combined with everything you've learned throughout this book, you can build different types of chatbots with various navigation schemes, using NLP and other AI services, with advanced conversational and voice interfaces that are easy for everyone to use.

Index

A

ActionTypes class 249
activities 11
   Activity class 67–68
   adding speech to 345–347
   code design 69–72
   handling 67–77
   listening for new 310–314
   pinging 76
   relationship changes 72–73
   sending 72, 316–318
   types 68
   typing indicators 77, 78–80
   user data deletion 75–76
Activity class 46, 59–62, 67–68
   custom message Activity 61–62
ActivityExtensions class 60–62
ActivityType class 68–69
ActivityType members 46
ActivityType.Message 46
Activity types 24–25
ActivityTypes class 249
Adaptive cards 267–276
   handling actions 274–275
   layout with containers 271–273
   using controls 273–274
AddRemainingFields method 175–176
AddTieAsync method 56
Analytics tab 280–283
AnalyzeImageAsync method 331
API Key 282
api/messages 36
Apple 6
Application ID 282
Application Insights 280–283

artificial intelligence (AI) 4–5, 183, 321
ASP.NET MVC applications 21
ASP.NET MVC Web API project 18
assembly references 19–20
attachments 112–113
   accepting from users 254–255
   sending 255–257
   working with 250–257
Audio cards 265–266
augmented/virtual reality (AR/VR) applications 7
Authenticate module 315
Autofac 19, 254
AutoFac 229
Azure
   deploying chatbot to 31–35
Azure Application Insights 280–283
Azure Table Storage 253

B

Bing
   searching with 321–330
Bing channel
   configuration 286–288
BotAuthentication attribute 23
Bot Builder 19
   Language.GenerateTerms method 172
Bot Builder dialogs. *See* dialogs
Bot Connector 9, 19, 24, 25, 35
   relationship changes and 72–73
   services 10–12
   state management service 47–59
Bot Emulator
   activities handling 67–77
   ConversationUpdate Activity 73–75
   features 65–67
   Typing Activity 80

Bot Framework

Bot Framework
  architecture 8–14
    Bot Connector 10–12
    channels 9–10
    chatbots 12–14
    components 8–9
  Direct Line API and 303–320
  InputHints class 353–354
  introduction to 3–14
  Natural Language Processing 183–202
  voice services and 345–360
Bot Framework Emulator
  communicating with chatbot 29–30
  configuration 27–29
  installing 26–27
  testing chatbot with 26–30
Bot Framework project 15–38
  default chatbot 22–25
  default code 18–25
  folder and file layout 21
  initial testing with emulator 26–30
  publishing 31–35
  registering 35–38
  starter template 15, 16–17
    assembly references 19–20
  starting new 17–18
  steps in building 15–16
  template installation 16–17
Bot Framwork State Service 47–59
bots. *See* chatbots
BotState 49–50
Bot State Service 102, 115, 253
browse dialog 243–244
BrowseDialog 326–329
BrowseDialog implementation 258–263
BuildForm method 127, 153, 163, 168
Build method 178
BuildWineDialog method 131–132

C

C# 11
Call method 209–210
camelCase 127
CancellationTokenSource 318, 319
cancelToken parameter 318
CardElement class 272–273

cards 237–278
  Adaptive 267–276
  attachments 250–257
  Audio 265–266
  BrowseDialog implementation 258–263
  building blocks 246–257
  creating 261–268
  displaying 257–267
  Hero 258, 261–262
  layout 261
    with containers 271–273
  Suggested Actions 246–250
  Thumbnail 258, 262
  using controls 273–274
Case 222
Chain class 213–225
  Chain.ContinueWith method 223–224
  Chain.From method 219–220
  Chain.Loop method 220–221
  Chain.Switch method 220–222
  LINQ statements 225–226
  posting and waiting methods 223–224
  WineBotChain 214–218
Chain.ContinueWith method 223–224
Chain.From method 219–220
Chain.Loop method 220–221
Chain.Switch method 220–222
ChannelId 46
Channel Inspector 289
channels 9
  Bing 286–288
  configuration 279–290
  Console 303–308
  Cortana 355–359
  custom 303–320
  Email 291–293
  overview of 9–10, 279–283
  Teams 284–286
  Twilio 293–294
  types of 279
  Webchat 295–300
chatbots
  activities handling 67–77
  adding voice services to 345–360
  advanced conversations 77–86
  analytics 280–283

benefits of 5–7
caching with multiple 48
channels 279–290
characteristics of 12–13
communications 13–14
conversation and 4–5, 8, 12–13, 39–64, 73–75
creating
    assembly references 19–20
    default chatbot 22–25
    folder and file layout 21
    initial testing with emulator 26–30
    starting new project 17–18
    template installation 16–17
custom information storage 11
defined 3
deploying to Cortana 355–359
deployment of 5–6
email 291–293
ending conversations with 318–320
FAQ 339–344
fine-tuning 65–88
FormFlow 124–132
identity 46–47
in Bot Framework 9
interruption handling 226–229
Music Chatbot 237–278
navigating between dialogs 204–207
pinging 76
platform independence of 6–7
publishing 31–35
registering 35–38
relationship changes 72–73
Rock, Paper, Scissors game bot 39–45, 51–56, 60–62
sending independent messages 80–86
sending user input to 316–318
steps to building 15–16
texting 293–294
typing indicators 77, 78–80
usage scenarios 7–8
use of LUIS by 183–202
using voice 6
Webchat control 295–300
WineBot 91–118
chatbotState variable 54
ChoiceStyleOptions enum 147–148
C# language 127

client-side coding
    Webchat control 296–297
CognitiveService class 322
Cognitive Services. *See* Microsoft Cognitive Services
command-line applications 4
commands
    configuration of FormFlow 160–161
communication 5
    chatbot 13–14
    multi-cultural 332–339
    out-of-band 226–229
    proactive 80–86
    voice 6
Compare method 42
condition parameter 164–165
configuration
    Application Insights 280–283
    Bing channel 286–288
    channels 279–290
    email channel 292–293
    FormFlow 154–161
        commands 160–161
        responses 156–157
        templates 157–159
    Teams channel 284–286
    Visual Studio 281
    Webchat channel 295–296
Configure class 315–316
ConfigureFormBuilder method 156–160
Confirm method 166–168
ConnectorClient class 25
console applications 4
Console channel
    code 305–308
    components 304–305
    ending conversations 318–320
    keeping conversation open 315–316
    listening for new activities 310–314
    overview of 303–308
    sending activities 316–318
    starting a conversation 308–309
ContactRelationUpdate Activity 72–73
containers
    with Adaptive cards 271–273
ContinueWith method 223–224, 224

controls
  using 273–274
conversation 39–64
  Activity class 59–62
  Activity state 46
  advanced 203–234
  advanced messages 77–86
  continuing 85–86
  criticality and 8
  defined 45
  designing 43
  dialogs and 105–107
  dialog stack 203–213
  elements of 45–47
  email 291–293
  ending 318–320
  FormFlow 132–137
  handling interruptions 226–229
  identifiers 45–46
  identity 47
  importance of 4–5
  keeping open 315–316
  management
    with chaining 213–225
  management of 12–13
  MessagesController class 44–45
  participating in 59–62
  responding to 59
  starting new 86
  starting, on custom channel 308–309
  state management 45–59
    bot state service capabilities and organization 49–51
    dialogs 106–108
    saving and retrieving state 47–59
  stray user input and 43
  text output formatting 230–232
  updates 73–75
  user and chatbot identities 46–47
  with Music Chatbot 237–246
conversational user interface (CUI) 3, 4, 13, 39, 183
Conversation property 47
ConversationReference.json file 83–84
ConversationUpdate Activity 73–75
Conversation.UpdateContainer method 229
Cortana 6, 345, 355–359
CreateReply method 25, 59, 232, 249
criticality 8

CUI. *See* conversational user interface
custom channels 303–320
  ending conversations 318–320
  keeping conversation open 315–316
  listening for new activities 310–314
  overview 303–308
  sending activities 316–318
  starting a conversation 308–309
customer support 339

D

DefaultCase 222
default.htm file 21
DeleteScoresAsync method 55
DeleteStateForUserAsync method 50
DeleteUserData Activity 75–76
dependencies parameter 164–165
deployment
  chatbots 5–6
Describe attribute 133–134, 138
DetectAndTranslateAsync method 336
DetectLanguageAsync method 334
dialogs
  browse 243–244
  building 91–118
  calling 115–116
  conversation flow 105–107
  dialog class creation 102–103
  finishing 211–214
    Done method 211–212
    Fail method 212–213
    Reset method 212–213
  FormFlow. *See* FormFlow
  IDialogContext parameter 103–105
  implementation of 98–114
  initialization and workflow 103
  LINQ statements 225–226
  playlist 244–245
  profile 243, 250–253
  PromptDialog class 108–114
  root 242, 246–248
  search 245–246
  searches 114–115

FormFlow

   serializable 102, 111
   WineBot 91–118
   WineForm 129–132
dialog stack 203–213
   defined 203–204
   navigating to other dialogs 204–207
   navigating via call 209–210
   navigating via forward 207–209
Direct Line API 11, 303–320
   ending conversations 318–320
   keeping conversation open 315–316
   listening for new activities 310–314
   sending activities 316–318
   starting a conversation 308–309
   Watermark property 307–308, 313
DisplayAsync method 255–256
distributed applications 14
DoChainLoopAsync method 220–221
DoChainSwitchAsync method 220–222
DoManageCase method 222, 223
Do method 224
DoneAsync method 229
Done method 211–212
DoSearchAsync method 114–115
DoSearchCase method 222, 223
DoSearch method 128, 176
duplicate messages 314

E

email accounts
   creating 291–292
Email channel 291–293
emulator
   communicating with chatbot 29–30
   configuration 27–29
   installing 26–27
   testing chatbot with 26–30
EndConversationAsync method 319
EndOfConversationCodes class 177–178
entities
   handling 196–200
   prebuilt 188–189
   simple 187–188
   specifying, in LUIS 187–189
ExtractEntities method 197

F

Fail method 212–213
FAQ chatbots 339–344
Field method 171–173
fields 168–175
   AddRemainingFields method 175–176
   dynamic field definition 171–172
   HasField method 175
   validation 173–175
FileService class 256
fluent interface
   FormFlow 153–154
formatting text 230–232
FormBuilder class 127–128
FormCanceledException 131
FormConfiguration class 156–160
FormFlow 119–152
   attributes 132–135
      Describe 133–134
      Numeric 134–135
      Optional 135
      Pattern 135
      Prompt 135–136
      Terms 136–137
   basic chatbot 124–132
   BuildForm method 127
   choosing between IDialog<T> and 150
   conversations 132–137
   customization 153–182
      Build method 178
      commands 160–161
      Configuration property 154–161
      Confirm method 166–168
      Message method 162–167
      OnCompletion method 176–178
      reponses 156–157
      templates 157–159
      working with fields 168–175
   features 119–124
   fluent interface 153–154
   help options 123
   initialization 178–181
   input error correction 121
   menu options 120
   serializable 127

Forward method

templates and patterns 137–150
  basic templates 139
  configuration 157–159
  pattern language 137–139
  template options 146–149
  TemplateUsage enum 139–146
using as dialog 129–132
WineForm 124–132
working with fields
  AddRemainingFields method 175–176
  dynamic field definition 171–172
  field validation 173–175
  HasField method 175
Forward method 207–209
Frequently Asked Questions (FAQs) 339–344
From method 219–220
From property 46

G

Game class 40–43
GameState class 51–56
generateMessage parameter 166–167
GetAudioCardsForPreview method 265–266
GetConversationReference 84–85
GetGenres method 241
GetMessage method 59
GetPreview method 241
GET requests 98
GetScoresAsync method 53–54, 56, 79–80
GetToken method 241
GetTracks method 241
GetUserImageAsync 98
GetUserInputAsync method 318, 319
GitHub 12
Google 6
graphical user interfaces (GUI) 4–5, 12
graphical user interfaces (GUIs) 237
Groove API 238–242
GrooveService class 238–241

H

HandleCanceledForm method 132–133
Handle method 71–72
HandleSystemMessageAsync method 116
HandleSystemMessages 24
Happy Path 237
HasField method 175

Help command 160
HelpScorable 226–229
Hero cards 258, 261–262
HttpOperationException 50
HTTP POST endpoint 22

I

IBotContext 104
IBotData 105
IBotDataStore<BotData> 253
IBotState 49–50
IBotToUser 105–106
IContactRelationUpdateActivity interface 72–73
IDialogContext 103–105, 177, 195
  Call method 209–210
  Done method 211
  Forward method 207–209
IDialog<object> 102–103
IDialogStack 104
IDialog<T> 130–131, 210
  choosing between FormFlow and 150
Id property 47
IFormBuilder<T> 154, 156
IFormDialog<T> 131–132
images
  interpreting 330–332
IMessageActivity 207
initialization
  FormFlow 178–181
input controls 274–275
input hints 353–355
InputHints class 353–354
installation
  Bot Framework Emulator 26–27
  project template 16–17
Instrumentation Key 282
intents
  adding 195–196
  building, in LUIS 186–187
interruptions
  handling 226–229
Inversion of Control (IoC) 229
IScorable 226–229
IsGroup property 47
IsRequired property 274
IStateClient 54

J

Json.NET 97
Json.NET package 49

K

keep-alive messages 313

L

Language.GenerateTerms method 172
language translation 332–339
Language Understanding Intelligence Service (LUIS) 14, 183–202, 321
  continuous model improvement 200–201
  essential concepts of 183–192
  publishing model 192
  setting up 185–188
    building intents 186–187
    model creation 185–186
    specifying entities 187–189
  training and deploying 189–192
    adding utterances 190–192
    model testing 191–192
  using in chatbots 193–200
    adding intents 195–196
    entity handling 196–200
Last-In First Out (LIFO) 203
LINQ statements 225
Linux operating systems 303
localization 66, 332
Loop method 220–221
LUIS. *See* Language Understanding Intelligence Service
LuisIntent attribute 195
LuisResult method 197

M

MainAsync method 306
Markdown 230–232
Material Design 6
Message class 307
Message method 127–128, 162–167
  condition parameter 164–165
  dependencies parameter 164–165
  generateMessage parameter 166–167
  prompt parameter 165

MessageReceivedAsync method 107–108, 115, 212, 271, 329–330
messages
  duplicate 314
  keep-alive 313
  POST 75
  receiving 310–314
  sending independent 80–86
  translating 334–339
MessagesController class 44–45, 57–59, 115–116, 129–132
MessagesController.cs file 21, 22
MessagesController.Post method 71–72
messaging apps 3–4, 5, 9–10
MicrosoftAppId 241
Microsoft App ID 36
Microsoft App password 36
MicrosoftAppPassword 241
Microsoft.Bot.Builder 19, 20
Microsoft.Bot.Connector 19
Microsoft Bot Framework. *See* Bot Framework
  platform independence and 6–7
Microsoft Cognitive Services 4–5, 6, 14, 321–344
  LUIS 183–202
  QnA Maker 339–344
  Search service 321–330
  Text Translation API 332–339
  Vision service 330–332
Microsoft Direct Line SDK 303, 308
multi-cultural communication 332–339
multi-threading 306–307, 308
Music Bot
  using search with 321–330
Music Chatbot 237–246
  attachments and 250–257
  Audio cards 265–266
  BrowseDialog 243–244, 258–261
  channels 279–290
  FileService class 256
  Groove API 238–242
  PlaylistDialog 244–245, 263–265
  ProfileDialog 243, 250–253
  PromptDialog 350–353
  RootDialog 242, 246–248
  RootMenuItem 248
  SearchDialog 245–246, 267–271

N

naming conventions 127
natural language processing (NLP) 4, 14
Natural Language Processing (NLP) 183–202, 321
   training models 185–192
navigation
   dialog stack and 203–213
   to other dialogs 204–207
   via call 209–210
   via forward 207–209
.NET Framework 310
NewsDialog 323–326, 330
ngrok 36, 37
NLP. *See* Natural Language Processing
None intent 195–196
NuGet 19, 20, 97, 303, 308
Numeric attribute 134–135

O

OnCompletion method 127–128, 176–178
Optional attribute 135
out-of-band communication 226–229

P

PascalCase 127
Pattern attribute 135
pattern language
   FormFlow 137–139
Ping Activity 76
platform independence
   chatbots and 6–7
Playlist Dialog 244–245
PlaylistDialog
   implementation 263–265
PlayList enum 43
Play method 42–43
PlayScore class 53
PlayType enum 39–40, 45
polling 310
port numbers 45
PostAsync method 177
POST messages 75
Post method 22, 23, 24, 44, 51, 57, 61, 71–72, 76, 130, 179, 204, 213, 218, 335, 347

PostToChain method 221
PostToUser method 222
prebuilt entities 188–189
proactive communication 80–86
ProfileDialog 243, 250–253
Program class 304
programming languages 11
PromptAsync method 347
Prompt attribute 135–136, 138, 139, 165
PromptAttribute 159
Prompt class 316–318
PromptDialog 350–353, 354
PromptDialog class 108–114
   Attachment method 112–113
   Choice method 109–110
   Confirm method 111–112
   Number method 110
   Text method 111–112
PromptDialog method 207
PromptOptions 353
promptOptions parameter 108
prompt parameter 165

Q

QnA maker 339

R

Rating property 164
RatingType enum 125–126
ReceiptCards 267
ReceiveAsync method 312
Recipient property 46
ReconnectToConversationAsync method 316
RefreshTokenAsync method 318
RefreshTokensAsync method 316
RegexCase 222–223
ReplyToId parameter 61
Representational State Transfer (REST) interface 22
Reset method 212–213
REST. *See* Representational State Transfer (REST)
REST API 11
REST endpoints 297
REST interface 303
ResumeAfterProfileAsync method 212

RetrieveMessagesAsync method 312, 319
Rock, Paper, Scissors game bot 39–45
   BuildMessageActivity method 60–62
   BuildTypingActivity method 78
   deploying to Cortana 355–359
   Game class 40–43
   GameState class 51–56
   GetScoresAsync method 79–80
   MessageController.Post method 71–72
   MessagesController class 44–45, 57–59
   PlayList enum 43
   PlayType enum 39–40
   SystemMessages class 69–71
RootDialog 205–207, 209, 218, 246
RootDialog.cs file 21
RootMenuItem 248
routing 10

S

SayAsync method 347–350, 354
ScorableBase class 227–228
SearchArguments class 275
SearchDialog 245–246, 267–271
SearchingIntent method 196–197
Search services 321–330
   BrowseDialog 326–329
   NewsDialog 323–326, 330
Select method 222
SendAsync method 116, 130, 204
Send System Activity menu 72
SendToConversationAsync method 257
SendToExistingConversation method 85–86
Serializable attribute 102
ServiceUrl 46
SetDefine 172
SetFieldDescription 172
SetType 172
SetValidation method 173–175
ShowMenuAsync method 249–251
SignInCards 267
Siri 6
SMS messages 293–294
Speak property 345, 347
speech capabilities
   adding to activities 345–347
   adding to PromptDialog 350–353
   adding with SayAsync 347–350
   Cortana 355–359
   input hints 353–355
speech recognition 357
Speech Synthesis Markup Language (SSML) 347
SSML. *See* Speech Synthesis Markup Language
StartAsync method 103, 105, 115, 210
StartConversationAsync method 309, 318
StartNewConversation method 86
StateClient 54
state management
   conversations 45–59, 106–108
StockingType enum 125–126
streaming 310–314
SubmitAction 274–275
Suggested Actions 246–250
Suggested Utterances tab 200–201
SuppressRepeatedActivities method 314–315
Switch method 220–222
SystemMessages class 69–71

T

Task Parallel Library (TPL) 312
Teams channel
   configuration 284–286
   using 286
Template attribute 139, 145
TemplateAttribute 157–159
TemplateBaseAttribute class 146–148
templates
   configuration 157–159
   FormFlow 137–150
   project 16–25
TemplateUsage enum 139–146
Terms attribute 136–137
text chatbots 293–294
text messaging 5
text output
   formatting 230–232
Text Translation API 332–339
Then method 224
Thumbnail cards 258, 262
Timestamp 46
TranslateResponseAsync method 338–339
TResult type parameter 103
Twilio channel 293–294

Typing Activity 77, 78–80
typing indicators 77, 78–80

U

Universal Windows Programs (UWP) 6
UnWrap method 222
UploadAsync method 254–255, 256
user data
  deletion of 75–76
user exits 319–320
user identity 46–47
user input
  interpretation of 199
  sending to chatbots 316–318
utterances 190–192

V

VideoCards 267
Vision services 330–332
Visual Studio
  configuration 281
  template 16–25
voice communication 6
voice services 345–360
  adding to PromptDialog 350–353
  Cortana 355–359
  input hints 353–355
  SayAsync method 347–350

W

Wait method 104, 210
Watermark property 307–308, 313
Webchat control 295–300
  client-side coding 296–297
  handling server requests 297–299
web pages
  adding chatbots to 295–300
Web Sockets 310, 312
Windows Subsystem for Linux (WSL) 303
WineApi class 96–97, 98, 124, 128
WineBot
  attachments 112–113
  calling dialog 115–116
  FormFlow chatbot 124–132
  implementation of 98–114

  introduction to 91–92
  MessagesController class 115–116, 129–132
  PromptDialog class 108–114
  WineApi class 96–97, 98, 124–132, 128
  WineCategories class 93
  Wine.com API 92–98
  WineForm class 126–127
  WineProducts class 93–96
  WineSearchDialog class 98–114, 111, 114–115
WineBotChain 214–225
WineBotDialog class 193–195
WineBotLuis 193–200
WineCategories class 93
Wine.com API 92–98
WineForm 124–132
  Confirm method 166–168
  FormFlow fluent interface 153–154
  Forward method 207–209
  OnCompletion method 176–178
  Rating property 164
  using as dialog 129–132
  working with fields 168–175
WineForm class 126–127
WineFormCompletedAsync method 231
WineProducts class 93–96
WineSearchDialog class 98–114, 111, 114–115
WineType enum 124–125, 198
WritePrompt 308

X

Xamarin 6

About the Author

JOE MAYO, Joe Mayo is an author and independent software consultant, specializing in Microsoft technologies. He has written several books, including *Programming the Microsoft Bot Framework: A Multiplatform Approach to Building Chatbots* by Microsoft Press. A long-time MVP with several years of awards. Joe can often be found tweeting (as @JoeMayo) about #BotFramework and #AI on Twitter.

Now that you've read the book...

Tell us what you think!

Was it useful?
Did it teach you what you wanted to learn?
Was there room for improvement?

Let us know at https://aka.ms/tellpress

Your feedback goes directly to the staff at Microsoft Press, and we read every one of your responses. Thanks in advance!

Microsoft

Hear about it first.

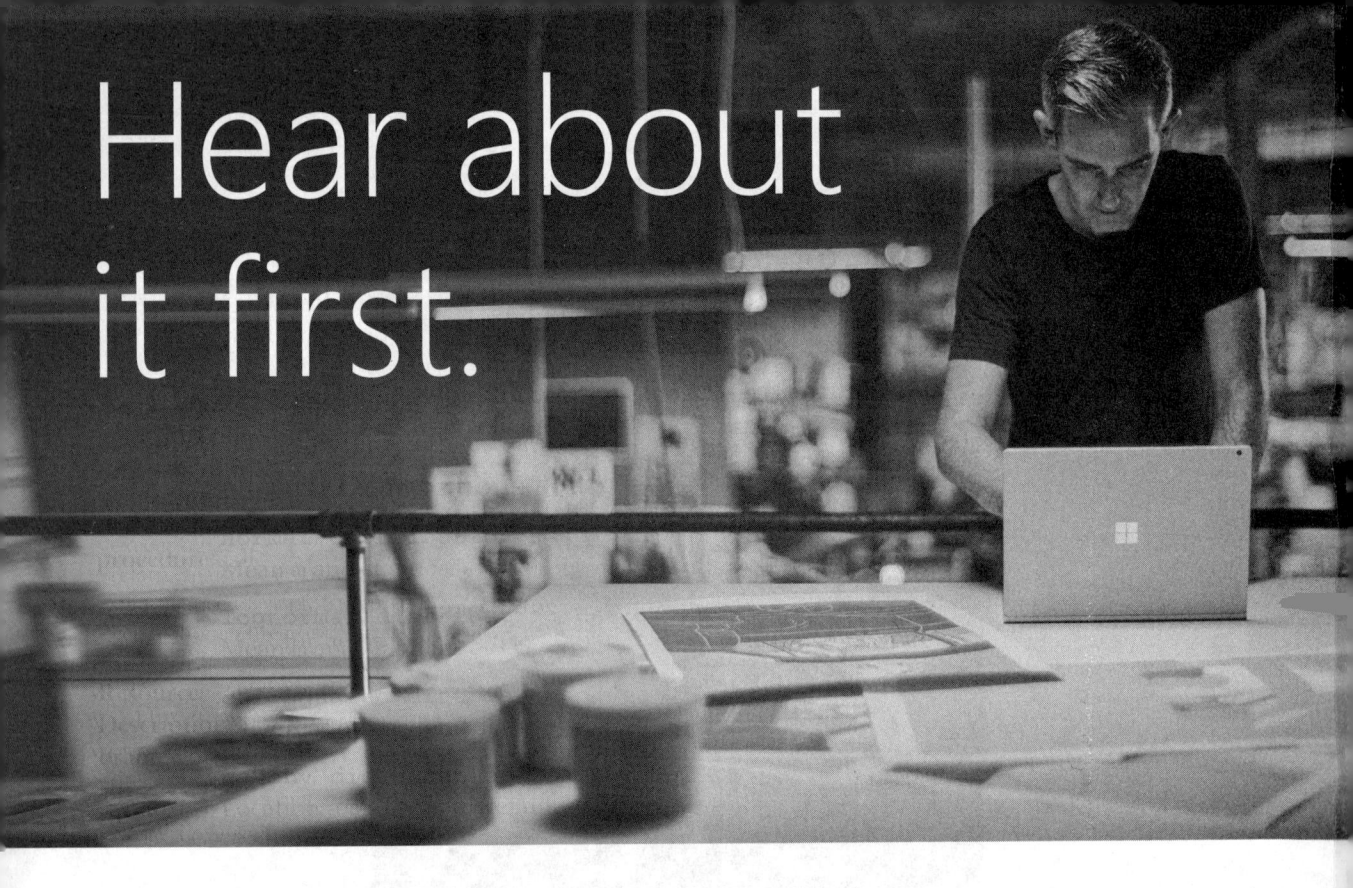

Get the latest news from Microsoft Press sent to your inbox.

- New and upcoming books
- Special offers
- Free eBooks
- How-to articles

Sign up today at MicrosoftPressStore.com/Newsletters

Microsoft